STOCHASTIC MODELLING OF ELECTRICITY AND RELATED MARKETS

ADVANCED SERIES ON STATISTICAL SCIENCE & APPLIED PROBABILITY

Editor: Ole E. Barndorff-Nielsen

Published

Vol. 1 Random Walks of Infinitely Many Particles
by P. Revesz

Vol. 2 Ruin Probabilities
by S. Asmussen

Vol. 3 Essentials of Stochastic Finance: Facts, Models, Theory
by Albert N. Shiryaev

Vol. 4 Principles of Statistical Inference from a Neo-Fisherian Perspective
by L. Pace and A. Salvan

Vol. 5 Local Stereology
by Eva B. Vedel Jensen

Vol. 6 Elementary Stochastic Calculus — With Finance in View
by T. Mikosch

Vol. 7 Stochastic Methods in Hydrology: Rain, Landforms and Floods
eds. O. E. Barndorff-Nielsen et al.

Vol. 8 Statistical Experiments and Decisions: Asymptotic Theory
by A. N. Shiryaev and V. G. Spokoiny

Vol. 9 Non-Gaussian Merton–Black–Scholes Theory
by S. I. Boyarchenko and S. Z. Levendorskiĭ

Vol. 10 Limit Theorems for Associated Random Fields and Related Systems
by A. Bulinski and A. Shashkin

Vol. 11 Stochastic Modelling of Electricity and Related Markets
by F E Benth, J S Benth and S Koekebakker

Advanced Series on Statistical Science & Applied Probability **Vol. 11**

STOCHASTIC MODELLING OF ELECTRICITY AND RELATED MARKETS

Fred Espen Benth

Jūratė Šaltytė Benth

University of Oslo, Norway

Steen Koekebakker

University of Agder, Norway

World Scientific

NEW JERSEY • LONDON • SINGAPORE • BEIJING • SHANGHAI • HONG KONG • TAIPEI • CHENNAI

Published by

World Scientific Publishing Co. Pte. Ltd.

5 Toh Tuck Link, Singapore 596224

USA office: 27 Warren Street, Suite 401-402, Hackensack, NJ 07601

UK office: 57 Shelton Street, Covent Garden, London WC2H 9HE

Library of Congress Cataloging-in-Publication Data

Benth, Fred Espen, 1969–

Stochastic modelling of electricity and related markets / by Fred Espen Benth, Jūratė Šaltytė Benth & Steen Koekebakker.

p. cm. -- (Advanced series on statistical science and applied probability ; v. 11)

Includes bibliographical references and index.

ISBN-13: 978-981-281-230-8 (hbk. : alk. paper)

ISBN-10: 981-281-230-X (hbk. : alk. paper)

1. Electric utilities--Mathematical models. 2. Energy industries--Mathematical models. 3. Stochastic models. I. Šaltytė Benth, Jūratė. II. Koekebakker, Steen. III. Title.

HD9685.A2B44 2008

333.793'20151922--dc22

2008002489

British Library Cataloguing-in-Publication Data

A catalogue record for this book is available from the British Library.

For Julia (F. E. B. & J. Š. B.)

For Anja (S. K.)

Preface

Since the early 1990s, the markets for electricity and related products have been liberalized worldwide. It all started off with the Nordic market Nord-Pool and the England & Wales market at around 1992, and over the last two decades trade in electricity and related products on all continents in the world has been liberalized. In 1999 the Chicago Mercantile Exchange organized a market for temperature derivatives that has gained momentum in recent years. Exchange-based markets for gas have emerged and are now actively traded at the New York Mercantile Exchange and the Intercontinental Exchange in London. As these markets mature, energy becomes increasingly more important asset class of investments, not only attracting the traditional actors in the markets, but also speculators like investment banks, hedge funds and pension funds.

The basic products in the electricity, gas and temperature markets are spot, futures and forward contracts and options written on these. With organized markets comes the need to have consistent stochastic models describing the price evolution of the products. Such models must reflect the stylised facts of the commodity prices we observe at the exchanges, but also lend themselves to analytical treatment like pricing of derivatives. Energy-related spot prices have several typical characteristics, with the most prominent being mean reversion towards a seasonally varying mean level, and frequently occurring spikes resulting from an imbalance between supply and demand. Electricity spot prices may, for instance, increase with several 100% over very short time intervals, before they come back to their normal levels. Such price volatilities are hardly found in any other markets than energy. Further, since the energy commodities are driven by the balance between demand and production, the prices tend to mean-revert. A natural class of stochastic models to describe such dynamics is the Ornstein-Uhlenbeck processes. We use these mean-reverting stochastic processes as our modelling tool throughout the book.

Contrary to more classical commodity markets like agriculture and metals, energy-related futures contracts deliver the underlying spot over a contracted period. The derivation of futures prices from spot is not straightforward, and depends on the choice of risk-neutral probability and the type of model. The delivery of the spot over a period creates technical problems when calculating the futures prices based on exponential spot models, while arithmetic models on the other hand are feasible for analytical pricing. The Heath-Jarrow-Morton approach suggests a direct modelling of the futures prices. However, again the existence of a delivery period creates problems since it turns out to be hard to propose arbitrage-free models which at the same time are tractable from statistical and theoretical points of view. All these challenges defend a study of energy markets on its own.

The background for writing this book is twofold. Over the recent years, we have worked on electricity and temperature modelling, and wanted to collect our results together in a consistent and general way. On the other hand, we also wished to provide a rigorous introduction to stochastic modelling of the energy markets. One of our goals is to present a mathematically sound foundation for the relevant models to energy-related products, being useful in both theory and practise.

In many practical examples we test our models on data collected from electricity, gas and temperature markets. However, there are many challenges related to the fitting of the relevant stochastic models in the energy markets, and it is outside the scope of this book to provide a detailed treatment of all the various estimation techniques and issues. The focus is on presenting a consistent and complete theoretical framework for energy market models with applications to derivatives pricing.

Acknowledgements: Many colleagues and friends have helped us out in the process of writing this book. One may say that "no one mentioned, no one forgotten". However, we feel that some persons have made a significant contribution to the contents, and it would be unfair not to thank them explicitly. Apart from being an enthusiastic source of information on the market, Fridthjof Ollmar has been the co-author on a scientific paper with us (resulting in Ch. 7) and provided electricity data (through Agder Energy). A crucial input from him has been the creation of a computer program for which smoothed forward curves can be constructed from observed electricity futures prices. This has been invaluable for a big part of the empirical work done for the electricity markets. Our supportive friend and colleague Alvaro Cartea has read parts of the manuscript and correcting some errors. He has provided us with data (through Oxford Economic Research Associates, Oxera) and shared his insight into the theoretical and practical aspects of electricity and gas markets.

Through their scientific collaboration, many colleagues have contributed (directly or indirectly) to the contents of the book. We are grateful to Roar Aadland, Kęstutis Dučinskas, Dennis Frestad, Petter E. Furrebøe, Pål Nikolai Henriksen, Paulius Jalinskas, Jan Kallsen, Paul C. Kettler, Rüdiger Kiesel, Rodwell Kufakunesu, Thilo Meyer-Brandis, Andrea Roncoroni, Sigbjørn Sødal and Valeri Zakamouline for all the fun in working together with you on problems related to the energy markets, as well as your friendship and support.

Stochastic models for energy markets are not very relevant if you do not have access to data to support (or dismiss) your theories. Data has been provided to us through several sources (some already mentioned). We greatly acknowledge the provision of gas and electricity data from Andre Damslora at PointCarbon, and Håvard Hvarnes and Bjarte Lima at Elkem. Temperature data was kindly made available to us by SMHI, the Swedish Meteorological Institute.

Finally, we thank our respective institutions for their support in this book project, and all our friends and colleagues there for creating such a great research environment and for all the every-day fun. We also feel indebted to the editor Ole E. Barndorff-Nielsen, who enthusiastically encouraged us to publish the book in this series. The staff at World Scientific is thanked for efficient handling.

Fred Espen Benth, Jūratė Šaltytė Benth and Steen Koekebakker
Oslo and Kristiansand, December 2007

Contents

Preface vii

1. A Survey of Electricity and Related Markets 1

1.1 The electricity markets 3
1.1.1 Electricity contracts with physical delivery 3
1.1.2 Financial electricity contracts 5
1.2 The gas market 8
1.2.1 Futures and options on gas 10
1.3 The temperature market 11
1.4 Other related energy markets 14
1.5 Stochastic modelling of energy markets 18
1.5.1 Spot price modelling 19
1.5.2 Forward and swap pricing in electricity and related markets 24
1.6 Outline of the book 32

2. Stochastic Analysis for Independent Increment Processes 37

2.1 Definitions 37
2.2 Stochastic integration with respect to martingales 41
2.3 Random jump measures and stochastic integration 43
2.4 The Lévy-Kintchine decomposition and semimartingales . . 45
2.5 The Itô Formula for semimartingales 48
2.6 Examples of independent increment processes 49
2.6.1 Time-inhomogeneous compound Poisson process . . . 49
2.6.2 Models based on the generalized hyperbolic distributions 51

2.6.3 Models based on the Variance-Gamma and CGMY distributions 55

3. Stochastic Models for the Energy Spot Price Dynamics 59

3.1 Introduction 59
3.2 Spot price modelling with Ornstein-Uhlenbeck processes . . 60
3.2.1 Geometric models 66
3.2.2 Arithmetic models 74
3.3 The autocorrelation function of multi-factor Ornstein-Uhlenbeck processes 78
3.4 Simulation of stationary Ornstein-Uhlenbeck processes: a case study with the arithmetic spot model 82

4. Pricing of Forwards and Swaps Based on the Spot Price 89

4.1 Risk-neutral forward and swap price modelling 89
4.1.1 Risk-neutral probabilities and the Esscher transform 95
4.1.2 The Esscher transform for some specific models . . . 99
4.2 Currency conversion for forward and swap prices 100
4.3 Pricing of forwards 104
4.3.1 The geometric case 104
4.3.2 The arithmetic case 114
4.4 Pricing of swaps 118
4.4.1 The geometric case 119
4.4.2 The arithmetic case 122

5. Applications to the Gas Markets 129

5.1 Modelling the gas spot price 129
5.1.1 Empirical analysis of UK gas spot prices 130
5.1.2 Residuals modelled as a mixed jump-diffusion process 136
5.1.3 NIG distributed residuals 139
5.2 Pricing of gas futures 142
5.3 Inference for multi-factor processes 146
5.3.1 Kalman filtering 147
5.3.2 Inference using forward and swap data 150

6. Modelling Forwards and Swaps Using the Heath-Jarrow-Morton Approach 155

6.1 The HJM modelling idea for forward contracts 156

6.2 HJM modelling of forwards 160
6.3 HJM modelling of swaps 164
6.3.1 Swap models based on forwards 168
6.4 The market models 172
6.4.1 Modelling with jump processes 176

7. Constructing Smooth Forward Curves in Electricity Markets 181

7.1 Swap and forward prices 183
7.1.1 Basic relationships 183
7.1.2 A continuous seasonal forward curve 184
7.2 Maximum smooth forward curve 187
7.2.1 A smooth forward curve constrained by closing prices 187
7.2.2 A smooth forward curve constrained by bid and ask spreads 190
7.3 Putting the algorithm to work 191
7.3.1 Nord Pool example I: A smooth curve 191
7.3.2 Nord Pool example II: Preparing a data set and analysing volatility 195

8. Modelling of the Electricity Futures Market 203

8.1 The Nord Pool market and financial contracts 205
8.2 Preparing data sets 206
8.3 Descriptive statistics 208
8.4 A market model for electricity futures 214
8.5 Principal component analysis 215
8.5.1 Principal component analysis of the total data set . . 217
8.5.2 Principal component analysis for individual market segments 220
8.6 Estimating a parametric multi-factor market model 224
8.6.1 Seasonal volatility 226
8.6.2 Maturity volatilities 227
8.7 Normalised logreturns and heavy tails 231
8.8 Final remarks 235

9. Pricing and Hedging of Energy Options 237

9.1 Pricing and hedging options on forwards and swaps 238
9.1.1 The case of no jumps – the Black-76 Formula 238
9.1.2 The case of jumps 247

9.2 Exotic Options 254
9.2.1 Spread options 254
9.2.2 Asian options 260
9.3 Case Study: Valuation of spark spread options – a direct approach 262
9.3.1 Modelling and analysis of spark spread options . . . 264
9.3.2 Empirical analysis of UK gas and electricity spread . 268

10. Analysis of Temperature Derivatives 277

10.1 Some preliminaries on temperature futures 277
10.2 Modelling the dynamics of temperature 280
10.2.1 The CAR(p) model with seasonality 281
10.2.2 A link to time series 283
10.3 Empirical analysis of Stockholm temperature dynamics . . . 285
10.3.1 Description of the data 285
10.3.2 Estimating the CAR(p) models 287
10.3.2.1 Fitting an AR(1) model 289
10.3.2.2 Fitting an AR(3) model 296
10.3.2.3 Identification of the parameters in the CAR(p) model 300
10.4 Temperature derivatives pricing 301
10.4.1 CAT futures 302
10.4.2 HDD/CDD futures 305
10.4.3 Frost Day index futures 312
10.4.4 Application to futures on temperatures in Stockholm 314

Appendix A List of abbreviations 319

Bibliography 321

Index 333

Chapter 1

A Survey of Electricity and Related Markets

In the beginning of the 1990s a liberalization of the electricity and gas markets started, resulting in the emergence of markets for spot and derivative products in numerous countries and regions spread over the world. The securitization of, for instance, weather, freight and greenhouse gas emission rights, contribute to a greater flexibility for risk control for both producers and consumers operating in the energy markets. In the present book, the aim is to develop tailor-made stochastic models for the various assets traded in electricity and related markets. These assets are in many ways distinct in nature and definition compared to what we find in the more "classical" commodity markets as oil, coal, metals and agriculture. Hence, new and challenging modelling problems appear.

Our main focus will be on stochastic modelling of the electricity market. In an arbitrage-free modelling framework, the spot price does not play the same important role as it does in other commodity markets. Electricity has very limited storage possibilities. Producers may store indirectly in water reservoirs (for hydro-based electricity production) and via gas, oil or coal (for thermal electricity production). However, the consumer of electricity cannot buy for storage. This has the implication that the cost-of-carry relationship between spot and forward prices breaks down. Further consequences of the lack of storeability are strong seasonality and possible spikes in prices. The spiky behaviour of spot electricity prices is a stylised feature of these markets, and appears when, for instance, a nuclear power plant must unexpectedly be closed down, or temperature drops significantly. Power prices may soar during short periods of time, and then fall back to more normal levels shortly after (giving a "spike" in the price path). Typically, the spot price volatilities may exceed the levels usually observed in stock markets by several orders. The limited storage possibility also means that electricity markets are regional. For instance, a difference in the price of electricity between the Nordic power exchange, Nord Pool, and the German-based European Power Exchange (EEX), does not necessarily imply an arbitrage opportunity. An arbitrageur cannot buy for storage and transportation, and therefore the spot asset cannot be used to set up dynamic hedging strategies exploiting price differentials.[1] The tradable assets in such markets are typically average-based forward contracts, that deliver electricity over a specified time period.

In the title of the book we refer to *related markets* of electricity. A technical approach to define a market as being related to electricity is to look for the same modelling characteristics, such as limited storeability of the spot, seasonally dependent prices with spikes, and where the tradeable assets are average based forward contracts. An economist, on the other hand, would typically have a different focus, defining related markets as those markets interacting more or less directly with the demand or supply side of the electricity market. In the application part of this book we study, in addition to the Nord Pool electricity market, UK natural gas and the temperature markets. Both temperature and natural gas share similarities with the electricity market from a modelling point of view. Temperature is obviously not possible to store. Natural gas on the other hand, can be stored, but most often it is quite costly.[2] Limited storage capacity makes the natural gas markets the less extreme sibling to electricity from a modelling point of view. We find average based forward type contracts in all these markets. From the economic point of view the temperature market is linked to the demand side of the electricity market. Electricity demand varies with temperature when power is needed for cooling in areas with warm summer temperatures, or heating in areas with cold winters. Temperature can also affect the electricity supply side, but this effect is typically weaker. In warm summer periods nuclear power plants in continental Europe have been forced to cut on production due to lack of cold water for cooling. High temperatures will also reduce hydro production because of excessive evaporation from water reservoirs. The gas market on the other hand, is mainly linked to the electricty market through the supply side of gas fired power plants.

Before starting our analysis, we provide a survey of the three markets we

[1]Many regional markets are interconnected through cables, however, these have limited capacity preventing a full exploitation of the potential arbitrage.

[2]There is limited storage capacity in the gas pipeline system, and big flexible underground storage facilities are typically major investments.

are mainly concerned about, namely electricity, gas and temperature. Our emphasis will be on how these markets function, with particular attention to the obstacles we face when trying to model the different products offered for trade. We also give an informal discussion on the models and techniques we are going to apply in this book, together with their relevance for the markets in question.

1.1 The electricity markets

Electricity is usually labeled a "commodity", although its non-storeability has a profound effect on the infrastructure and the organization of the electricity market compared with other commodity markets.[3] Electrical power is only useful for practical purposes if it can be delivered during a period of time. This is why electricity has been called a *flow* commodity. Deregulated power markets have market mechanisms to balance supply and demand, where electricity is traded in an auction system for standardized contracts. All contracts guarantee the delivery of a given amount of power for a specified future time period. Some contracts prescribe physical delivery, while others are financially settled.

Financial power contracts are linked to some reference electricity spot price, and they are settled in cash. The market for such contracts is open to speculators, since it is not required to have consumption or production of electricity to participate in the market. We will focus our modelling efforts on the Nordic power market Nord Pool, but our results can be applied to contracts traded at other power exchanges. For instance, the base load financial contracts traded on the EEX and the French exchange Powernext are more or less identical to the Nord Pool contracts. In the following subsections we will describe both the physical and the financial electricity contracts traded at Nord Pool, along with a brief review of some of the relevant literature connected to the modelling of electricity prices.

1.1.1 *Electricity contracts with physical delivery*

By physical electricity contracts we mean contracts with actual consumption or production as part of contract fulfillment. Since capacity is restricted, and the supply and demand must balance, these markets must be

[3] [Stoft (2002)] provides a unique treatment of the interplay between economics and engineering in deregulated electricity markets. [Wolak (1997)] gives a description of worldwide electricity market organization after deregulation.

supervised by a transmission system operator (TSO). Moreover, typically, the players in these markets are restricted to those with proper facilities for production or consumption. The contracts for physical delivery are usually organized in two different markets, the real time and day ahead market.[4] This is known as the two-settlement system.

The *real-time* market (henceforth RT market) is organized by a system operator for short-term upward or downward regulation. The auction specifies both load and time period for generation or consumption. Bids in the RT market are submitted to the TSO. Bids may be posted or changed close to the operational time, in accordance with agreed rules. RT market bids are for upward regulation (increased generation or reduced consumption) and downward regulation (decreased generation or increased consumption). Both demand and supply side bids are posted, stating prices and volumes. Market participants must be able to commit significant power volumes on short notice. In the Nordic market the TSOs are Statnett (Norway), Svenska Kraftnät (Sweden), Fingrid (Suomen Kataverkko Oyj) (Finland), Energinet.dk (Denmark). TSOs list bids for each hour in priority order, according to price (merit order), and the merit order for each hour is used to balance the power system. Upward regulation is applied to resolve a grid power deficit. Then the RT market price is set at the highest price of the units called upon from the merit order. In the case of grid power surplus, downward regulation is applied, and the lowest price of the units called upon from the participation list sets the RT market price. The auctions in each country are in effect Walrasian auctions, but the specific rules for determining the hourly price of power imbalances, based on the RT market price, differ among the Nordic TSOs. In addition to the RT auction market the national TSOs have established markets for various necessary ancillary services providing balance power to market actors and securing the operational reserves needed for the system balance.

There also exists a *day-ahead* market (henceforth DA) in most deregulated electricity markets. In the Nordic area, the DA market is a non-mandatory market called Elspot and it is organized by Nord Pool. The UK Power Exchange (UKPX), Powernext and EEX are also examples of non-mandatory DA markets, contrary to, for instance, Omel in Spain. On Elspot, hourly power contracts are traded daily for physical delivery in the next day's 24-hour period (midnight to midnight). On Nord Pool's spot

[4]See Part 3 in the book by [Stoft (2002)] for a detailed description of alternative market architectures for both day-ahead and real time-markets.

market, Norwegian, Swedish, Finnish and Danish players trade in hourly contracts for each of the 24 hours of the coming day. Each morning, the players submit their bids for purchasing or selling a certain volume of electricity for the different hours of the following day. Once the spot market is closed for bids, at noon each day, the DA price is derived for each hour next day. The DA price is called the *system price*, and is common to all Nordic countries. In case of congestion due to capacity constraints, the Nordic market is divided into different bidding areas, resulting in area (or zonal) prices. Each contract is assigned a specific load for a given future delivery period. This means that, strictly speaking, the DA market is trading in electricity forward contracts with delivery over a specified hour the next day. Figure 1.1 shows a time series of weekly averages of the system price in the Nordic market.

There also exists a market that somewhat closes the gap between the DA and RT markets. This is called the Elbas market. The time span between the Elspot price fixing round and the actual delivery hour of the concluded contracts is quite long (36 hours at the most). The Elbas market supplements the Elspot and the national Nordic RT markets, as it allows a market player to adjust the market exposure between the DA auction on Elspot at noon and the actual hour of delivery the following day. Elbas is run by Nord Pool Finland Oy (formerly EL-EX Electricity Exchange Ltd.). It has been in operation since 1999. This market provides continuous power trading 24 hours a day covering individual hours in the same way as the Elspot market (one hour delivery period with 1 MW load). However, the contracts are open for trade only after the Elspot auction, so 24 new contracts are introduced daily. Just like Elspot, Elbas is a physical market for power trading in hourly contracts. The products can be traded up to one hour prior to delivery (two hours prior to delivery in Eastern Denmark). It only covers the trading areas of Finland, Sweden and Eastern Denmark, and the amount of power traded is limited by the free cross border transmission capacity. Unlike the Elspot market, Elbas allows participants to buy and sell the same physical contract several times before delivery, and the position can be closed prior to delivery, so that no actual physical delivery is necessary.

1.1.2 *Financial electricity contracts*

Specifications and rules of trading for financial electricity contracts vary among the different power exchanges. The fact that these contracts are

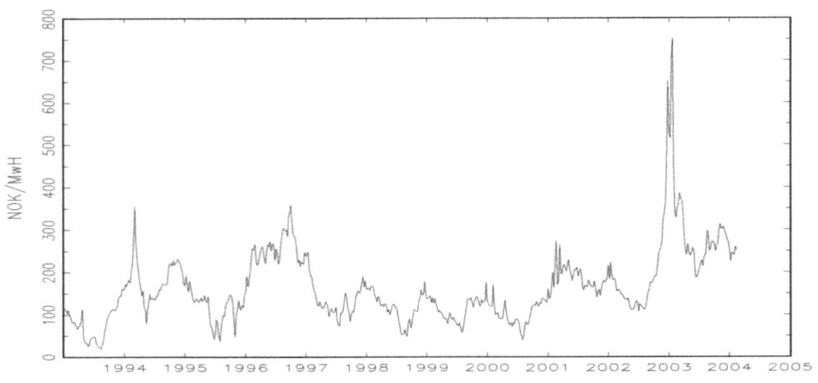

Fig. 1.1 Time series of spot prices from Nord Pool in the period 1993-2004 (weekly average of DA prices).

settled financially against a reference price, implies that the market place for financial electricity contracts does not require central coordination. They can be considered as side bets on the physical system. We will describe contracts traded on Nord Pool only.

Although contracts for future delivery of power are called futures or forwards, this denomination may be misleading. The basic exchange traded contracts at Nord Pool are written on the (weighted) average of the (hourly) system price over a specified *delivery period*. During the delivery period the contract is settled in cash against the system price, hence, financial electricity contracts are in fact swap contracts, exchanging a floating spot price against a fixed price. However, to be in line with the market jargong, we call these contracts *electricity futures* in this book.[5] When we refer to the spot price in our mathematical modelling, this should be interpreted as whatever reference price which a given exchange has tied its financial contracts to. The specified reference price is typically the DA price described in the previous section. In this way the financial electricity contracts are not the relevant risk management vehicles for hedging RT electricity price risk. Contracts on Nord Pool are not traded during the delivery period, and market participants typically close their position prior to the delivery period. We shall consider only electricity futures dynamics in the trading period in our mathematical models. The trading period is the time period the contract is available for trading. The term "time to maturity" used for

[5]We will use the term swap, but then as a common reference to electricity futures and gas futures.

fixed maturity forward contracts is replaced by *time to delivery*.

Nord Pool has facilitated trading in financial electricity contracts since 1995. Since the contracts are settled against hourly DA prices (the Nord Pool system price), the underlying amount of electrical energy is determined by

$$\mathrm{DP} \times 24\,\mathrm{MWh}\,,$$

with DP being the "delivery period" measured in days. These are base load contracts. To be able to compare contracts with different delivery periods, prices are listed in Euros (EUR) for 1 MWh of power delivered as a constant flow during the delivery period.

Since the start in 1995, contract specifications have changed several times. Peak load contracts were available the first couple of years, but were taken from the market due to low liquidity. However, in the summer of 2007 they were reintroduced. There has also been a change of delivery periods for monthly and seasonal contracts. Block contracts with delivery periods of exactly four weeks, have been replaced by monthly contracts with delivery period equal to the respective calendar month. Quarterly contracts have replaced the former three-season regime. The quarterly contracts were first introduced for the year 2005.

In the first trading day in January each year, four new quarterly contracts (Q1, Q2, Q3 and Q4) are listed. The new Q1 contract trades for two years, the new Q2 contract trades for two years and three months, etc. A new yearly contract that trades for three years is also introduced. Thus, it is possible to hedge the exposure to electricty prices in the Nordic market three to four years into the future at all times. Every month a monthly contract is unlisted, and a new one is introduced that trades for six months. Each week one weekly contract is unlisted, and a new one is introduced that trades for eight weeks. New daily contracts are introduced every Thursday. The Friday contract only trades for one day. All contracts trade until the last trading day prior to the delivery period.

The contracts differ when it comes to how settlement is carried out during the trading period. Daily and weekly contracts are futures contracts. The value of such a contract is calculated daily, reflecting changes in the contract's market price. These changes are settled on a margin account for each participant. The electricity futures with monthly, quarterly and yearly delivery are forward-style contracts.

Nord Pool's financial market also includes option contracts and Contracts for Differences (CfD). Call and put options are written on the electricity futures contracts, and they are of European type. Exercise day is set as the third Thursday in the month before the delivery period of the underlying contract starts. The options are traded on quarterly and yearly contracts. The activity on the option market on Nord Pool is, at the time of writing this book, rather low. Hence, it is not easy to derive implied volatilities from this market. The EEX is also offering options written on electricity futures. Asian options written directly on the system price are frequently traded in the bilateral over-the-counter (OTC) market. As mentioned above, area prices may differ from the system price in case of congestion. CfDs are defined as the area price minus the system price. The different tradeable area prices are Oslo, Stockholm, Copenhagen, Aarhus, Helsinki and northern Germany. The CfDs are defined similar to the electricity futures contracts with identical delivery periods. However, delivery periods shorter than one month do not trade. Using CfDs in combination with electricity futures allows a market participant to effectively remove the price risk associated with congestion.[6]

1.2 The gas market

Natural gas is an important fuel for heating and when generating electricity. For instance, in 2002 one-third of the electricity production in the UK came from gas fired power plants, with a prospect of 60% by 2020 (see [Geman (2005)]). The figure for US is that 14% of gas demand comes from electricity generation. The gas markets, foremost in UK/Europe and the US, have been liberalized over the years, with some structural differences and similarities with the electricity markets. In this section we give a brief overview of the specifics of the gas markets actively traded in the US and UK.

The gas markets are located around different hubs, which are connection and arrival points for gas transportation systems and where there are infrastructure capabilities like, for instance, storage and a concentration of buyers and sellers. Two important hubs are Henry Hub located in Louisiana (US) at the Mexico Gulf and the National Balancing Point (NBP) in the UK. The latter is a *notional* hub without any physical location, where all UK gas flows through. The market for short-term delivery of gas is usually

[6]See http://www.nordpool.no/nordpool/financial/index.html for details on the different financial contracts traded on Nord Pool. The interested reader is recommended to read [Kristiansen (2004)] for more on the pricing of CfDs.

referred to as the *spot market*, and the trading is mostly OTC. Futures contracts ensure the delivery of gas over longer time periods like weeks, months, quarters, or even years, and the settlement of these resembles closely electricity futures. Although the largest portion of the trade in futures takes place in the OTC market, some exchanges also offer futures with physical delivery of gas through a hub. We shall refer to futures contracts in the gas market as *gas futures*, following the terminology used in the industry.

Gas prices, very much like electricity prices, exhibit sudden spikes during periods of high demand or shortage of production (or low storage), as can be seen in Fig. 1.2 presenting gas spot prices at the NBP. This gives

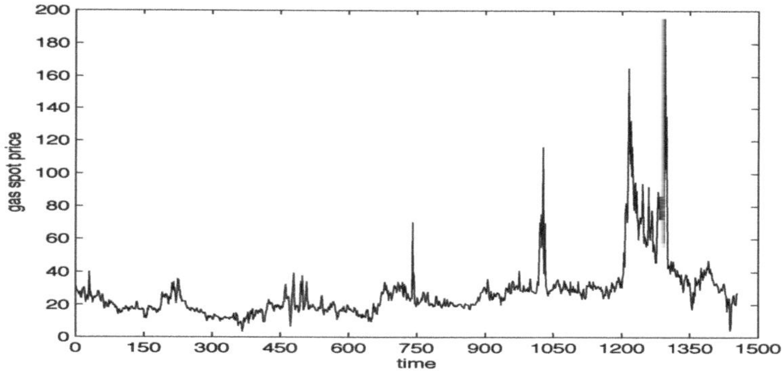

Fig. 1.2 Gas spot prices (Heren DA index) at the NBP for the period 6 February 2001 till 24 October 2006.

rise to a price dynamics having a higher volatility than what is normally observed in other commodity markets (like oil, say). Furthermore, especially in the UK market, the gas prices are seasonal since demand is very much dependent on temperature. Due to capacity constraints, one may even observe negative prices of gas from time to time (however, naturally rather infrequent). On the other hand, gas can be stored, which makes it possible to use for hedging. This links the analysis of the gas markets closer to more "classical" commodities like oil.

In the market place, the energy content of gas is measured in units of 'therms' or 'British thermal units' (Btu). By definition there are 100,000 Btu in 1 therm, whereas 1 therm is the equivalent of 105.5 MJ. Since there are 3.6 GJ per MWh, we have the relation

$$1\text{therm} = 105.5\text{MJ} \cdot \frac{\text{MWh}}{3.6 \cdot 1000\text{MJ}} = 0.029306\text{MWh}\,.$$

In the US, gas transactions are denominated in Dollars per MBtu, while in the UK pence per therm is used as the unit.

1.2.1 *Futures and options on gas*

Although most of the trading of futures on gas takes place bilaterally, there exist organized markets as well. For instance, New York Mercantile Exchange (NYMEX) offers trading in standardized futures contracts with physical delivery of gas at Henry Hub over a specified month. Altogether 72 contracts are offered for trade at all times, covering the nearest consecutive months. The participants of this market have thus access to futures covering six years ahead. In addition, European options are written on the gas futures contracts.

At the Intercontinental Exchange (ICE), structured UK natural gas futures are traded. The contracts deliver gas at a fixed rate through the NBP over a specified period of time. The delivery periods are the first 10-12 consecutive months, 11-12 quarters and six seasons. There are two seasons, a summer season ranging from April to September, and a winter season from October to March. It is worth noticing the similarities with the electricity futures traded at Nord Pool, where the market is also separated into contracts with longer and shorter delivery periods, with only long delivery contracts in the long end of the curve.

There is no organized market for gas spot trading, in the sense of standardized spot contracts traded on an open exchange leading to publicly available prices. This raises the question how a gas futures contract can be benchmarked against the spot market. The lack of an objective reference price for the futures is resolved by objective indexes provided for the market. For the ICE gas futures, the Heren NBP DA index is used as a measurement of the spot price of gas at the NBP. This index is provided by Heren Energy,[7] and is published daily in the European Spot Gas Markets report.[8] The index for a specific day is the volume-weighted average of transaction prices for gas to be delivered at NBP the next day. This index constitutes the counterpart to the system price at Nord Pool, except that

[7] www.heren.com
[8] www.energypublishing.com

the latter is set for each hour the next day, and is a real trading price. The index gives, of course, just the average level of all relevant trades agreed on for the next day in the short-term market. There also exist similar indices for Henry Hub, which are used to settle NYMEX gas futures.

An important class of derivatives contracts is the so-called spark spread options. These are call and put options written on the difference between electricity and gas prices. A typical example may be a European put option on the futures price spread of the two energies, where the futures have a specified monthly or quarterly delivery period. Such options may be used for risk management of a gas fired power plant, giving the plant owner a possibility to hedge undesirable moves in the electricity and gas prices. These options are rather popular, and traded to a siginificant extent in the OTC market. Spark spreads may also be used for real option valuation of gas fired power plants.

1.3 The temperature market

In recent years the trade in contracts written on weather has emerged and become a new and interesting asset class for risk management. There are obviously close connections between energy and weather, like, for instance, an increase in power consumption during cold and warm periods. This means that both consumers and producers of energy may be interested in financial contracts that can be used to manage weather risk. Markets for weather derivatives are thus of importance for the energy industry. A thorough account on weather derivatives markets and valuation is given in [Geman (1999)] and [Jewson and Brix (2005)]. In this Section we shall concentrate on the market for temperature derivatives found at the Chicago Mercantile Exchange (CME).

From a modelling point of view, there is a close resemblance between weather and energy markets. The markets are incomplete, in the sense that hedging using the underlying is impossible. Further, there are clear evidences of mean reversion. In the energy markets this is due to the balance between demand and supply, while for temperature it may be explained by conservation of energy. Further, the typical temperature futures contracts are written on some temperature index measured over a period of time, which one may think of as a "delivery period". In this respect, the temperature futures have "delivery" of the underlying "asset" over a period rather than at a fixed point in time. To be in line with the industry terminology, we shall refer to such contracts as *temperature futures.*

At the time of writing this book, the market for weather derivatives is steadily increasing. The OTC market offers a wide range of different weather deals, while the volumes for temperature futures and options at the CME have experienced constant growth. New weather contracts like futures and options written on the amount of snowfall in New York and the frost days in Amsterdam, have emerged, and it is expected that even more weather related derivatives will be introduced at the exchange in the future.

In our discussion we shall focus on the temperature based products actively traded at the CME. CME organizes trade in futures contracts based on four different temperature indices. These indices measure the aggregation of daily mean temperature or its deviation from a fixed level, in 18 US, two Japanese and nine European cities. In addition, there is a Frost Day index based on the temperature in Amsterdam. We refer the reader to *www.cme.com* for a detailed description of all these temperature products, which we are going to discuss.[9]

For the US cities, the contracts are written on the aggregated amount of *heating-degree days* (HDD) and *cooling-degree days* (CDD). The amount of CDD on a particular day is defined as the difference between mean temperature and 65°F (18°C), whenever this is positive. In mathematical terms, the CDD on day t is defined to be

$$\mathrm{CDD}(t) = \max\left(T(t) - c, 0\right) ,$$

where $T(t)$ is the mean temperature on day t. The mean temperature is interpreted as the average of the maximum and minimum temperature over the 24 hours of the day in consideration. The contracts are written on the accumulated amount of CDD over a month or a season.[10] The constant c denotes the threshold 65°F (or 18°C). Since most air conditioners are switched on when temperatures are above c, the CDD gives a measurement of how much air conditioning it is required, and therefore is linked to the demand for power. The temperature futures contracts written on the CDD index is for the warmer half of the year, ranging from April to October. The CDD futures are settled financially in terms of \$20 per unit for the US cities. If the contract is specified as the accumulated CDD over a period

[9]We are not going to consider the snowfall contracts for New York, since these are not directly temperature linked.

[10]A season being two to seven months long.

$[\tau_1, \tau_2]$, the amount of money to be paid to the buyer of the contract is

$$\sum_{t=\tau_1}^{\tau_2} \mathrm{CDD}(t) \times \$20\,.$$

The HDD index measures the amount of degrees below the threshold c, which is an index of how much heating it is required. It is defined as

$$\mathrm{HDD}(t) = \max\left(c - T(t), 0\right)\,,$$

and the futures contracts are written on accumulated HDD over a month or a season in the colder part of the year, lasting from October to April.

For the European cities, there is trade in HDD contracts in the winter season, while in the summer season the underlying temperature index is the so-called *cumulative average temperature* (CAT). The CAT over a period $[\tau_1, \tau_2]$ is defined as

$$\sum_{t=\tau_1}^{\tau_2} T(t)\,.$$

The currency is British pounds for the European futures contracts, which are also settled in units of 20. Finally, the Japanese futures are settled against the index called *Pacific Rim* (PRIM), being the average temperature over a period

$$\frac{1}{\tau_2 - \tau_1 + 1} \sum_{t=\tau_1}^{\tau_2} T(t)\,.$$

The currency is Japanese yen and the settlement is in units of 250,000. The contracts are listed for all the 12 months of the year.

A different class of futures and options traded at the CME is based on an index measuring the days where there is a danger of icy conditions on the runway of Schiphol airport in Amsterdam. The Frost Day index is defined as the accumulated number of days in a month or season when there is frost conditions observed at Schiphol airport. Each day counting as a frost day in the measurement period gives a contribution 1 to the index. If the temperature during a day satisfies one or more of the following three conditions, it is defined as a frost day:

(1) The temperature at 7 a.m. is less than or equal to $-3.5°$C,
(2) The temperature at 10 a.m. is less than or equal to $-1.5°$C

(3) The temperatures at 7 a.m. and 10 a.m. are less than or equal to $-0.5°$C.

At the CME, futures are listed on monthly November to March Frost Day indices. The seasonal Frost Day index is listed for the whole period of November to March. The trade unit is EUR 10,000 times the index. Note that the Frost Day index over a measurement period becomes a natural number including zero. The upper limit is the number of measurement days for the index in question. A frost day on day t is mathematically defined as

$$\begin{aligned}\mathrm{FD}(t) = 1\Big(&\{T(t+7/24) \leq -3.5\} \cup \{T(t+10/24) \leq -1.5\} \\ &\cup \{\{T(t+7/24) \leq -0.5\} \cap \{T(t+10/24) \leq -0.5\}\}\Big)\,.\end{aligned}$$

Here $1(\cdot)$ is the indicator function. Moreover, we use the convention that the time t is measured in days, implying that $t + 7/24$ is at 7 a.m., and $t + 10/24$ is at 10 a.m. of the day in question. The Frost Day index over a measurement period $[\tau_1, \tau_2]$ is then defined as

$$\mathrm{EUR}10,000 \times \sum_{t=\tau_1}^{\tau_2} \mathrm{FD}(t)\,.$$

In the market place, only weekdays are taken into account when finding the Frost Day index. Thus, in the summation above, we should disregard weekends. To avoid the introduction of a new (and more messy) notation we consider here a summation over all the days in the measurement period.

The temperature market at CME also includes options written on the different futures. The options are plain vanilla European call and put options, with a temperature futures as the underlying asset.

1.4 Other related energy markets

Natural gas and temperature are not the only two markets related to electricity, there are others not discussed in detail in this book. Different oil products are certainly important fuels for electricity generation. However, since storage is easier, and since the forward market for oil has been studied quite extensively over the past decades (see, for instance, [Schwartz (1997)]), we do not discuss the oil market in this book. We only briefly comment on the market for coal.

The economic link between coal and electricity is strong, as coal fired plants represent an important part of total power supply in many electricity markets. The market for coal has historically been a physical market, with big players on both the supply and the demand side. The contractual agreements have typically been long-term and bilateral. In recent years there has been increasing London-based OTC trading activity in forward type contracts. The forward contracts bear close resemblence with the average-based contracts which we describe in this book. Heren Energy collects fair prices from different market players for coal delivered at certain landing points around the world. This is done each week. The OTC traded forward contracts are settled financially on average price fixings during the settlement month. Exchange traded coal contracts have been around for a while, but so far this market has yet to see a big trading activity. The link to the electricity market has made the EEX to launch exchange traded coal futures.

A market with a more indirect link to the electricity market is the growing financial shipping market. The development of the freight derivatives market has spanned more than two decades, starting with the Baltic International Freight Futures Exchange (BIFFEX) market in 1985 and, since 1992, gradually developing into an active OTC Forward Freight Agreement (FFA) market. Towards the end of 2001, the development came full circle, with the emergence of electronic trading of route-specific cleared tanker derivatives on the Oslo-based IMAREX platform. Since 2005, large global clearing houses such as London Clearing House and NYMEX have also offered clearing of traditional voice-brokered FFA contracts. However, both contracts are cash settled against the average spot freight rate for a particular route, as published daily by the Baltic Exchange, over a specified future period of time and for a specified quantity of cargo and vessel type.

From a modelling point of view, the contracts are identical to the electricity and gas futures contracts studied in detail in this book. Research in the area of freight derivatives has been primarily concerned with various aspects of market efficiency, first in the BIFFEX futures market and later in the OTC FFA market. The research topics include the applicability of the unbiasedness hypothesis, hedging effectiveness for ship operators, and the interaction between the spot and forward markets. [Aadland and Koekebakker (2007)] and [Koekebakker, Aadland and Sødal (2007)] are early attempts to study freight rate dynamics in a continuous time model similiar to the modelling framework advocated in this book. The dry bulk shipping market has been very volatile in recent years, and freight rates have peaked. In some periods freight has been a significant cost factor for coal fired power plants with short-term contracts. This linkage between shipping and electricity markets may become even more important in the future.

A market segment that has increased in size over the years is the freight market for natural gas. Gas can be liquified and compressed through a cooling process (liquified natural gas is known as LNG). LNG carriers are very expensive, but the new building activity has been high for this vessel type in recent years due to the increase in gas fired power plants around the world. As storage facilities for natural gas are both expensive and require a certain infrastructure, storage capacity is scarce. Recently it has been speculated that LNG ships are hired also as floating storage devices to exploit 'LNG arbitrage' in natural gas pipeline systems. If these trends escalate in the future, it will make sense to include certain shipping markets as integral parts of the electricity market.

Finally, we comment on the emerging CO2 emission market. In 1997, many governments adopted the Kyoto protocol accepting mandatory constraints on reduction of greenhouse gases emission. The Kyoto protocol contains different mechanisms to reduce emissions: International Emission Trading, Joint Implementation, and Clean Development Mechanism. Carbon dioxide is by far the most important greenhouse gas, the runner up is methane. The protocol promotes permit trading as the champion mechanism to reduce CO2 emissions.

There is a close economic link between prices on CO2 emissions and electricity. Increased cost of pollution increases costs for many power producers. Producers can either buy enough emissions and keep on polluting, or they can invest in cleaning technology (or both). From a modelling point of view, the issue of non-storability is a shared characteristic with the electricity market. The basic 'commodity' is overall CO2 emissions. Since it is not possible to trade physical emissions, spot certificates have been introduced as tradeable assets. According to [Daskalakis *et al.* (2006)] there are four active emission allowance markets: the European Union Emission Trading Scheme (EU ETS), the UK Emission Trading System (UK ETS), the New South Wales GHG Abatement Scheme and the voluntary Chicago Climate Exchange (CCX). The EU ETS is dominant with a share of approximately 97% of the total transactions during the first three quarters of 2006 ([Daskalakis *et al.* (2006)]).

Financial research in this market is still scarce, and the research conducted has to our knowledge been focusing exclusively on EU ETS. An early discussion of emission trading and reports from an expert survey is given in [Uhrig-Homburg and Wagner (2006)]. The EU ETS is a joint effort by EU member states to efficiently reach their Kyoto goals. The EU ETS breaks down the emissions trading to the company level. Companies in industries with big emission needs, are allocated a certain amount of EU Allowances (EUAs). One EUA gives the holder the right to emit one tonne of CO2. If a company does not use all its allocated permits, due, for instance, to new environmental friendly technology, the surplus EUAs can be sold. Other companies, where new technology is more expensive or perhaps does not exist, can buy additional permits if needed. Spot EUAs are sold at Powernext, EEX, Nordpool, Energy Exchange Austria (EXAA), European Climate and Exchange (ECX) and Climex (see [Uhrig-Homburg and Wagner (2007)]). Some research has examined spot price dynamics of EUAs, see [Benz and Trück (2006)] and [Paolella and Taschini (2006)] for econometric studies. In the paper [Fehr and Hinz (2006)], the authors build a microeconomic equilibrium model for price formation of carbon emission rights.

Forward contracts on EUAs are also traded. The EUAs can only be used within a particular trading period. The first trading period is 2005 – 2007, the second trading period coincides with the Kyoto commitment period spanning from 2008 to 2012. From a modelling point of view, we observe the interesting feature that the traditional cost-of-carry relationship should hold for forwards that mature within a trading period. But since there exists no EUAs for the second trading period, there can be no spot-forward relationship (see [Uhrig-Homburg and Wagner (2007)], [Borak *et al.* (2006)] and [Daskalakis *et al.* (2006)] for discussions and empirical results). The trading period system suggests that the price dynamics of EUAs changes over time depending in particular on the total emissions. If towards the end of a trading period, cumulative emissions in the period are high, then EUAs would be in high demand, supplies of EUAs are scarce, and we would expect very high prices. In the case of low cumulative emissions, we would expect the opposite, and the prices of EUAs would plummet. [Seifert, Uhrig-Homburg and Wagner (2006)] propose an equilibrium model consistent with such predictions.

1.5 Stochastic modelling of energy markets

The energy related markets consist in general of three different segments, a market for physical spot trading, futures contracts on the spot with either physical or financial settlement over a period, and an option market with the futures contracts as underlying. The exception is the market for temperature, where there is obviously no trading in the "spot". Thus, modelling of the energy markets can be separated into three tasks: spot price modelling, derivation or modelling of futures, and pricing of options. In this section we discuss the different modelling issues, to establish a common foundation for the theoretical and empirical analyses which will be our focus in the subsequent chapters. The discussion here will be kept at an informal level, to leave space for fixing the ideas and highlight the approaches we are going to use.

We emphasise that in this book we refer to swaps being futures contracts with delivery over a period. This will be used as a general reference including electricity and gas futures. Wherever it is natural, we use the terminology "electricity futures" and "gas futures" instead of the general notion "swaps". Temperature futures will be discussed separately. By forwards we understand exclusively contracts with a fixed delivery time. We will be consistent in this separation throughout the book.

A dynamics for the spot price evolution is desirable for several reasons. Models describing the uncertainty in the spot price is of interest for traders operating in these markets. However, they are also used as the reference point for settlement of forward and futures contracts, and thus is a basic input in understanding the dynamics of these derivatives. The spot dynamics will be based on Ornstein-Uhlenbeck (OU) processes, which model mean reversion in a natural way. The stochastic driver may allow for jumps, where we can explain spikes in electricity prices, for example. Finally, it is paramount to allow for seasonal variations, since the demand after electricity and gas vary with temperatures, which are highly dependent on season. In the literature, one usually connects stationarity properties to OU processes. When we include seasonlity (in, for instance, the jump occurrence and size), the traditional notion of stationarity breaks down. From this point of view, the terminology "OU process" may not be natural. However, we keep the name "OU process" in this book, which seems to be the standard use. We interpret it as a dynamic model with certain mean reversion properties.

The next modelling point is to establish the connection between the spot and futures/forward price dynamics. In markets like oil, say, there exists an extensive theory for the spot-forward relation, including storage costs and convenience yields. For electricity and temperature, and perhaps gas, the connection is not at all clear since the underlying spot is not storable. One may explain the relation through a market price of risk, which essentially is the specification of a risk-neutral probability. The existence of a delivery period for the electricity and gas futures (swaps) puts restrictions on the class of spot models feasible for analytical pricing.

Alternatively to explaining the forward and swap prices by the underlying spot, one may adopt the Heath-Jarrow-Morton (HJM) approach from interest rate theory (see [Heath, Jarrow and Morton (1992)]). Rather than trying to establish a spot-futures/forward relation via the specification of a risk-neutral probability, the HJM approach suggests to directly assume a dynamics for the forward and swap price evolution. This can be done in terms of market dynamics, or under the risk-neutral measure. Modelling the swap price dynamics, where the energy delivers over a period, creates challenges that are not present in the fixed income markets theory (see, for example, [Musiela and Rutkowski (1998)] for a discussion of the HJM approach in fixed income markets).

Having a forward and swap price dynamics, our final task is to look at how to price options. With a risk-neutral dynamics available for the forward and swap price, this entails in an exercise of calculating a conditional expectation of the pay-off from the option, which for many of our models can be done more or less explicitly by using Fourier techniques. The question of hedging will also be analysed, however, leaving out a significant part related to incomplete markets.

We discuss now these modelling aspects in more detail trying to explain our choice of models and approaches that we are going to consider in the following chapters.

1.5.1 *Spot price modelling*

In mathematical finance, the traditional models are based on stochastic processes driven by a Brownian motion $B(t)$, also called a Wiener process. The most frequently encountered model for the price dynamics $S(t)$ of a financial asset is the geometric Brownian motion (see [Samuelson (1965a)]), being the exponential of a drifted Brownian motion. It is defined as

$$S(t) = S(0) \exp\left(\mu t + \sigma B(t)\right) ,$$

with μ and $\sigma > 0$ being constants. Brownian motion is a process with independent and stationary increments, where the increments are normally distributed. This implies that the logarithmic returns (or logreturns, for short), defined as logarithmic price changes over a time interval Δt

$$\ln S(t + \Delta t) - \ln S(t) ,$$

will become independent and stationary, a reasonable property in view of the market efficiency hypothesis (see, for example, [Fama (1970)]). In addition, logreturns are normally distributed.

A natural and frequently used generalization of the geometric Brownian motion is the exponential of a Lévy process (see, for example, [Barndorff-Nielsen (1998)] and [Eberlein and Keller (1995)]),

$$S(t) = S(0) \exp(L(t)) .$$

Lévy processes $L(t)$ open for the possibility to model price jumps and leptokurtic behaviour of asset prices on small time scales. These processes have independent and stationary increments, with Brownian motion being a special case. With these models at hand, we may incorporate the possibility of large price variations, and even skewness in the price fluctuations. However, due to stationarity, the variation in prices is homogeneous over the year, and we cannot allow for more variable prices during winter than summer.

Energy markets, and in particular electricity markets, are seasonally varying markets. By appropriate modelling of the mean level of energy prices, one may remove much of the seasonal features observed in prices, however, there are still distinctive characteristics which call for models that may vary with time. In the electricity market we observe seasonality in the jump size and frequency. For instance, in the Nord Pool market spikes are most frequent in the winter period. Further, the temperature dynamics underlying weather derivatives turns out to have seasonal features like a time-dependent volatility. In other markets, like gas, we see similar seasonal variations in the dynamics, explained by demand being weather dependent.

The classical model for commodity markets is the Schwartz model (see [Schwartz (1997)]), which is an extension of the geometric Brownian motion allowing for mean reversion. In the simplest case, it may be defined as

$$S(t) = S(0) \exp(X(t)) , \tag{1.1}$$

where

$$dX(t) = \alpha(\mu - X(t))\,dt + \sigma\,dB(t)\,. \tag{1.2}$$

Extending to Lévy process innovations, we still preserve the homogeneity in jump size and frequency, and we will not be able to explain the observed seasonal features. A reasonably flexible class of models are provided by the *independent increment* (II) processes, which generalize Lévy processes in a way that the increments are independent, but not necessarily stationary. This opens up for multi-factor models of the Schwartz type which may have one or more factors with seasonally dependent jump frequencies and sizes in addition to mean reversion. In other words, the logarithmic spot price is represented as a sum of OU processes driven by II processes. In this way we may model the typical spikes observed for electricity spot prices by having an OU process with big, but rare, jumps coupled with a strong mean reversion. The II processes allow for a stochastic analysis which can be utilised for calculating derivatives prices. This class of models is a reasonable compromise between modelling flexibility and analytical tractability, and will be our stochastic driver in the spot price dynamics. Further, by substituting $S(0)$ in (1.1) by a deterministic funtion $\Lambda(t)$, we can model explicitly a seasonally varying mean level.

We may argue in favour of arithmetic models rather than geometric ones for the spot price evolution, that is, we may assume that the spot price dynamics is represented as a sum of OU processes directly. This makes analytical pricing of swap contracts feasible for a large class of models, a possibility not shared with most geometric models. In this book we shall introduce a class of arithmetic models where we ensure positivity of spot prices, using the specific choice of increasing II processes as stochastic drivers.

The question of estimating such models on data is not an easy one. For some simple one-factor models, this may be a straightforward task, as we shall demonstrate in many examples. However, if the jumps are seasonal we immediately face problems when trying to fit the stochastic model to spot data. For multi-factor models this may be an even more challenging problem, involving highly sophisticated estimation techniques. It is outside the scope of this book to give an exhaustive presentation and application of the different estimation approaches. We shall present many examples, where we can use simple estimation techniques. Emphasis is put on transparency and on showing the connection between data and model without having to implement advanced estimation procedures. Admittedly, to apply our models at full strength, this is not satisfactory. We will indicate possible estimation approaches along the way for the convenience of the reader.

The traditional models in mathematical finance belong to the class of semimartingale processes. The reason for this is the existence of so-called equivalent (local) martingale measures, being probability measures equivalent to the objective (or market) P probability, and such that the discounted price dynamics is a (local) martingale. Existence of such probabilities, which are often coined *risk-neutral probability measures*, leads to markets where there are no arbitrage possibilities, since the martingale property of the discounted prices makes it impossible to create portfolios with a sure win (we refer the reader to [Bjørk (1998)] and [Bingham and Kiesel (1998)] for excellent accounts on this theory). In markets like temperature or electricity the underlying spot (being either temperature itself, or the spot electricity), is not tradeable in the sense of being an asset that can be liquidly bought or sold, and kept in a portfolio over time. Hence, the spot is not accounted for as being a tradeable asset, and will not be a part of the definition when fixing a martingale measure. Therefore, any probability measure Q being equivalent to the objective probability P is also an equivalent martingale measure. This has the implication that we do not need to restrict the class of spot price models to be semimartingales. However, all our models will be semimartingales, since this is a convenient class of processes from an analytical point of view.

On the other hand, the swap and temperature futures markets are liquid, and these contracts have to be priced so that arbitrage opportunities do not exist. Thus, it is required that the dynamics of the forward and swap price possesses the semimartingale property in order to ensure the existence of risk-neutral probabilities. Connecting the spot dynamics with the forward and swap price leads to formulas representable in terms of conditional expectations of the spot dynamics. We will discuss this in more detail in the following subsection.

Let us elaborate on the spot price dynamics for the electricity market. As we discussed in Sect. 1.1, the spot markets of electricity quote prices on an hourly basis (or half-hourly in some markets). This means that strictly speaking, the spot price should be modelled as a time series. Thus, it will not make sense to talk about the spot price of electricity at any time t. On the other hand, we know that electricity futures are settled against the hourly spot prices, but are traded in a continuous market in the sense that the actors can buy or sell at any time as long as they find a counterpart in the market. Hence, contrary to most other commodity markets where there is liquid trading in both spot and futures/forwards, we face the situation of a discrete-time spot and a continuous-time futures market.

Let us introduce a continuous-time stochastic process $\widetilde{S}(t)$ being the *unobserved* instantaneous spot price of electricity, that is the price of electricity at time t with delivery in the interval $[t, t+dt)$. Associated to the process is a filtration $\mathcal{F}_t$ modelling the stream of information. We can think of the process $\widetilde{S}(t)$ as the price market participants know they would have to pay *if* they could buy electricity at time t with infinitesimal delivery time (that is, like a shock of electricity).

What we do *observe* in the market is the price of electricity with delivery over a specified hour. Let us say that the hour is the time interval $[t_i^d, t_{i+1}^d)$, with $i = 0, 1, \ldots, 23$ denoting the hour and d the day. Thus, t_i^d is the start of hour i on day d. Entering a spot contract will then give us the following expenses

$$\int_{t_i^d}^{t_{i+1}^d} \widetilde{S}(u)\,du\,, \tag{1.3}$$

if we would know the instantaneous spot price. The hourly spot price in the market is set before the delivery takes place. Thus, a natural assumption is that the hourly spot price is the best prediciton of (1.3), *given* the information up to start of delivery. Hence, the hourly spot price should be

$$S_i^d = \mathbb{E}\left[\int_{t_i^d}^{t_{i+1}^d} \widetilde{S}(u)\,du \,|\, \mathcal{F}_{t_i^d}\right]. \tag{1.4}$$

This definition makes the time series S_i^d measurable with respect to $\mathcal{F}_{t_i^d}$, meaning that the hourly spot price contains all market information up to the start of delivery, but not into the delivery period.

Let us approximate the integral inside the conditional expectation in (1.4) with

$$\int_{t_i^d}^{t_{i+1}^d} \widetilde{S}(u)\,du \approx \widetilde{S}(t_i^d)\,,$$

using the convention that time is measured in hours and thus $t_{i+1}^d - t_i^d = 1$. From the measurability of $\widetilde{S}(t_i^d)$ it follows that

$$S_i^d \approx \mathbb{E}\left[\widetilde{S}(t_i^d) \,|\, \mathcal{F}_{t_i^d}\right] = \widetilde{S}(t_i^d)\,.$$

This argues in favour of defining a spot price process in the market as

$$S(t) = \widetilde{S}(t)\,, \tag{1.5}$$

where we observe the spot price at time moments t_i^d, that is, $S(t_i^d)$ are the observations of an underlying continuous-time spot price process of electricity. This is the assumption usually made (implicitly) in the literature when modelling a spot price of electricity. Note that this connection makes it possible to estimate the parameters of the unobserved process $\widetilde{S}$ directly.

The above arguments for linking the unobserved process $\widetilde{S}$ to the electricity spot price may be questioned from several different angles. The spot price is determined in the market from bids in an auction, which results in prices for all hours the next day. It is not clear how the available information is taken into account in this price determination. It is therefore not simple to understand the connections linking the assumed continuous-time process, the filtration $\mathcal{F}_t$ and the electricity spot price, if such a modelling approach is at all valid. To keep matters simple, we assume a continuous-time stochastic process for the electricity spot price, and relate it to the observed spot via (1.5). As we discuss in the next subsection, this setup will also lead to natural connections between the spot and electricity futures price. Note that gas can in principle be purchased OTC at any time (continuously), thus we may assume that the gas spot price is a continuous-time process. Temperature may obviously be viewed as a continuous-time process as well.

1.5.2 *Forward and swap pricing in electricity and related markets*

The key driving factor for the swap price is the underlying spot. The relation between spot and swap prices are of crucial importance to the players in the energy market, and one of the central topics in this book. In the current subsection we aim at illustrating some of the ideas and problems encountered when deriving swap prices for the energy markets.

Suppose $S(t)$ is a stochastic process[11] defining the price dynamics of the spot, and $r > 0$ is the constant risk-free interest rate. For simplicity, let us discuss forward contracts first. Assume that we have entered a forward contract delivering the spot at time τ. Denoting $f(t, \tau)$ the forward price

[11]Since we are not going to give a rigorous treatment of the forward-spot relation, we do not go into details on the exact dynamics of the spot and the hypotheses required. We leave the mathematical details for later chapters.

at the time $t \leq \tau$ of entry of the contract, the payoff from the position is

$$S(\tau) - f(t, \tau)$$

at delivery time τ. From the theory of mathematical finance (see, for example, [Duffie (1992)]), we know that the value of any derivative is given as the present expected value of its payoff, where the expectation is taken with respect to a risk-neutral probability Q. Hence, since a forward contract is entered at no cost,

$$e^{-r(\tau-t)}\mathbb{E}_Q\left[S(\tau) - f(t, \tau) \mid \mathcal{F}_t\right] = 0 .$$

Here, $\mathcal{F}_t$ is the filtration containing all market information up to time t, and $\mathbb{E}_Q$ is the expectation operator with respect to the risk-neutral measure.

The forward price is set at time t, and therefore cannot include any more information about the market than given by $\mathcal{F}_t$, which therefore implies that it must be adapted to this filtration. Thus, we obtain the following formula for the spot-forward relationship

$$f(t, \tau) = \mathbb{E}_Q\left[S(\tau) \mid \mathcal{F}_t\right] . \tag{1.6}$$

This definition yields an arbitrage-free dynamics of the forward price process $t \mapsto f(t, \tau)$, since this process is a martingale under Q. In effect, the relation (1.6) implies that the forward price is the best risk-neutral prediction at time t of the spot price $S(\tau)$ at delivery. In this book we exclusively consider the situation where the interest rate r is constant. We know then that forward and futures prices coincide. We will not make any distinction between the two, and reserve the terminology "forwards" for these contracts.

Suppose now that the spot can be liquidly traded in a market (like a stock, say). Then we can perfectly hedge a short position in the forward contract by a long position in the spot, financed by borrowing at the risk-free rate r. This hedging strategy is known as the *buy-and-hold strategy*, and uniquely defines the forward price. Since Q is a risk-neutral probability, it follows by definition that the discounted spot price $S(t)e^{-rt}$ is a martingale under Q, and thus we get

$$f(t, \tau) = S(t)e^{r(\tau-t)} . \tag{1.7}$$

This is the well-known connection between a forward contract and the underlying spot in a market where the two assets can be traded frictionless (a complete market). From (1.7) we easily see that the forward price converges to the underlying spot price when time to delivery $\tau - t$ approaches zero.

When running a buy-and-hold strategy in a commodity market, the commodity must be stored. Thus, the hedger will be incurred additional costs reflected in the forward price (1.7) as an increased interest rate to be paid. On the other hand, holding the commodity has a certain advantage over being long a forward contract due to the greater flexibility. For instance, the access to a gas storage facility means that one can sell gas when prices are high, and store if prices are low. Furthermore, if you run a gas fired power plant, you ensure production with such a storage facility. These opportunities are lost when holding a forward contract instead. The notion of *convenience yield* is introduced to explain this additional benefit accrued to the owner of the physical commodity. If it is assumed that the convenience yield comes at a constant rate, the hedging argument leading to (1.7) is modified exactly as if the spot would be a dividend paying stock. Hence, letting the convenience yield rate be c, and the storage costs be measured at a rate s, we get

$$f(t,\tau) = S(t)\mathrm{e}^{(r+s-c)(\tau-t)}\,. \tag{1.8}$$

Obviously, to measure the convenience yield is a more delicate task than the dividends paid from a stock. Note that the relation (1.8) can be derived from (1.6) via an appropriate choice of a risk-neutral measure for reasonable spot price models. Therefore, it may be more convenient and give more flexibility to start out with (1.6) as the definition of the forward price. Choosing Q will correspond, loosely speaking, to specifying the convenience yield. We will choose this approach for gas. We refer to [Geman (2005)] and [Eydeland and Wolyniec (2003)], and the references therein, for more details on the convenience yield and storage in commodity markets.

In the electricity market, these considerations break down since electricity is a non-storable commodity. For temperature, it does not make sense to talk about any trading in the underlying, which also makes the hedging arguments senseless. However, since the forward contracts need to have a price dynamics being arbitrage-free, we use (1.6) as a definition of the forward price, but now based on *any* equivalent probability measure Q. Recall that a risk-neutral probability turns all *tradeable* assets into martingales after discounting. Since both electricity spot and temperature are not tradeable in the usual sense, we are left with the bank account, which trivially becomes a martingale under any equivalent measure Q after discounting. In conclusion, we cannot pin down a unique forward price dynamics based on arbitrage arguments.

The *rational expectation hypothesis* in interest rate theory has also been considered in relation to forward prices in commodity markets. In this context, it says that the forward price is the best prediction of the spot price at delivery, or, in mathematical terms,

$$f(t,\tau) = \mathbb{E}\left[S(\tau) \,|\, \mathcal{F}_t\right] . \tag{1.9}$$

In view of (1.6), the rational expectation hypothesis means choosing $Q = P$ as the risk-neutral probability. In reality, it is not to be expected that the rational expectation hypothesis holds. The theory of normal backwardation argues that producers of a commodity will wish to hedge their revenues by selling forwards, and thereby willing to accept a discount on the expected spot price. Thus, in normal backwardation, we should have $f(t,\tau) < \mathbb{E}[S(\tau) \,|\, \mathcal{F}_t]$, saying that the hedgers are willing to pay a premium for getting rid of the spot price risk. The risk premium is defined as

$$\mathrm{RP}(t,\tau) \triangleq f(t,\tau) - \mathbb{E}\left[S(\tau) \,|\, \mathcal{F}_t\right] , \tag{1.10}$$

which is negative when the market is in normal backwardation. [Geman and Vasicek (2001)] find evidence of a positive risk premium in the Pennsylvania-New Jersey-Maryland (PJM) electricity market for contracts with a short time to maturity, and explain this by the market's aversion for the high volatility and thereby willingness to pay high prices to ensure delivery. For longer matured contracts, the sign of the risk premium changes in their study. [Longstaff and Wang (2004)] perform a non-parametric study of the PJM market, obtaining evidence of significant positive risk premium for the short-term contracts. Their study is extended by [Diko, Lawford and Limpens (2006)], who analyse risk premia in the three markets EEX, Powernext, and Dutch market APX. A term structure for the risk premium is found, which varies significantly from the short- to the long-term segment of the market. [Benth, Cartea and Kiesel (2006)] present a framework for explaining the sign of the risk premium in terms of the certainty equivalent principle and jumps in the spot price dynamics.

If the forward price is set under a risk-neutral probability Q as in (1.6), then the risk premium measures exactly the difference between the risk-neutral and the "market probability" predicitions. The choice of Q determines the risk premium, and opposite, having knowledge of the risk premium determines the choice of the risk-neutral probability. It is common to select a parametric class of risk-neutral probabilities, to explain the risk premium. These risk-neutral probabilities introduce a parametric change of the drift of the spot. To explain this further, suppose for simplicity that the spot is defined as a drifted Brownian motion

$$S(t) = \mu t + \sigma B(t)\,,$$

with $\sigma > 0$. Consider a change of measure given by the Girsanov transformation (see, for example, [Bjørk (1998)]). For a constant θ, there exists a probability Q equivalent to P such that

$$B^\theta(t) = B(t) - \theta t$$

is a Brownian motion under Q. Hence, we find

$$\begin{aligned}\mathrm{RP}(t,\tau) &= f(t,\tau) - \mathbb{E}[S(\tau)\,|\,\mathcal{F}_t] \\ &= (\mu+\sigma\theta)\tau + \sigma\mathbb{E}_Q[B^\theta(\tau)\,|\,\mathcal{F}_t] - \mu\tau - \sigma\mathbb{E}[B(\tau)\,|\,\mathcal{F}_t] \\ &= \sigma\theta\tau + \sigma B^\theta(t) - \sigma B(t) \\ &= \sigma\theta(\tau - t)\,.\end{aligned}$$

We see that the risk premium is positive if and only if θ is positive. It seems to be a common view that the risk premium is modelled as a change in the drift of the spot dynamics, or implictly, a Girsanov-type change of probability (see, for example, [Clewlow and Strickland (2000)]).

In this book we use the Esscher transform as the way to select risk-neutral probabilities. The Esscher transform is a parametric structure-preserving change of measure which generalizes the Girsanov transform for Brownian motion to a general II process. The drift of the spot dynamics will be changed using the Esscher transform, along with a change in the jump frequency and size, through possibly time-dependent parameters. These parameters are called the *market prices of risk*, and are closely linked to the risk premium. In normal backwardation, the risk premium is negative, equivalent to a negative market price of risk in the above context. However, for power commodities, the sign of the market price of risk may change depending on the time horizon in question. For instance, [Cartea and Williams (2006)] show that in the gas market, in the long-term the sign is positive, whereas in the short-term sign may change. [Weron (2005)] finds a changing sign of the market price of risk in the Nord Pool market, when considering Asian-style options, whereas [Cartea and Figueroa (2005)] argue for a negative market price of risk in the UK electricity market.

A complication in the electricity, gas and temperature markets is the fact that we have swap contracts traded, and not forwards that deliver the underlying energy at a fixed maturity time. Gas and electricity are flow commodities, in the sense that being long a swap ensures receiving a flow of the commodity over a specified time period. Similarly, temperature futures are contracts written on different temperature indices measured over specified periods like months or quarters of a year. To be able to calculate expressions for swap prices, we must constrain the class of models seriously if we want to avoid simulation-based pricing. As we shall see in a moment, the swap price is expressed through the average (or a weighted average) of the spot price over the delivery period. For exponential models like geometric Brownian motion or the Schwartz model, this may be difficult to calculate analytically. Arithmetic models may resolve this problem, and we shall discuss a class of such models for which the spot price is restricted to be positive. Interestingly, we cannot any longer expect to have the convergence of swap prices to the spot price when time to delivery approaches zero.

We discuss the pricing of electricity futures in more detail. As discussed in the previous subsection, the electricity spot price is strictly speaking *not* a continuous-time process. The reference price for the electricity futures contracts is given as the hourly price for electricity in the spot market, and therefore a time series model for the spot price dynamics should be used in determining the electricity futures price. Let us explore the consequences of this view by starting with the spot-forward relation in (1.6).

Consider an electricity futures contract with finanical delivery over a time interval $[\tau_1, \tau_2]$. The payoff from being long such a contract entered at time t is

$$\sum_{t_i=\tau_1}^{\tau_2-1} S(t_i) - (\tau_2 - \tau_1)F(t, \tau_1, \tau_2),$$

where the electricity futures price is denoted $F(t, \tau_1, \tau_2)$ at $t \leq \tau_1$. In the electricity market, the futures price is customarily denominated in terms of currency per MWh, which means that the total amount paid is $F(t, \tau_1, \tau_2)$ times the length of the delivery period. The hourly prices between τ_1 and τ_2, with $\tau_2 - 1$ being the last hour before end of the delivery period, are denoted by $S(t_i)$. If we suppose that the electricity futures is settled at the end of the delivery period, the price will be defined through the relation

$$e^{-r(\tau_2-t)}\mathbb{E}_Q\left[\sum_{t_i=\tau_1}^{\tau_2-1} S(t_i)-(\tau_2-\tau_1)F(t,\tau_1,\tau_2)\,|\,\mathcal{F}_t\right]=0\,,$$

yielding

$$F(t,\tau_1,\tau_2)=\mathbb{E}_Q\left[\frac{1}{\tau_2-\tau_1}\sum_{t_i=\tau_1}^{\tau_2-1} S(t_i)\,|\,\mathcal{F}_t\right]\,. \tag{1.11}$$

If we consider an hourly spot price model as in (1.4), we need to calculate

$$F(t,\tau_1,\tau_2)=\mathbb{E}_Q\left[\frac{1}{\tau_2-\tau_1}\sum_{t_i=\tau_1}^{\tau_2-1}\mathbb{E}\left[\int_{t_i}^{t_{i+1}}\widetilde{S}(u)\,du\,|\,\mathcal{F}_{t_i}\right]\,|\,\mathcal{F}_t\right]\,,$$

or, by appealing to $\mathcal{F}_t\subset\mathcal{F}_{t_i}$,

$$F(t,\tau_1,\tau_2)=\mathbb{E}_Q\left[\frac{1}{\tau_2-\tau_1}\int_{\tau_1}^{\tau_2}\widetilde{S}(u)\,du\,|\,\mathcal{F}_t\right]\,. \tag{1.12}$$

This integral formulation of the electricity futures price will be used throughout the book. Note that by interchanging the expectation and integration in (1.12), the electricity futures price can be viewed as an average of forward prices maturing over the delivery period.

Let us discuss briefly the consequences of modelling the hourly electricity spot prices directly as a time series without an underlying continuous-time process. Suppose that $S(t_i)$ is a time series defined at the hourly time moments t_i, $i=0,1,2,\dots,n$, where $t_0=0$ and $t_n=\tau_2$. Next, we assume that we have a discretely defined filtration $\mathcal{F}_{t_i}$ associated to the spot process. This is naturally enlarged to all times t by setting $\mathcal{F}_t=\mathcal{F}_{t_i}$ for $t\in[t_i,t_{i+1})$, which means that there is no new information coming from the spot price process before next time instance t_{i+1}. This implies that

$$\begin{aligned}F(t,\tau_1,\tau_2)&=\mathbb{E}_Q\left[\frac{1}{\tau_2-\tau_1}\sum_{t_i=\tau_1}^{\tau_2-1} S(t_i)\,|\,\mathcal{F}_t\right]\\&=\mathbb{E}_Q\left[\frac{1}{\tau_2-\tau_1}\sum_{t_i=\tau_1}^{\tau_2-1} S(t_i)\,|\,\mathcal{F}_{t_i}\right]\\&=F(t_i,\tau_1,\tau_2)\,.\end{aligned}$$

Hence, the electricity futures price becomes constant over each hour, that is, it becomes a time series process rather than a continuous-time stochastic process. This behaviour is not what we observe in the actual market, where electricity futures prices move according to trades taking place also within every hour. We mend this by introducing a continuous-time unobserved spot price process as above, which then introduces more information in the model.

We remark that temperature is naturally a continuous-time process, even though the indices used as underlying for the temperature futures contracts are discretely monitored. The gas spot can also be viewed as a process in continuous time, since one can buy a delivery of gas on short notice at a desired (in principle) time. This makes electricity as a rather particular case for the above discussion on discrete- vs. continuous-time models.

The HJM approach in the interest rate markets proposes to model the forward rates directly rather than the spot rates. This approach has been suggested to be used for modelling the forward price dynamics in commodity markets. In particular, [Bjerksund, Rasmussen and Stensland (2000)], [Keppo *et al.* (2004)], [Benth and Koekebakker (2005)] and [Kiesel, Schindlmayer and Börger (2006)] have done this for the contracts in the Nord Pool and EEX electricity markets, while a discussion of the approach to general energy markets can be found in [Clewlow and Strickland (2000)]. Note that both [Bjerksund, Rasmussen and Stensland (2000)] and [Clewlow and Strickland (2000)] suggest to use the HJM approach to model forward contracts, while in [Benth and Koekebakker (2005)] electricity futures, the actual contracts traded in the market, are considered. A large portion of this book is devoted to the application of the HJM approach, with a particular view towards the electricity markets. Some issues arise when trying to apply the HJM theory to electricity (and gas) futures.

First of all, what kind of contracts should the HJM approach be used on. Following the interest rate method directly, one may be tempted to model the forwards contracts, as it is done by [Bjerksund, Rasmussen and Stensland (2000)] and [Clewlow and Strickland (2000)]. However, in the electricity market we do not have data for such contracts, and the question of how to estimate the model to market observations arises. One way out is to smoothen the observed electricity futures prices, in order to transform the data to forward prices. Alternatively, one may integrate up the forward prices to get an implied dynamics for the observed futures. Instead of using the HJM technique on forwards that is not traded in the market, one may instead consider modelling the electricity futures directly.

In the Nord Pool market, electricity futures contracts with overlapping delivery periods are traded. For example, you can enter a yearly contract, or four quarterly contracts covering the entire year. Hence, certain consistency conditions need to be satisfied for the price dynamics of the contracts in order to avoid arbitrage opportunities. In the strict sense, the HJM method models the electricity futures price dynamics for arbitrary delivery periods. As we shall see, it is difficult to state models satisfying the arbitrage conditions and at the same time being analytically tractable. Furthermore, the condition rules out models like geometric Brownian motion.

To resolve this problem, we follow the path given by the LIBOR[12] models in interest rate theory (see, for example, [Brigo and Mercurio (2001)]). We model exclusively those contracts that are traded in the market, and in addition have delivery periods which cannot be decomposed into other traded contracts. With this way of modelling, we are much more free to state reasonable stochastic dynamical models which can easily be estimated on data and used for risk management analysis.

A possibly undesirable consequence of the HJM approach for electricity futures price modelling is the loss of a connection with the underlying spot price. Given an electricity futures price dynamics, one cannot trace back a spot price dynamics except in trivial and not relevant cases. This is a serious matter on one hand, since the spot is namely the reference index for the futures. On the other hand, one may view the electricity spot market as itself being a futures market, where the contracts have hourly settlement periods throughout the day.

1.6 Outline of the book

The basic modelling tools in this book are II processes and mean-reverting stochastic processes driven by these. In order to understand the models, and price products like swaps and options, we need a stochastic analysis for the II processes. The theory on stochastic integration and differentiation (for example, Itô's Formula) for such processes is surveyed in Chapter 2. The purpose of the chapter is to provide the reader an easy reference for the fundamental results and notions which will be useful in the modelling and pricing analysis of energy markets. The chapter is *not* meant to give a complete theory, for which the reader is referred to the existing literature in the area, for instance [Jacod and Shiryaev (1987)]. For us, the most useful results will be the Lévy-Kintchine representation, Itô's Formula, the

[12]London interbank offer rate.

stochastic Fubini Theorem and Bayes' Formula. However, to have a complete theoretical foundation for the analysis, we also need to understand stochastic integration with respect to II processes. The chapter includes examples of some of the most used stochastic processes in finance, and in particular energy markets. The reader being eager to process to the modelling and analysis of electricity and related markets, may skip reading this chapter and only use it for reference.

In Chapter 3 we model spot prices in energy markets based on OU processes. We analyse both geometric and arithmetic models, and present in particular an arithmetic model which preserves positivity of prices. The models are multi-factor, driven by both Brownian motion and pure jump processes, with possible seasonally dependent jump size and intensity. Stochastic simulation of these models is discussed in a case study of the arithmetic model.

Based on these spot models, we derive the forward and swap price dynamics in Chapter 4. We recall here that forwards in our use of the terminology are contracts with a fixed maturity time, whereas swaps are used as a general reference to electricity and gas futures. After some general considerations, we apply the Esscher transform to construct risk-neutral probabilities. The Esscher transform preserves the distributional properties of the jump processes, and can be thought of as a generalization of the Girsanov transform used for Brownian motions. Forward prices for the arithmetic and geometric spot models introduced in Chapter 2 are derived. For the swaps, the geometric models do not in general admit any explicit formulas for the price dynamics. Choosing an arithmetic spot model, we can derive an explicit swap dynamics. The issue of currency conversion, being relevant, for instance, on the Nord Pool market, is discussed in detail.

Our spot models and derived swap price dynamics are applied to the UK gas market in Chapter 5. A simple one-factor model with both Brownian motion and jump-driven increments are considered, a frequently used dynamics for energy spot prices. Recursive filtering is implemented for identifying the jumps in the spot price series. The heavy-tailed normal inverse Gaussian distribution for the spot price innovations is considered and estimated on data as well. The different spot price models are next used as a basis for deriving gas futures prices. We analyse the theoretical prices in view of the observed gas futures term structure in the UK market, and discuss the market price of risk, that is, the choice of the parameters in the Esscher transform. The chapter also contains discussions on how multi-factor spot models can be estimated on data, incorporating, for instance, OU processes having different speeds of mean reversion.

The HJM approach to the modelling of forward and swap prices is presented in Chapter 6. The different modelling issues regarding forward prices and swaps are investigated in detail, along with the incorporation of jump processes. As we show, the no-arbitrage condition for the term structure dynamics of the swap price rules out most of the relevant models. To resolve this issue, we introduce market models for the swaps, much in the spirit of LIBOR models for fixed income markets.

When applying the HJM approach to electricity markets, one may base the electricity futures price dynamics on a model for non-traded forwards. To estimate such models, one needs to derive forward data from the observed electricity futures prices. An algorithm for the derivation of smooth forward curves in electricity markets is presented in Chapter 7. The algorithm may be applied to gas markets as well. We demonstrate the algorithm at work on Nord Pool electricity futures data, and further apply it to study the term structure of volatility of electricity.

The smoothing algorithm is also applied in Chapter 8, where we empirically analyse the Nord Pool electricity futures market using HJM-based models. The smoothing algorithm enables us to derive a data set which is structured and more easy to use in an empirical investigation of the market. A principal component analysis reveals certain structures for the short- and long-term market, and motivate a parametric multi-factor market model, including seasonal volatility with maturity effect. The model is fitted to market data.

Following is a more theoretical chapter dealing with the pricing and hedging of options traded for energies. Chapter 9 presents pricing formulas for call and put options based on the various proposed spot, forward and swap models. The option prices become generalizations of the Black-76 formula when the underlying models are depending on Brownian motions only. For models with jumps, we use a Fourier approach to derive formulas for the prices. Issues of hedging are discussed for these options. The pricing of spread and Asian options are analysed for arithmetic multi-factor models, where reasonably explicit formulas are available based on the cumulant functions of the jump processes. A case study on the pricing of spark spread options in the UK market is presented, based on a direct modelling approach for the spread between electricity and gas.

The final Chapter 10 is devoted to the market for temperature futures. We present continuous-time mean reversion models being generalizations of autoregressive moving average time series. Applying these to temperature data, we find that the "volatility" of temperature has a clear seasonal pattern. The temperature models allow for rather explicit pricing of the typical futures traded on CME. The chapter includes a thorough empricial analysis of Stockholm temperature data in view of the proposed models.

Chapter 2

Stochastic Analysis for Independent Increment Processes

The purpose of this chapter is to give the necessary background in stochastic analysis for independent increment (II) processes that we need in order to model the electricity and related markets. We refrain from giving proofs, since most of the results require a much deeper theoretical foundation than we aim for. We have collected results from [Ikeda and Watanabe (1981)], [Jacod and Shiryaev (1987)], [Sato (1999)] and [Shiryaev (1999)], all containing the background and proofs that the interested reader is encouraged to explore in more detail.

2.1 Definitions

Let $(\Omega, \mathcal{F}, \{\mathcal{F}_t\}_{t\geq 0}, P)$ be a *complete filtered probability space*. A mapping $X : \Omega \mapsto \mathbb{R}^d$ is said to be a *random variable* if it is $\mathcal{F}$-measurable, whereas a *stochastic process* is a family of random variables $\{X(t)\}_{t\geq 0}$ parametrized over the time t. The time t will usually be defined on a finite interval $[0, \mathcal{T}]$, but sometimes we shall work with an infinite time horizon $[0, \infty)$. This will always be clear from the context. Frequently, we use the notation $X(t)$ for the process, and not only for the random variable at time t. A stochastic process $X(t)$ is said to be $\mathcal{F}_t$*-adapted* if every $X(t)$, $t \geq 0$ is measurable with respect to the σ-algebra $\mathcal{F}_t$.

A process $X(t)$ is said to be *RCLL* if its paths $t \mapsto X(t,\omega)$ are *r*ight-continuous for $t \geq 0$, and has *l*eft-*l*imits for $t > 0$, *a.s.*[1] RCLL processes are often coined *càdlàg* in the literature, being an acronym of the french "continu à droite avec des limites à gauche". If the paths are continuous

[1] The abbreviation *a.s.* stands for *almost surely*, and means that a property holds for all $\omega \in \Omega$ except for a subset with probability zero.

a.s., we simply refer to $X(t)$ as a continuous stochastic process. We write

$$\Delta X(t) = X(t) - X(t-), \tag{2.1}$$

to denote the jump at time t of an RCLL process $X(t)$, with

$$X(t-) = \lim_{s \uparrow t} X(s), \tag{2.2}$$

being the limit from the left of the process.

A *stopping time* τ is a random variable with values in $[0, \infty]$ and with the property

$$\{\omega \in \Omega \,|\, \tau(\omega) \leq t\} \in \mathcal{F}_t,$$

for every $t \geq 0$. The notion of a (local) martingale is often useful.

Definition 2.1 An adapted RCLL stochastic process $M(t)$ is a *martingale* if it is in $L^1(P)$ for all $t \geq 0$, and for every $t \geq s \geq 0$

$$\mathbb{E}\left[M(t) \,|\, \mathcal{F}_s\right] = M(s). \tag{2.3}$$

$M(t)$ is said to be a *local* martingale if there exists a sequence of stopping times $\tau_n < \infty$, where $\tau_n \uparrow \infty$, such that $M(t \wedge \tau_n)$ is a martingale.

In later chapters we will need to know how to calculate conditional expectations with respect to different probabilities. This is done by Bayes' Formula. Consider a finite time horizon $[0, \mathcal{T}]$, and let Q be a probability measure equivalent to P. Denote the density process of the Radon Nikodym derivative by $Z(t)$, for $t \in [0, \mathcal{T}]$, that is,

$$\frac{dQ}{dP}|_{\mathcal{F}_t} = Z(t)\,.$$

Suppose that $Z(t)$ is a martingale. Then the Bayes' Formula says

$$\mathbb{E}_Q\left[X \,|\, \mathcal{F}_t\right] = Z^{-1}(t)\mathbb{E}\left[XZ(\mathcal{T}) \,|\, \mathcal{F}_t\right]\,, \tag{2.4}$$

where X is an integrable random variable and $\mathbb{E}_Q[\cdot]$ is the expectation operator with respect to Q. We refer to [Liptser and Shiryaev (1977)] for more on the Bayes' Formula.

We introduce the class of stochastic processes that we are going to use as the basic model to describe the random evolution of energy prices.

Definition 2.2 An adapted RCLL stochastic process $I(t)$ starting in zero is an *II process* if it satisfies the following two conditions:

(1) The increments $I(t_0), I(t_1)-I(t_0), \ldots, I(t_n)-I(t_{n-1})$ are independent random variables for any partition $0 \le t_0 < t_1 < \cdots < t_n$, and $n \ge 1$.
(2) It is continuous in probability, that is, for every $t \ge 0$ and $\varepsilon > 0$, it holds

$$\lim_{s \to t} P\left(|I(s) - I(t)| > \varepsilon\right) = 0. \tag{2.5}$$

The main characteristic of an II process is the first property. If we add the property that increments are stationary, in the sense that the distribution of $I(t) - I(s)$, $t > s \ge 0$, is only dependent on $t - s$, and not on s and t separately, $I(t)$ is called a *Lévy process*. We encounter Lévy processes quite frequently in this book, and often use the notation $L(t)$ to denote such processes. Furthermore, in the special case of a Lévy process with increments $L(t) - L(s)$ being normally distributed with zero mean and variance $t - s$, we have a *Brownian motion*, usually denoted $B(t)$. Lévy processes which are increasing, that is, having only positive jumps, are often called subordinators.

Definition 2.2 is adapted from [Sato (1999)], where the II processes are called *additive processes*. We prefer to use the name "independent increment process" since it is more descriptive, although, the name introduced by Lévy in [Lévy (1965)] was *processus additif*. In [Jacod and Shiryaev (1987)], Property 2 in Definition 2.2 is substituted with *fixed times of discontinuities* $t \ge 0$, being times where the II process has a positive probability of making a jump, that is,

$$P(\Delta I(t) \neq 0) > 0.$$

Note that a process which is continuous in probability cannot have any fixed times of discontinuities.

The characteristic function of an II process is defined in the following Proposition.

Proposition 2.1 *The characteristic function of the II process $I(t)$ is*

$$\mathbb{E}\left[\exp\left(i\theta(I(s) - I(t))\right)\right] = \exp(\psi(s, t; \theta)) \tag{2.6}$$

for $0 \le s < t$, $\theta \in \mathbb{R}$ and

$$\begin{aligned}\psi(s, t; \theta) = i\theta(\gamma(t) - \gamma(s)) - \frac{1}{2}\theta^2(C(t) - C(s)) \\ + \int_s^t \int_{\mathbb{R}} \left(e^{i\theta z} - 1 - i\theta z 1_{|z| \le 1}\right) \ell(dz, du).\end{aligned} \tag{2.7}$$

The function $\psi(s,t;\theta)$ is called the cumulant function of the process $I(t)$, and the generating triplet of the II process is $(\gamma(t), C(t), \ell)$, with the properties

(1) $\gamma : \mathbb{R} \mapsto \mathbb{R}$ is a continuous function with $\gamma(0) = 0$,
(2) $C : \mathbb{R} \mapsto \mathbb{R}$, is non-decreasing and continuous, with $C(0) = 0$,
(3) ℓ is a σ-finite measure on the Borel σ-algebra of $[0,\infty) \times \mathbb{R}$, with the properties

$$\ell(A \times \{0\}) = 0\,, \quad \ell(\{t\} \times \mathbb{R}) = 0, \quad \textit{for } t \geq 0 \textit{ and } A \in \mathcal{B}(\mathbb{R}_+)$$

and

$$\int_0^t \int_{\mathbb{R}} \min(1, z^2)\, \ell(ds, dz) < \infty\,.$$

The measure ℓ relates to the jumps of the II process, while C is the covariance of the continuous martingale part of the process. The drift is given by γ. We shall come back to this when introducing random jump measures and semimartingales.

If the measure ℓ can be written as

$$\ell(ds, dz) = ds\, \widetilde{\ell}(dz),$$

and $\gamma(t) = \gamma t$ and $C(t) = ct$ for two constants γ and $c \geq 0$, we have $\psi(t,s;\theta) = (t-s)\widetilde{\psi}(\theta)$ with

$$\widetilde{\psi}(\theta) = \mathrm{i}\theta\gamma - \frac{1}{2}\theta^2 c + \int_{\mathbb{R}} \left(\mathrm{e}^{\mathrm{i}\theta z} - 1 - \mathrm{i}\theta z 1_{|z|\leq 1}\right) \widetilde{\ell}(dz)\,.$$

The function $\widetilde{\psi}$ is then the cumulant function of a Lévy process L. From now on we write $\psi(\theta)$ for the cumulant function in the stationary case, and $\psi(t,s;\theta)$ for the general non-stationary case.

The generating triplet $(\gamma(t), C(t), \ell)$ is related to a representation of the II process in terms of jump processes and continuous processes with finite and infinite variation. It is usually referred to as the Lévy-Kintchine decomposition, and we introduce it in terms of random jump measures. To do so, we need some theoretical foundation on integration with respect to such measures. In the next two sections we discuss stochastic integration from a general point of view, returning to the Lévy-Kintchine decomposition in Sect. 2.4.

2.2 Stochastic integration with respect to martingales

The class of square integrable martingales are suitable as stochastic integrators.

Definition 2.3 Define $\mathcal{M}_2$ to be the set of martingales $M(t)$ which are square integrable, and $\mathcal{M}_2^{\rm loc}$ to be the set of *local* martingales $M(t)$, where the sequence of stopped processes $M^{\tau_n}(t) \in \mathcal{M}_2$ for every n.

The Doob-Meyer decomposition theorem connects a square-integrable (local) martingale to a so-called *natural integrable increasing* processes. First, let us introduce a natural integrable increasing process.

Definition 2.4 A one-dimensional stochastic process $A(t)$ is called a *natural integrable increasing* process if

(1) $A(t)$ is adapted,
(2) $A(0) = 0$, $A(t)$ is nondecreasing and right continuous *a.s.*,
(3) $\mathbb{E}[A(t)] < \infty$ for every $t \in [0, \infty)$,
(4) $\mathbb{E}[\int_0^t M(s)\, dA(s)] = \mathbb{E}[\int_0^t M(s-)\, dA(s)]$, for all $t \in [0, \infty)$ for every bounded martingale $M(t)$.

The Doob-Meyer decomposition theorem states the following.

Theorem 2.1 *If $M(t) \in \mathcal{M}_2$ then there exists a unique natural integrable increasing process $A(t)$ such that $M^2(t) - A(t)$ is a martingale. If $M(t), N(t) \in \mathcal{M}_2$, then there exists a unique process $A(t)$ which can be expressed as the difference of two natural integrable increasing processes such that $M(t)N(t) - A(t)$ is a martingale.*

The process $A(t)$ in the Doob-Meyer theorem is called the *quadratic variation process* of the martingales $M(t)$ and $N(t)$ (or only of $M(t)$, as in the first part of the theorem). Customarily, it is denoted as $\langle M, N\rangle$, or simply $\langle M\rangle$ in the case of $\langle M, M\rangle$. Remark in passing that $\langle B\rangle(t) = t$ and $\langle N\rangle(t) = at$, when $B(t)$ is the Brownian motion and $N(t)$ is the Poisson process with intensity a. If $M(t), N(t) \in \mathcal{M}_2^{\rm loc}$, we obtain from the uniqueness in the Doob-Meyer theorem that there exists a unique quadratic variation process such that for a localising sequence $\{\tau_n\}$,

$$\langle M, N\rangle(t \wedge \tau_n) = \langle M^{\tau_n}, N^{\tau_n}\rangle(t \wedge \tau_n)\,.$$

The integrands with respect to a square integrable (local) martingale will be collected from the class of *predictable* processes.

Definition 2.5 A stochastic process $X(t)$ is called *predictable* if it is measurable with respect to the smallest σ-algebra on $[0,\infty)\times\Omega$ such that all left continuous processes are measurable. This σ-algebra is called the *predictable* σ-algebra.

We remark that the quadratic variation processes are predictable because of the last condition in the definition of a natural integrable increasing process. For the sake of completeness, we also introduce *optional* processes.

Definition 2.6 A stochastic process $X(t)$ is called *optional* if it is measurable with respect to the smallest σ-algebra on $[0,\infty)\times\Omega$ such that all right-continuous processes are measurable. This σ-algebra is called the *optional* σ-algebra.

The set of integrands with respect to $M(t)\in\mathcal{M}_2^{(\mathrm{loc})}$ are defined as follows.

Definition 2.7 We say that the stochastic process $X(t)$ is integrable with respect to $M(t)\in\mathcal{M}_2$, if,

(1) $X(t)$ is predictable, and
(2) for every $t>0$,

$$\mathbb{E}\left[\int_0^t X(s)^2\,d\langle M\rangle(s)\right]<\infty\,.$$

If $M(t)\in\mathcal{M}_2^{\mathrm{loc}}$, the last condition is substituted with

$$\int_0^t X(s)^2\,d\langle M\rangle(s)<\infty\,,a.s.,$$

for every $t>0$.

If $X(t)$ is integrable with respect to $M(t)\in\mathcal{M}_2$, we define the stochastic integral of $X(t)$ with respect to $M(t)$ via approximating the integrand with step functions, multiplying with the increments of $M(t)$ and passing to the limit in variance. The resulting stochastic integral, denoted $\int_0^t X(s)\,dM(s)$, becomes a square integrable martingale as well (that is, belonging to the set $\mathcal{M}_2$). Moreover,

$$\mathbb{E}\left[\left(\int_0^t X(s)\,dM(s)\right)^2\right]=\mathbb{E}\left[\int_0^t X(s)^2\,d\langle M\rangle(s)\right]\,. \tag{2.8}$$

This relation is known as the *Itô isometry* for stochastic integrals with respect to Brownian motion. When $M(t)\in\mathcal{M}_2^{\mathrm{loc}}$, the same construction holds after introducing a localising sequence of stopping times $\{\tau_n\}$. It is easily observed that the stochastic integral is linear in the integrands and in the integrators, as long as the necessary integrability conditions are satisfied. Moreover, the stochastic integral has the following properties.

Proposition 2.2 *If $X(t)$ is integrable with respect to $M(t) \in \mathcal{M}_2^{(loc)}$, then the stochastic integral $\int_0^t X(s)\,dM(s)$ has the following properties:*

(1) $\int_0^0 X(s)\,dM(s) = 0$, *a.s.,*
(2) for a stopping time τ,

$$\int_0^{t\wedge\tau} X(s)\,dM(s) = \int_0^t X(s) 1_{s\le\tau}\,dM(s)\,,$$

(3) if $Y(t)$ is an integrable process with respect to $N(t) \in \mathcal{M}_2$, then,

$$\mathbb{E}\Big[\int_0^t X(s)\,dM(s) \int_0^t Y(s)\,dN(s)\Big] = \mathbb{E}\Big[\int_0^t X(s)Y(s)\,d\langle M,N\rangle(s)\Big]\,.$$

The last property holds after introducing a localising sequence when $M(t), N(t) \in \mathcal{M}_2^{loc}$.

Stochastic integrals are important in defining the price dynamics of assets.

2.3 Random jump measures and stochastic integration

For an II process $I(t)$, define for each $t > 0$ and U a Borel subset in $\mathbb{R}\setminus\{0\}$

$$N((0,t]\times U) = \sum_{s\le t} 1_{\Delta I(s)\in U}\,. \tag{2.9}$$

Since the process $I(t)$ has RCLL paths, the sum above is finite. Hence, N defines a counting measure on the Borel subsets of $(0,\infty)\times\mathbb{R}\setminus\{0\}$, which we call the *random jump measure* associated to the process I. Moreover, from the independent increment property of I, the process $t \mapsto N((0,t]\times U)$ is also an II process. Define the σ-finite measure on the Borel sets of $(0,\infty)\times\mathbb{R}\setminus\{0\}$ as

$$\nu((0,t]\times U) = \mathbb{E}[N((0,t]\times U)]\,, \tag{2.10}$$

for each Borel set $U \in \mathbb{R}\setminus\{0\}$. Then it holds that the process $t \mapsto N((0,t]\times U) - \nu((0,t]\times U)$ is a local martingale, and we call ν the *compensator measure* of N. It turns out that ν coincides with ℓ, defined in the characteristic triplet of I. From now on we use the notation ℓ for the compensator measure, and remark that this is called the Lévy measure when the process is a Lévy process. It measures the expected number of jumps of a certain size occurring over a time interval.

To establish the link with semimartingales, and to prepare for Itô's Formula, we need to introduce some definitions related to stochastic integration with respect to random jump measures. We focus on real-valued random fields $X(t,z,\omega)$ defined on $[0,\infty)\times\mathbb{R}\times\Omega$, being *predictable*, as the integrands.

Definition 2.8 A real-valued random field $X(t,z,\omega)$ defined on $[0,\infty)\times\mathbb{R}\times\Omega$, is said to be *predictable* if

(1) for each $t>0$, the mapping

$$(z,\omega)\mapsto X(t,z,\omega)\,,$$

is $\mathcal{B}(\mathbb{R})\times\mathcal{F}_t$-measurable.

(2) For each (z,ω), the mapping

$$t\mapsto X(t,z,\omega)\,,$$

is left-continuous.

For predictable integrands, we consider stochastic integrals of the form

$$\int_0^t\int_{\mathbb{R}\backslash\{0\}} X(s,z)\,N(ds,dz)\,, \tag{2.11}$$

and

$$\int_0^t\int_{\mathbb{R}\backslash\{0\}} X(s,z)\,\widetilde{N}(ds,dz)\,. \tag{2.12}$$

The notation $\widetilde{N}$ stands for the compensated random jump measure, $\widetilde{N}=N-\ell$. These two integrals are defined in different manners according to certain integrability properties of the integrand.

First, consider the integral (2.11). If $X(t)$ satisfies the integrability property

$$\int_0^t\int_{\mathbb{R}\backslash\{0\}} |X(s,z)|\,N(ds,dz)<\infty\,,a.s., \tag{2.13}$$

we have that (2.11) is well-defined *a.s.* in the Lebesgue-Stieltjes sense, and the following equality holds

$$\int_0^t \int_{\mathbb{R}\setminus\{0\}} X(s,z)\,N(ds,dz) = \sum_{s\le t} X(s,\Delta I(s))\,. \tag{2.14}$$

Concerning the integral (2.12), we have several possibilities. Supposing

$$\mathbb{E}\left[\int_0^t \int_{\mathbb{R}\setminus\{0\}} |X(s,z)|\ell(ds,dz)\right] < \infty\,, \tag{2.15}$$

one has that X satisfies the condition in (2.13), and we define (2.12) by

$$\begin{aligned}\int_0^t \int_{\mathbb{R}\setminus\{0\}} X(s,z)\,\widetilde{N}(ds,dz) &= \int_0^t \int_{\mathbb{R}\setminus\{0\}} X(s,z)\,N(ds,dz)\\ &\quad - \int_0^t \int_{\mathbb{R}\setminus\{0\}} X(s,z)\,\ell(ds,dz)\,.\end{aligned} \tag{2.16}$$

In this case, the process

$$t \mapsto \int_0^t \int_{\mathbb{R}\setminus\{0\}} X(s,z)\,\widetilde{N}(ds,dz)\,,$$

is a martingale. It is possible to define the integral for random fields satisfying

$$\mathbb{E}\left[\int_0^t \int_{\mathbb{R}\setminus\{0\}} |X(s,z)|^2\ell(ds,dz)\right] < \infty\,, \tag{2.17}$$

by a similar procedure as with the Itô integral. By considering step functions, and constructing a Cauchy sequence in an appropriate Hilbert space, we get that there exists a square integrable martingale process which we denote by (2.12). One may further generalize this to localised predictable integrands, in the sense that for a sequence of finite stopping times $\tau_n \uparrow \infty$, $X(t\wedge\tau_n, z)$ satisfies condition (2.17). The stochastic integral then becomes a local martingale.

2.4 The Lévy-Kintchine decomposition and semimartingales

It holds that

$$\sum_{s\le t} \Delta I(s) 1_{\Delta I(s)\ge 1} = \int_0^t \int_{|z|\ge 1} z\,N(ds,dz)\,,$$

and the process

$$t \mapsto \int_0^t \int_{|z|<1} z \, \widetilde{N}(ds, dz)$$

is a square integrable local martingale. The following representation of I can be derived

$$I(t) = \gamma(t) + M(t) + \int_0^t \int_{|z|<1} z \, \widetilde{N}(ds, dz) + \int_0^t \int_{|z|\geq 1} z \, N(ds, dz) \,, \quad (2.18)$$

where $M(t)$ is a local square integrable continuous martingale with quadratic variation equal to $C(t)$. This representation is often called the Lévy-Kintchine decomposition of I. The Lévy-Kintchine decomposition is the bridge connecting to semimartingales, which we discuss next. We call $I(t)$ a *pure jump* II process if the continuous martingale part $M(t)$ is identically equal to zero. If we have two independent pure jump II processes $I(t)$ and $J(t)$, then they cannot jump at the same time *a.s.*

II processes are closely related to semimartingales. We introduce semimartingales, since they are heavily used in mathematical finance to model prices of financial assets. The class of semimartingale processes is closed under stochastic integration, differentiation (Itô's Formula) and measure change, among other things, making it a tractable tool for analysis. We shall exploit their properties in the context of energy markets.

We define a semimartingale as in [Ikeda and Watanabe (1981)].

Definition 2.9 An adapted RCLL stochastic process $S(t)$ is a semimartingale if it has the representation

$$S(t) = S(0) + A(t) + M(t) + \int_0^t \int_{\mathbb{R}\setminus\{0\}} X_1(t,z) \, \widetilde{N}(ds, dz) + \int_0^t \int_{\mathbb{R}\setminus\{0\}} X_2(t,z) \, N(ds, dz) \,, \quad (2.19)$$

where $A(t)$ is an adapted continuous stochastic process having paths of finite variation on finite time intervals, and $M(t)$ is a continuous square integrable local martingale. Further, $S(0)$ is an $\mathcal{F}_0$-measurable random variable, the random fields X_1 and X_2 satisfy the conditions (2.17) (in a localised sense) and (2.13), respectively, with $X_1(t, z, \omega) X_2(t, z, \omega) = 0$.

This definition is more restrictive than the conventional one, assuming that a semimartingale is decomposable into an RCLL finite variation process

$A(t)$ and a local martingale. However, the definition in [Ikeda and Watanabe (1981)] suits our purposes and we do not need to go into further generality.

In general, an II process is not a semimartingale, albeit very close to being so. Theorem 5.1 in [Jacod and Shiryaev (1987)] states that an II process $I(t)$ may be represented as the sum of a deterministic RCLL function and a semimartingale, the latter being an II process

$$I(t) = \widetilde{A}(t) + \widetilde{I}(t). \tag{2.20}$$

Hence, $I(t)$ is a semimartingale if and only if the function $\widetilde{A}(t)$ is equal to zero and the drift function $\gamma(t)$ in the Lévy-Kintchine decomposition (2.18) of I is of finite variation on finite intervals. In this book, we only consider II processes $I(t)$ which are semimartingales.

From Sects. 2.2-2.3 we know how to integrate with respect to a semimartingale $S(t)$. We simply decompose the integrator $S(t)$ into a pathwise integral with respect to a finite variation process $A(t)$, a local martingale integration and an integration with respect to random jump measures. In the sequel, it will be convenient to have the following stochastic Fubini Theorem (this is the second version in [Protter (1990), page 160]). Denote by $(U,\mathcal{U})$ a measure space equipped with a finite measure $m(du)$, and $\mathcal{P}$ the predictable σ-algebra.

Theorem 2.2 *Let $X(t)$ be a semimartingale, and $H(u,t,\omega)$ be $\mathcal{U}\otimes\mathcal{P}$-measurable. Assume*

$$\left\{\int_U H^2(u,t,\cdot)\,m(du)\right\}^{1/2}$$

is integrable with respect to $X(t)$. Letting

$$\int_0^t H(u,s,\cdot)\,dX(s)$$

be $\mathcal{U}\otimes\mathcal{B}(\mathbb{R}_+)\otimes\mathcal{F}$ measurable and RCLL for each u, then

$$\int_U\int_0^t H(u,s,\cdot)\,dX(s)\,m(du)$$

exists and is a RCLL version of

$$\int_0^t\int_U H(u,s,\cdot)\,m(du)\,dX(s)\,.$$

In all our applications, U will be an interval in $\mathbb{R}_+$, with $\mathcal{U}$ being the Borel sets and $m(du)$ the Lebesgue measure.

2.5 The Itô Formula for semimartingales

We formulate the Itô Formula for semimartingales (see [Ikeda and Watanabe (1981)]). Let $\mathbf{S}(t) \triangleq (S_1(t), \ldots, S_n(t))$ be n semimartingales, each having a dynamics as defined in Def. 2.9,

$$S_i(t) = S_i(0) + A_i(t) + M_i(t) + \int_0^t \int_{\mathbb{R}\setminus\{0\}} X_{1,i}(t,z)\,\widetilde{N}_i(ds,dz) + \int_0^t \int_{\mathbb{R}\setminus\{0\}} X_{2,i}(t,z)\,N_i(ds,dz)\,.$$

Here, we assume that the random jump measures N_i are independent. Further, let $f(t,x)$ be a real-valued function on $[0,\infty) \times \mathbb{R}^n$ which is once continuously differentiable in t and twice continuously differentiable in x. Then $f(t,\mathbf{S}(t))$ is again a semimartingale, with the following representation:

$$\begin{aligned}
f(t,\mathbf{S}(t)) &= f(0,\mathbf{S}(0)) + \int_0^t \partial_t f(u,\mathbf{S}(u))\,du \\
&+ \sum_{i=1}^n \int_0^t \partial_{x_i} f(u,\mathbf{S}(u))\,dA_i(u) + \sum_{i=1}^n \int_0^t \partial_{x_i} f(u,\mathbf{S}(u))\,dM_i(u) \\
&+ \frac{1}{2}\sum_{i,j=1}^n \int_0^t \partial_{x_i x_j} f(u,\mathbf{S}(u))\,d\langle M_i, M_j\rangle(u) \\
&+ \sum_{i=1}^n \int_0^t \int_{\mathbb{R}\setminus\{0\}} f(u,\mathbf{S}(u-) + X_{1,i}(u,z)\mathbf{e}_i) - f(u,\mathbf{S}(u-))\,\widetilde{N}_i(du,dz) \\
&+ \sum_{i=1}^n \int_0^t \int_{\mathbb{R}\setminus\{0\}} f(u,\mathbf{S}(u-) + X_{2,i}(u,z)\mathbf{e}_i) - f(u,\mathbf{S}(u-))\,N_i(du,dz) \\
&+ \sum_{i=1}^n \int_0^t \int_{\mathbb{R}\setminus\{0\}} f(u,\mathbf{S}(u) + X_{1,i}(u,z)\mathbf{e}_i) - f(u,\mathbf{S}(u)) \\
&\qquad - X_{1,i}(u,z)\partial_{x_i} f(u,\mathbf{S}(u))\,\ell_i(du,dz)\,.
\end{aligned}$$

Here, $\partial_t f$ and $\partial_{x_i} f$ denote the first derivatives with respect to t and x_i of $f(t,x)$, respectively, and $\partial_{x_i x_j} f$ is the second derivative of f with respect to x_i and x_j. Further, we have denoted by $\mathbf{e}_i$, $i = 1, \ldots, n$, the ith basis vector in $\mathbb{R}^n$, with 1 on coordinate i and zeros otherwise.

Note that this formula is a reformulation of the one that can be found in [Ikeda and Watanabe (1981)]. In their case, there is only one random jump measure, in which all semimartingales are represented. Based on the multi-dimensional formula in [Protter (1990)], and the fact that jumps are independent for the semimartingale processes, we obtain the Itô Formula above.

2.6 Examples of independent increment processes

In this section we discuss typical II processes, including Brownian motion, compound Poisson processes and Lévy processes of different kinds. For Lévy processes, we concentrate on two main classes which are relevant in energy price modelling, the normal inverse Gaussian (NIG) and the CGMY Lévy processes.

The standard model to describe the stochastic fluctuations of asset prices is the Brownian motion. A Brownian motion $B(t)$ is an II process with stationary increments being normal random variables, that is, $B(t) - B(s)$ is normally distributed with zero expectation and variance given by $t - s$ for $0 \leq s < t$. The Brownian motion has continuous sample paths *a.s.*, although, not of finite variation. It constitutes the "simplest" II process, in the sense that it is the only infinite variation II process with continuous sample paths.

We proceed to more sophisticated models for the stochastic fluctuations of energy prices, starting with the compound Poisson process.

2.6.1 *Time-inhomogeneous compound Poisson process*

A Poisson process $N(t)$ with intensity λ is a one-dimensional stochastic process which has stationary and independent increments, and where $N(t)-N(s)$ is Poisson distributed with intensity $\lambda(t-s)$, $0 \leq s < t$, that is, for every $k = 0, 1, 2, \ldots$,

$$P(N(t) - N(s) = k) = \frac{\lambda^k (t-s)^k}{k!} \mathrm{e}^{-\lambda(t-s)} . \tag{2.21}$$

We call $N(t) - \lambda t$ the *compensated* Poisson process, which is a martingale. A Poisson process jumps with size one at exponentially distributed jumping times, and remain constant between jumps.

A popular way to introduce jumps in a spot price dynamics is by a so-called *compound Poisson process*. Let $\{X_i\}_{i=1}^{\infty}$ be a sequence of independent and identically distributed (i.i.d) random variables. Define

$$L(t) = \sum_{i=1}^{N(t)} X_i \,,$$

where $N(t)$ is a Poisson process with intensity λ independent of X_i. The compound Poisson process $L(t)$ is a Lévy process, where jumps occur at an intensity λ, and jump size is determined by X_i. The cumulant function of $L(1)$ is

$$\psi(\theta) = \lambda \left(\mathrm{e}^{\psi_X(\theta)} - 1 \right) \,, \tag{2.22}$$

where ψ_X is the cumulant function of X_i. The process $L(t)$ has Lévy measure given by

$$\widetilde{\ell}(dz) = \lambda \, F_X(dz) \,,$$

where F_X denotes the distribution of X_i. Note that the Lévy measure in this case integrates over zero, and thus the compound Poisson process is of finite variation.

In [Geman and Roncoroni (2006)], a *time-inhomogeneous* version of the compound Poisson process is considered as the basic noise process of electricity prices. The Poisson process $N(t)$ is assumed to have a time-varying intensity $\lambda(t)$ in order to account for the seasonality in the arrival of spikes frequently observed in the electricity markets. More specifically, they assume

$$\lambda(t) = c \left(\frac{2}{1 + |\sin(\pi(t-\tau)/k)|} - 1 \right)^d \,,$$

and interpret it as a seasonally varying intensity function with k controlling the concentration of price shocks occuring in a multiple of k years. The exponent d is adjusting the dispersion of jumps around peaking times, and the constant c is the maximum expected number of jumps per unit time. [Geman and Roncoroni (2006)] give the example of $k = 0.5$, meaning a concentration of price shocks twice a year. Letting the phase change be $\tau = 7/12$, we will have the maximum number of jumps concentrated in January and August every year.

Observe that $N(s) - N(t)$ is a Poisson random variable with intensity

$\int_t^s \lambda(u)\,du$. Hence, we find the cumulant function of $L(t)$ to be

$$\psi(s,t;\theta) = \int_s^t \lambda(u)\,du\left(e^{\psi_X(\theta)} - 1\right).$$

Further, the compensator measure ℓ is

$$\ell(dz, ds) = \lambda(s) F_X(dz)\,ds.$$

Hence, the jump process $L(t)$ is an II process. The compensator measure is time-varying, and we therefore lose the stationarity we had for the compensated Poisson process. On the other hand, this II process is nothing but a compensated Poisson process where the jumps arrive at a rate which depends on time.

2.6.2 Models based on the generalized hyperbolic distributions

The *generalized hyperbolic* (GH) distributions were first introduced in [Barndorff-Nielsen (1977)] in connection with the study of wind-blown sand. This family of distributions has later been proven to be useful for financial modelling, as demonstrated by [Eberlein and Keller (1995)], [Eberlein, Keller and Prause (1998)], [Keller (1997)], [Rydberg (1997)], [Barndorff-Nielsen (1998)], [Prause (1999)], [Raible (2000)], [Lillestøl (2000)] and [Bølviken and Benth (2000)] on various financial data series. In [Benth and Šaltytė-Benth (2004)], the NIG Lévy process was used to model the evolution of gas and oil spot prices, while in [Benth and Šaltytė-Benth (2005)] it was argued that temperature dynamics could be modelled reasonably well by a GH distribution. For Value at Risk computations, [Eberlein and Stahl (2003)] argue in favour of the GH distributions when modelling the electricity spot price on the EEX.

The family of GH distributions is infinitely divisible, and has density function

$$f_{\text{gh}}(x;\lambda,\mu,\alpha,\beta,\delta) = c\big(\delta^2+(x-\mu)^2\big)^{(\lambda-\frac{1}{2})/2}\exp\big(\beta(x-\mu)\big)\times K_{\lambda-\frac{1}{2}}\big(\alpha\sqrt{\delta^2+(x-\mu)^2}\big), \tag{2.23}$$

where K_s is the modified Bessel function of the third kind with index s (see, for example, [Johnson, Kotz and Balakrishnan (1994)], Section A12, or [Abramowitz and Stegun (1972)], Section 9.6[2]) and

$$c = \frac{(\alpha^2 - \beta^2)^{\lambda/2}}{\sqrt{2\pi}\alpha^{\lambda-\frac{1}{2}}\delta^\lambda K_\lambda\big(\delta\sqrt{\alpha^2-\beta^2}\big)},$$

being the normalising constant. Here, α is controlling the steepness of the distribution, μ is the location parameter, β is related to skewness and δ is the scaling parameter. If $\beta = 0$ the GH distribution is symmetric. The parameter λ is identifying the sub-family of distributions within the GH class. The moment generating function of a random variable X distributed according to a GH distribution is explicitly given by

$$\mathbb{E}\big[e^{uX}\big] = e^{\mu u}\Big(\frac{\alpha^2-\beta^2}{\alpha^2-(\beta+u)^2}\Big)^{\lambda/2}\frac{K_\lambda\big(\delta\sqrt{\alpha^2-(\beta+u)^2}\big)}{K_\lambda\big(\delta\sqrt{\alpha^2-\beta^2}\big)}, \tag{2.24}$$

whenever $|\beta + u| < \alpha$. Hence, this family of distributions have finite moments of all orders (except for some asymptotic cases like the Student t, see below).

We say that $L(t)$ is a *GH Lévy process* if $L(t)$ is a Lévy process with $L(1)$ being a GH distributed random variable. The Lévy measure is given by

$$\widetilde{\ell}_{\mathrm{GH}}(dz) = |z|^{-1}e^{\beta z}\left\{\frac{1}{\pi^2}\int_0^\infty \frac{\exp(-\sqrt{2y+\alpha^2}|z|)}{J_\lambda^2(\delta\sqrt{2y})+Y_\lambda^2(\delta\sqrt{2y})}\,\frac{dy}{y} + \lambda e^{-\alpha|z|}\right\}dz, \tag{2.25}$$

when $\lambda \geq 0$, and

$$\widetilde{\ell}_{\mathrm{GH}}(dz) = |z|^{-1}e^{\beta z}\frac{1}{\pi^2}\int_0^\infty \frac{\exp(-\sqrt{2y+\alpha^2}|z|)}{J_{-\lambda}^2(\delta\sqrt{2y})+Y_{-\lambda}^2(\delta\sqrt{2y})}\,\frac{dy}{y}\,dz, \tag{2.26}$$

when $\lambda < 0$. Here J_λ and Y_λ are the Bessel functions of the first and second kind, respectively, with index (or order) λ (see, for example, [Johnson, Kotz and Balakrishnan (1994)], Section A5 and [Abramowitz and Stegun (1972)], Section 9.1.). Another name for Y_λ is the Weber function.

The GH Lévy processes $L(t)$ are pure jump processes, that is, the continuous martingale part in their Lévy-Kintchine representation is zero. Furthermore, the Lévy measure $\ell_{\mathrm{GH}}(dz)$ does not integrate $|z|$ in the neighbourhood of zero, which implies that the paths of $L(t)$ are not of finite variation.

[2]Note that the class of functions K_s is termed the modified Bessel functions of the *second kind* in Matlab.

Two special cases of the GH distribution are the hyperbolic and NIG distributions. Choosing $\lambda = 1$, we get

$$f_{\mathrm{H}}(x) = \frac{\sqrt{\alpha^2 - \beta^2}}{2\alpha\delta K_1(\delta\sqrt{\alpha^2 - \beta^2})} \exp\left(-\alpha\sqrt{\delta^2 + (x-\mu)^2} + \beta(x-\mu)\right). \quad (2.27)$$

This family of hyperbolic distributions was studied by [Eberlein and Keller (1995)] as a model for financial logreturns, see also [Eberlein, Keller and Prause (1998)], [Prause (1999)], [Keller (1997)] and [Raible (2000)]. Later, [Rydberg (1997)] and [Barndorff-Nielsen (1998)] considered the case $\lambda = -1/2$, corresponding to the NIG family of distribution. The density for this family is

$$f_{\mathrm{NIG}}(x) = \frac{\alpha}{\pi} \exp\left(\delta\sqrt{\alpha^2 - \beta^2} + \beta(x-\mu)\right) \frac{K_1\left(\alpha\delta\sqrt{1 + \left(\frac{x-\mu}{\delta}\right)^2}\right)}{\sqrt{1 + \left(\frac{x-\mu}{\delta}\right)^2}}, \quad (2.28)$$

and the Lévy measure simplifies to

$$\widetilde{\ell}_{\mathrm{NIG}}(dz) = \frac{\alpha\delta}{\pi|z|} K_1(\alpha|z|)\mathrm{e}^{\beta z}\, dz. \quad (2.29)$$

A Lévy process with hyperbolic or NIG distributed increments will be a *hyperbolic* or *NIG* Lévy process, respectively. The NIG Lévy process will be a pure jump Lévy process (as the hyperbolic), having a drift given as $\gamma(t) = \gamma t$ with

$$\gamma = \frac{2\delta\alpha}{\pi} \int_0^1 \sinh(\beta x) K_1(\alpha|x|)\, dx\,. \quad (2.30)$$

Using the fact that

$$K_{-\frac{1}{2}}(x) = \sqrt{\frac{\pi}{2}} x^{-1/2} \mathrm{e}^{-x}\,,$$

we derive from (2.24) that the logarithm of the moment generating function of a NIG variable is

$$\ln \mathbb{E}\left[\mathrm{e}^{uX}\right] = \mu u + \delta\left(\sqrt{\alpha^2 - \beta^2} - \sqrt{\alpha^2 - (\beta + u)^2}\right).$$

The cumulant function is therefore

$$\psi_{\mathrm{NIG}}(\theta) = \mathrm{i}\theta\mu + \delta\left(\sqrt{\alpha^2 - \beta^2} - \sqrt{\alpha^2 - (\beta + \mathrm{i}\theta)^2}\right). \quad (2.31)$$

Although the NIG distribution has a complicated density function, the cumulant and moment generating function have simple expressions. These are properties which will become useful later.

Sometimes it is convenient to represent the four NIG parameters in the so-called shape triangle. Transform the NIG parameters by the following relations

$$\xi = \left(1 + \delta\sqrt{\alpha^2 - \beta^2}\right)^{-1/2}, \quad \chi = \frac{\xi\beta}{\alpha}. \tag{2.32}$$

Then, the coordinates (χ, ξ) will lie within a triangle (turned upside down), defined by $0 \leq |\chi| < \xi < 1$. The parameter χ measures the skewness of the NIG distribution, with $\chi > 0$ being a positively skewed distribution and $\chi < 0$ negatively. Obviously, $\chi = 0$ is a symmetric NIG distribution since this corresponds to $\beta = 0$. The ξ parameter measures the departure from normality, or tail heaviness of the NIG distribution. In fact, the limiting case $\xi = 0$ (and by implication $\chi = 0$), corresponds to a normal distribution. Hence, having parameters which give $(\xi, \chi) \approx (0, 0)$, means that the NIG distribution is close to being normal. On the other hand, the limit $\xi = 1$ gives the heavy-tailed Cauchy distribution. Often, in empirical studies, the shape triangle is used for testing departure from normality.

Another popular subclass of the GH distributions is the Student t distribution. It was suggested as a model for asset logreturns by [Praetz (1972)], and later studied in more detail by [Blattberg and Gonedes (1974)]. Both [Praetz (1972)] and [Blattberg and Gonedes (1974)] compared empirically the performance of a Student t distribution to the stable laws on stock price data. The centered Student t distribution (with scale 1) has a density function given by

$$f_{\text{Student}}(x) = \frac{\sqrt{\nu^\nu}\,\Gamma\left(\frac{1}{2}(1+\nu)\right)}{\sqrt{\pi}\,\Gamma(\frac{1}{2}\nu)} \left(\nu + x^2\right)^{-\frac{1}{2}(1+\nu)},$$

where $\nu > 0$ denotes the degrees of freedom. We obtain this distribution by choosing $\lambda = -\nu/2$, $\alpha = \beta = 0$ and $\delta = \sqrt{\nu}$ as parameters in the GH family. Observe that when $\nu = 1$ we obtain the Cauchy distribution, while the normal distribution is recovered as the limit $\nu \to \infty$. The Student t distribution has finite moments of all orders up to $n < \nu$, and has higher kurtosis than the normal distribution.

The GH distributions can be described as mean-variance mixtures of a normal distribution with a generalized inverse Gaussian (GIG). Let σ^2 be GIG distributed with parameters λ, δ and $\gamma = \sqrt{\alpha^2 - \beta^2}, 0 \leq |\beta| < \alpha$.

That is, assume that the density of σ^2 is given by

$$f_{\mathrm{GIG}}(x;\lambda,\delta,\gamma) = \left(\frac{\gamma}{\delta}\right)^{\lambda} \frac{x^{\lambda-1}}{2K_\lambda(\delta\gamma)} \exp\left(-\frac{1}{2}\left(\frac{\delta^2}{x} + \gamma^2 x\right)\right), x > 0\,. \quad (2.33)$$

Assume that X conditioned on σ^2 is a normally distributed random variable with expectation $\mu + \beta\sigma^2$ and variance σ^2, then X will be a GH distributed random variable with parameters $\lambda, \mu, \alpha, \beta$ and δ. Note that the GIG distribution has several interesting special cases. For instance, choosing $\lambda = -1/2$ yields the inverse Gaussian (IG) law with parameters δ and γ, having density

$$f_{\mathrm{IG}}(x;\delta,\gamma) = \frac{\delta}{\sqrt{2\pi}} e^{\delta\gamma} x^{-3/2} \exp\left(-\frac{1}{2}(\delta^2 x^{-1} + \gamma^2 x)\right).$$

We recall the connection to the NIG distribution. Further, the more classical Gamma distribution is obtained by letting $\delta = 0$ and restricting $\lambda > 0$. By letting $\alpha = \gamma^2/2$ in the GIG distribution, we obtain the distribution of a $\Gamma(\lambda,\alpha)$ random variable as,

$$f_\Gamma(x;\lambda,\alpha) = \frac{\alpha^\lambda}{\Gamma(\lambda)} x^{\lambda-1} \exp(-\alpha x)\,. \quad (2.34)$$

A straightforward calculation shows that

$$\psi_\Gamma(\theta) = \lambda \ln\left(\frac{\alpha}{\alpha - \mathrm{i}\theta}\right) \quad (2.35)$$

is the cumulant function of the Gamma distribution.

2.6.3 *Models based on the Variance-Gamma and CGMY distributions*

The CGMY distribution and the associated Lévy process was introduced by [Carr *et al.* (2002)]. It is a four-parameter family of distributions[3] used for modelling logreturns of financial assets. In [Geman (2005)] it is also mentioned in connection with electricity spot prices as a potential model for Value at Risk calculations. The CGMY distribution is a generalization of the Variance-Gamma (VG) distribution, and we introduce the two families here using the theory and notation in [Carr *et al.* (2002)].

[3] The name of the distribution is after the authors' surnames.

The CGMY distribution with parameters C, G, M and Y is defined through its cumulant function as

$$\psi_{\mathrm{CGMY}}(\theta) = C\Gamma(-Y)\left[(M - \mathrm{i}\theta)^Y + (G + \mathrm{i}\theta)^Y - G^Y\right]. \tag{2.36}$$

It turns out that the CGMY distribution is infintely divisible and therefore a Lévy process $L(t)$, where $L(1)$ is CGMY distributed, can be constructed. The Lévy measure for this process is absolutely continuous with respect to the Lebesgue measure, explicitly given as

$$\widetilde{\ell}_{\mathrm{CGMY}}(dz) = \begin{cases} C|z|^{-1-Y}\exp(-G|z|)\,dz\,, & \text{for } z < 0\,, \\ C|z|^{-1-Y}\exp(-M|z|)\,dz\,, & \text{for } z > 0\,. \end{cases} \tag{2.37}$$

Here, $Y < 2$ in order to have a Lévy measure which integrates $|z|^2$ around zero.

Letting $Y = 0$, the CGMY distribution reduces to the VG distribution. Consider the Gamma distribution $\Gamma(1/\nu, 1/\nu)$, where the parameter ν is called the variance rate. The cumulant function is given by (2.35)

$$\psi_\Gamma(\theta) = -\frac{1}{\nu}\ln(1 - \mathrm{i}\nu\theta)\,.$$

One may construct the VG distribution from a time-changed Brownian motion, where time is following a $\Gamma(1/\nu, 1/\nu)$ process, that is, $L_{\mathrm{G}}(t)$ where $L_{\mathrm{G}}(1)$ is $\Gamma(1/\nu, 1/\nu)$ distributed. The process $L_{\mathrm{G}}(t)$ becomes a subordinator. Let $B(t)$ be a Brownian motion and let δ and σ be constants. Then,

$$L_{\mathrm{VG}}(t) = \delta L_{\mathrm{G}}(t) + \sigma B(L_{\mathrm{G}}(t))$$

is a VG Lévy process, where the cumulant function of $L_{\mathrm{VG}}(t)$ is

$$\psi_{\mathrm{VG}}(\theta) = \ln(1 - \mathrm{i}\delta\nu\theta + \sigma^2\nu\theta^2/2)^{-1/\nu}\,.$$

This is the cumulant function of the VG distribution. Another representation of the VG distribution is through the difference of two independent Gamma processes. This follows from the factorisation of the characteristic function as

$$\frac{1}{1 - \mathrm{i}\delta\theta + \sigma^2\nu\theta^2/2} = \left(\frac{1}{1 - \mathrm{i}\eta_p\theta}\right)\left(\frac{1}{1 + \mathrm{i}\eta_n\theta}\right),$$

with constants η_p and η_n satisfying $\eta_p - \eta_n = \delta\nu$ and $\eta_p\eta_n = \sigma^2\nu/2$. The solutions are

$$\eta_p = \sqrt{\frac{\delta^2\nu^2}{4} + \frac{\sigma^2\nu}{2}} + \frac{\delta\nu}{2}\,,$$

$$\eta_n = \sqrt{\frac{\delta^2\nu^2}{4} + \frac{\sigma^2\nu}{2}} + \frac{\delta\nu}{2}.$$

Let

$$L_{\mathrm{VG}}(t) = L_{\mathrm{G},p}(t) - L_{\mathrm{G},n}(t),$$

with $L_{\mathrm{G},p}, L_{\mathrm{G},n}$ being Gamma processes with mean and variance rates μ_p, μ_n and ν_p, ν_n, respectively. Here $\mu_p = \eta_p/\nu, \mu_n = \eta_n/\nu$ and $\eta_p = \mu_p^2\nu, \eta_n = \mu_n^2\nu$. The Lévy measure of the VG Lévy process is then

$$\widetilde{\ell}_{\mathrm{VG}}(dz) = \begin{cases} \frac{\mu_n^2}{\nu_n}|z|^{-1}\exp(-\frac{\mu_n}{\nu_n}|z|)\,dz\,, \text{ for } z < 0\,, \\ \frac{\mu_p^2}{\nu_p}|z|^{-1}\exp(-\frac{\mu_p}{\nu_p}|z|)\,dz\,, \text{ for } z > 0\,. \end{cases} \tag{2.38}$$

The CGMY Lévy process for $Y = 0$ is a VG Lévy process with $C = 1/\nu$, $G = 1/\eta_n$ and $M = 1/\eta_p$. We refer to [Carr *et al.* (2002)] for more theory on the CGMY Lévy process and an empirical study of logreturns using these models.

Chapter 3

Stochastic Models for the Energy Spot Price Dynamics

In this Chapter we study different types of stochastic processes for modelling energy spot prices. Motivated by the nature of commodity prices, general classes of mean reversion dynamics will be analysed. The models will include jumps to describe typical features likes spikes of energy spot prices.

3.1 Introduction

The classical stochastic process for the spot dynamics of commodity prices is given by the so-called Schwartz' model (see, for instance, [Schwartz (1997)]). It is defined as the exponential of an OU process, and has become the standard model for energy prices possessing mean-reverting features. The Schwartz model is feasible for calculating explicit forward prices in an arbitrage-free way for markets where hedging is (partly) possible.

We present in this chapter spot price models based on a sum of non-Gaussian OU processes. We will consider geometric models of the kind proposed by Schwartz, but also arithmetic models which may be more analytically tractable in markets where the forward contracts have a delivery period. Representing the logarithmic prices, or the prices itself, by a series of OU processes allows us to model different speeds of mean reversion, and to incorporate a mixture of jump and diffusional behaviour of the prices. Spikes may be modelled by an OU process having a low frequency of big jumps, with a fast mean reversion, while more "normal" price variations are represented by a slower mean-reverting process driven by Brownian motion. Seasonality in jumps is captured using an II process in the jump model. Arithmetic models usually lead to prices which may become negative. We introduce a class of models which ensure positive prices, still being of an arithmetic type.

Throughout this chapter we suppose that the market trades in products which have a finite time horizon $\mathcal{T} < \infty$.

3.2 Spot price modelling with Ornstein-Uhlenbeck processes

The basic process to model the spot price behaviour of energies is the OU process. It is also the basic model for the temperature dynamics that we study in Chapter 10.

Denote by $I(t)$ an II process with Lévy-Kintchine representation

$$\psi(t,s;\theta) = \mathrm{i}\theta(\gamma(s)-\gamma(t)) - \frac{1}{2}\theta^2(C(s)-C(t)) \\ + \int_t^s \int_{\mathbb{R}} \left\{ \mathrm{e}^{\mathrm{i}z\theta} - 1 - \mathrm{i}z\theta 1_{|z|<1} \right\} \ell(dz,du) \,, \qquad (3.1)$$

where γ is of finite variation. Thus, $I(t)$ is a semimartingale process. Let us define an OU process with non-Gaussian innovations.

Definition 3.1 A RCLL process $X(s)$, $t \le s \le \mathcal{T}$, is called a (non-Gaussian) OU process if it is the unique strong solution of the stochastic differential equation

$$dX(s) = (\mu(s) - \alpha(s)X(s))\,ds + \sigma(s)\,dI(s), \quad X(t) = x. \qquad (3.2)$$

Here, μ, α and σ are all real-valued continuous functions on $[0, \mathcal{T}]$.

In order for this definition to be well-posed, we need to know whether there exists a unique solution X of the stochastic differential equation (3.2). This is indeed the case, as demonstrated by the following Proposition.

Proposition 3.1 *The unique strong solution $X(s)$, $s \ge t$, of (3.2) is given by*

$$X(s) = x\exp\left(-\int_t^s \alpha(v)\,dv\right) + \int_t^s \mu(u)\exp\left(-\int_u^s \alpha(v)\,dv\right)du \\ + \int_t^s \sigma(u)\exp\left(-\int_u^s \alpha(v)\,dv\right)dI(u). \qquad (3.3)$$

Proof. First, we consider the question of uniqueness. Let X_1 and X_2 be two solutions to (3.2), and define the process $Y(s) = X_1(s) - X_2(s)$.

$Y(s)$ will be RCLL from the properties of X_1 and X_2. Appealing to their dynamics we find

$$Y(s) = -\int_t^s \alpha(u)Y(u)\,du,$$

with $Y(t) = 0$. By induction we have

$$Y(s) = (-1)^n \int_t^s \alpha_n(u)Y(u)\,du, \quad n = 1, 2, \ldots,$$

for

$$\alpha_n(u) = \alpha(u)\int_u^s \alpha(u_{n-1})\int_{u_{n-1}}^s \alpha(u_{n-2})\cdots\int_{u_2}^s \alpha(u_1)\,du_1\cdots du_{n-1}.$$

From the continuity of α, we know that it is bounded on $[s,t]$ by a constant, denoted $c > 0$. Hence,

$$\sum_{n=1}^{\infty} |\alpha_n(u)| \le \sum_{n=1}^{\infty} \frac{c^n}{(n-1)!}(s-u)^{n-1} < \infty,$$

and therefore $(-1)^n\alpha_n(u)$ converges to zero as $n \to \infty$ uniformly on $u \in [t,s]$. This implies that $Y(s) = 0$, and uniqueness of solution is established.

The existence follows from Itô's Formula. Let

$$Z(s) = \exp\left(-\int_t^s \alpha(v)\,dv\right)\int_t^s \sigma(u)\exp\left(\int_t^u \alpha(v)\,dv\right)dI(u).$$

Then, it follows that

$$dZ(s) = -\alpha(s)Z(s)\,ds + \sigma(s)\,dI(s).$$

This shows that $X(s)$ is a strong solution to the stochastic differential equation (3.2). Hence, the proposition follows. □

An equivalent way of defining an OU process would of course be to say that it is given by (3.3), and next use Itô's Formula to show that it indeed satisfies the dynamics in (3.2). However, the definition based on the stochastic differential equation (3.2) seems to be the preferred way of introducing an OU process when $I = B$, a Brownian motion, and we stick to this tradition here when generalizing it to II processes driving the noise.

The characteristic function of $X(s)$ provides us with all the probabilistic information about the process that we need for further calculations.

Proposition 3.2 *The characteristic function of an OU process is given by*

$$\mathbb{E}\left[e^{i\theta X(s)} \mid X(t) = x\right] = \exp\Big(i\theta\{xe^{-\int_t^s \alpha(v)\,dv} + \int_t^s \mu(u)e^{-\int_u^s \alpha(v)\,dv}\,du\}$$
$$+\psi(t,s;\theta\sigma(\cdot)e^{-\int_\cdot^s \alpha(v)\,dv})\Big) \tag{3.4}$$

for any $\theta \in \mathbb{R}$, where $\psi(t,s;\cdot)$ is the cumulant function of I and $\psi(t,s;g(\cdot))$ is defined to be

$$\psi(t,s;g(\cdot)) = i\int_t^s g(u)\,d\gamma(u) - \frac{1}{2}\int_t^s g^2(u)\,dC(u)$$
$$+\int_t^s \int_{\mathbb{R}} \{e^{ig(u)z} - 1 - ig(u)z1_{|z|<1}\}\,\ell(dz,du)\,, \tag{3.5}$$

for a continuous function $g : [0,T] \mapsto \mathbb{R}$.

Proof. Let f be a step function on $[t,s]$ given as

$$f(u) = \sum_{k=1}^n a_k 1_{(u_{k-1},u_k]}(u), \quad \text{with}\, t = u_0 < u_1 < \cdots < u_n = s.$$

Then, by the independent increment property of I we have

$$\begin{aligned}
\mathbb{E}\left[\exp\left(\mathrm{i}\theta\int_t^s f(u)\,dI(u)\right)\right] &= \mathbb{E}\left[\exp\left(\mathrm{i}\theta\sum_{k=1}^n a_k(I(u_k) - I(u_{k-1})\right)\right] \\
&= \prod_{k=1}^n \mathbb{E}\left[\exp\left(\mathrm{i}\theta a_k(I(u_k) - I(u_{k-1}))\right)\right] \\
&= \prod_{k=1}^n \exp\left(\psi(u_{k-1},u_k;\theta a_k)\right) \\
&= \exp\left(\sum_{k=1}^n \psi(u_{k-1},u_k;\theta a_k)\right) \\
&= \exp\left(\psi(t,s;\theta f(\cdot))\right).
\end{aligned}$$

Now, every continuous function on $[t,s]$ may be approximated by a step function, and passing to the limit we find by bounded convergence that the proposition follows. □

We find the expectation of $X(t)$ in the following Lemma.

Lemma 3.1 *Suppose that*

$$\int_t^s \int_{|z|\geq 1} |z|\, \ell(dz, du) < \infty\,.$$

Then

$$\begin{aligned}\mathbb{E}\left[X(s)\,|\,X(t)=x\right] &= x\exp\left(-\int_t^s \alpha(v)\,dv\right) \\ &+ \int_t^s \mu(u)\exp\left(-\int_u^s \alpha(v)\,dv\right) du \\ &+ \int_t^s \sigma(u)\exp\left(-\int_u^s \alpha(v)\,dv\right) d\gamma(u) \\ &+ \int_t^s \int_{|z|\geq 1} z\sigma(u)\left(-\int_u^s \alpha(v)\,dv\right) \ell(dz, du)\,.\end{aligned}$$

Proof. Consider the expectation of $\int_t^s f(u)\,dI(u)$ where f is a continuous function. From the integrability condition on the compensator measure, we find by appealing to the characteristic function of I that

$$\begin{aligned}&\mathbb{E}[\int_t^s f(u)\,dI(u)] \\ &= (-\mathrm{i})\frac{d}{d\theta}\mathbb{E}\left[\exp\left(\mathrm{i}\theta\int_t^s f(u)\,dI(u)\right)\right]\Big|_{\theta=0} \\ &= (-\mathrm{i})\frac{d}{d\theta}\exp\left(\psi(t,s;\theta f(\cdot))\right)\Big|_{\theta=0} \\ &= \int_t^s f(u)\,d\gamma(u) + \int_t^s \int_{\mathbb{R}} zf(u) - zf(u)1_{|z|<1}\,\ell(dz,du)\,.\end{aligned}$$

Thus, the Lemma follows. □

Let us consider the special case when the II process $I(t)$ is a Brownian motion $B(t)$. Then it turns out that $X(t)$ is a Gaussian process with a stationary distribution.

Lemma 3.2 *Suppose $I(t) = B(t)$. Then, at each time instance $s \geq t \geq 0$, the random variable $X(s)$ conditioned on $X(t) = x$ is normally distributed, with mean*

$$\begin{aligned}\mathbb{E}[X(s)\,|\,X(t)=x] &= x\exp\left(-\int_t^s \alpha(v)\,dv\right) \\ &+ \int_t^s \mu(u)\exp\left(-\int_u^s \alpha(v)\,dv\right) du\,,\end{aligned}$$

and variance

$$Var[X(s) \mid X(t) = x] = \int_t^s \sigma^2(u) \exp\left(-2\int_u^s \alpha(v)\,dv\right) du\,.$$

Proof. This is a straightforward calculation using either the known properties for the Itô integral, or appealing to the characteristic function in Prop. 3.2 with $\psi(t, s, g(\cdot)) = -0.5\int_t^s g^2(u)\,du$. □

We discuss the stationarity of the OU process, and for a moment we let $\mathcal{T} = \infty$. Obviously, stationarity of $X(s)$ is reflected in properties of the parameter functions and the II process $I(t)$. We restrict our attention to the case of constant coefficients $\mu(s) \equiv \mu$, $\alpha(u) \equiv \alpha$ and $\sigma(u) \equiv \sigma$, where α is strictly positive. Considering the logarithm of the characteristic function of $X(t)$ in Prop. 3.2, we see that the first term converges to zero when $s \to \infty$. In the case $I(t) = B(t)$, we find from the Lemma above that

$$\lim_{t\to\infty} X(t) = X_\infty\,,$$

where X_∞ is a normally distributed random variable with mean given by μ/α and variance $\sigma^2/2\alpha$.

Consider $I(t) = L(t)$, a Lévy process. Then if

$$\int_{|z|>2} \ln|z|\,\widetilde{\ell}(dz) < \infty\,,$$

for the Lévy measure $\widetilde{\ell}(dz)$, we have that the OU process has a limit distribution with cumulant function given as

$$\mathrm{i}\theta\frac{\mu}{\alpha} + \int_0^\infty \psi(\theta \mathrm{e}^{-\alpha s})\,ds\,. \tag{3.6}$$

Here, $\psi(\theta)$ is the cumulant function of $L(1)$. This result is stated and proven by [Sato (1999)], Thm. 17.5.

As an example, let for simplicity $\mu = 0$ and consider $L(t) = \sum_{k=1}^{N(t)} Z_k$, where $N(t)$ is a Poisson process with intensity λ. The jumps Z_k are i.i.d. exponentially distributed random variables, with density given by

$$f_{\mathrm{Exp}}(z) = \frac{1}{\mu_J}\exp(-z/\mu_J)\,. \tag{3.7}$$

The mean jump size is μ_J. The cumulant function of Z_k is easily calculated to be

$$\psi_Z(\theta) = -\ln(1 - \mathrm{i}\theta\mu_J)\,.$$

It follows from (2.22) that the cumulant function of $L(1)$ is

$$\psi(\theta) = \lambda\left(\mathrm{e}^{\psi_Z(\theta)} - 1\right).$$

Hence, from (3.6) we can calculate the cumulant function for the stationary distribution of $X(t)$ driven by $L(t)$ as

$$-\frac{\lambda}{\alpha}\ln(1 - \mathrm{i}\theta\mu_J).$$

We recognise this as the cumulant function of the Gamma distribution $\Gamma(\lambda/\alpha, 1/\mu_J)$ from (2.35).

These stationarity properties can effectively be used in modelling, as discussed by [Barndorff-Nielsen and Shephard (2001)] for the case of stochastic volatility. In many situations one may have access to, or at least an idea of, the stationary distribution D of the OU process (from data analysis, say). The question is then if there exists a Lévy process $L(t)$, denoted the *background driving Lévy process*, such that the solution $X(t)$ of

$$dX(t) = (\mu - \alpha X(t))\,dt + dL(t)$$

has the desired stationary distribution D. The answer is affirmative, whenever the distribution D is so-called *self-decomposable.* A distribution is self-decomposable if its cumulant function has the property

$$\psi(\theta) = \psi(c\theta) + \psi_c(\theta),$$

for all $\theta \in \mathbb{R}$ and $c \in (0,1)$, with ψ_c being some family of cumulant functions. In [Barndorff-Nielsen and Shephard (2001)], these issues (with references to relevant literature) are discussed in more detail. We note in passing that [Halgreen (1979)] proved the self-decomposability of the GIG distribution, which implies that, for instance, the Gamma distribution is self-decomposable. The latter fact also follows, of course, from the calculation above where we see that the background driving Lévy process is a compound Poisson process with exponentially distributed jumps. We shall apply this distribution in some examples later (see Sect. 3.4).

Note that when the parameter functions of the OU process are time-dependent, we need to impose integrability conditions to have stationarity. Also, going from a Lévy process to an II process introduces essentially a time-inhomogeneity in the associated random jump measure, which complicates the stationarity question. We leave the issues on stationarity, and move on to introduce more specific spot price models based on the OU process.

3.2.1 *Geometric models*

Next we introduce n pure jump semimartingale II processes $I_j(t), j = 1, \ldots, n$, where $I_j(t)$ and $I_k(t)$ are independent of each other for all $j \neq k$. We can represent each process via its associated random jump measure $N_j(dt, dz)$ by the Lévy-Kintchine representation

$$I_j(t) = \gamma_j(t) + \int_0^t \int_{|z|<1} z\, \widetilde{N}_j(dz, du) + \int_0^t \int_{|z|\geq 1} z\, N_j(dz, du)\,.$$

We recall the assumption that γ_j has bounded variation and that the compensator measure is denoted by $\ell_j(dz, du)$. Note that we do not suppose that the processes are martingales, but semimartingales. They may, thus, impose a drift in the overall dynamics. Further, assume that B_k, $k = 1, \ldots, p$ are p independent Brownian motions.

Let the stochastic process $S(t)$ be defined as

$$\ln S(t) = \ln \Lambda(t) + \sum_{i=1}^{m} X_i(t) + \sum_{j=1}^{n} Y_j(t)\,, \tag{3.8}$$

where, for $i = 1, \ldots, m$,

$$dX_i(t) = (\mu_i(t) - \alpha_i(t) X_i(t))\, dt + \sum_{k=1}^{p} \sigma_{ik}(t)\, dB_k(t)\,, \tag{3.9}$$

and, for $j = 1, \ldots, n$,

$$dY_j(t) = (\delta_j(t) - \beta_j(t) Y_j(t))\, dt + \eta_j(t) dI_j(t)\,. \tag{3.10}$$

The deterministic (seasonal) price level is modelled by the function $\Lambda(t) : [0, \mathcal{T}] \mapsto (0, \infty)$, which is assumed to be continuously differentiable. It captures the seasonality in mean (log-)prices, and sometimes it is referred to as the seasonal function. Further, the coefficients $\mu_i, \alpha_i, \delta_j, \beta_j, \sigma_{ik}$ and η_j are all continuous functions. From a modelling perspective it is natural to choose $\mu_i(t) = \delta_j(t) = 0$ since the OU processes should ideally revert towards zero in order to have the seasonality function as the mean price level. Further, it is also reasonable to have constant speeds of mean reversions. However, for the sake of generality we include the time-dependent case since it does not imply any additional mathematical complications (except, possibly, more notation). We do, on the other hand, assume that the speeds of mean reversion $\alpha_i(t)$ and $\beta_j(t)$ are non-negative for all $t \in [0, \mathcal{T}]$. To have the flexibility of correlating different factors X_i, we model the random fluctuations based on a common set of independent Brownian motions B_k, $k = 1, \ldots, p$. In many models (for example, the electricity spot model by [Lucia and Schwartz (2002)]), the factors represent the long- and short-term fluctuations of the spot dynamics which may be correlated. We suppose that the jump components I_j are independent, which is an obvious restriction of generality. In Chapter 6 we will briefly discuss how to make them dependent using the theory of copulas.

From the conditions we have that $\ln S(t)$, and thus $S(t)$ is a semimartingale process. We assume that the initial conditions of X_i and Y_j are such that

$$\sum_{i=1}^{m} X_i(0) + \sum_{j=1}^{n} Y_j(0) = \ln S(0) - \ln \Lambda(0) . \tag{3.11}$$

From (3.3), an explicit representation of $S(t)$ is given by

$$S(t) = \Lambda(t) \exp\left(\sum_{i=1}^{m} X_i(t) + \sum_{j=1}^{n} Y_j(t) \right) , \tag{3.12}$$

where, for $i = 1, \ldots, m$,

$$\begin{aligned} X_i(t) = X_i(0) \exp\left(- \int_0^t \alpha_i(v)\, dv \right) \\ + \int_0^t \mu_i(u) \exp\left(- \int_u^t \alpha_i(v)\, dv \right) du \\ + \sum_{k=1}^{p} \int_0^t \sigma_{ik}(u) \exp\left(- \int_u^t \alpha_i(v)\, dv \right) dB_k(u) , \end{aligned} \tag{3.13}$$

and, for $j = 1, \ldots, n$,

$$\begin{aligned} Y_i(t) = Y_j(0) \exp\left(- \int_0^t \beta_j(v)\, dv \right) \\ + \int_0^t \delta_j(u) \exp\left(- \int_u^t \beta_j(v)\, dv \right) du \\ + \int_0^t \eta_j(u) \exp\left(- \int_u^t \beta_j(v)\, dv \right) dI_j(u) . \end{aligned} \tag{3.14}$$

Observe that the jump parts may impose an additional drift, since we do not assume them to be martingale processes. It is highly natural to let them have a drift because they represent the jumps in the market. For instance, the occurrence of price spikes should add to the overall expected spot price in excess of the seasonal function. Hence, for a model with jumps, the mean level of the log-spot is not the same as the log of the seasonal function, modified by the terms stemming from mean reversion levels given by μ_i and δ_j. The difference is measured by the expected value of the jump parts. However, note that any big price jump (in positive or negative direction), will be smoothed out after a time depending on the strength of the mean reversion β_j.

We continue with stating the dynamics of $S(t)$.

Proposition 3.3 *The dynamics of $S(t)$ is given by*

$$
\begin{aligned}
\frac{dS(t)}{S(t-)} &= \left\{\frac{\Lambda'(t)}{\Lambda(t)} + \frac{1}{2}\sum_{k=1}^{p}\sum_{i_1,i_2=1}^{m}\sigma_{i_1k}(t)\sigma_{i_2k}(t) + \sum_{i=1}^{m}(\mu_i(t)-\alpha_i(t)X_i(t))\right\}dt \\
&\quad + \left\{\sum_{j=1}^{n}(\delta_j(t)-\beta_j(t)Y_j(t))\right\}dt + \sum_{j=1}^{n}\eta_j(t)d\gamma_j(t) \\
&\quad + \sum_{j=1}^{n}\int_{|z|<1}\{e^{\eta_j(t)z}-1-\eta_j(t)z\}\,\ell_j(dz,dt) \\
&\quad + \sum_{k=1}^{p}\sum_{i=1}^{m}\sigma_{ik}(t)\,dB_k(t) \\
&\quad + \sum_{j=1}^{n}\int_{|z|<1}(e^{\eta_j(t)z}-1)\,\widetilde{N}_j(dz,dt) \\
&\quad + \sum_{j=1}^{n}\int_{|z|\geq 1}(e^{\eta_j(t)z}-1)\,N_j(dz,dt)\,.
\end{aligned}
$$

Proof. This follows directly from Itô's Formula since $X_i(t)$ and $Y_j(t)$, $i=1,\ldots,m, j=1,\ldots,n$ are independent semimartingales. □

In the case of $m=1, p=1$ and $n=0$, the process $S(t)$ specialises to

$$
\begin{aligned}
\frac{dS(t)}{S(t)} &= \left\{\frac{\Lambda'(t)}{\Lambda(t)} + \alpha(t)\ln\Lambda(t) + \frac{1}{2}\sigma^2(t) + (\mu(t)-\alpha(t)\ln S(t))\right\}dt \\
&\quad + \sigma(t)\,dB(t)\,, \qquad (3.15)
\end{aligned}
$$

which is the Schwartz one-factor model, see [Schwartz (1997)].

Let us have a look at other, more general and flexible models proposed in the literature fitting into the framework of the spot price process defined in (3.8). The simplest extension of the one-factor model of Schwartz (3.15) is to include jumps. Hence, one can consider the following stochastic process

$$d\ln S(t) = d\ln \Lambda(t) - \alpha(t)\left(\ln S(t) - \ln \Lambda(t)\right) dt + \sigma(t)\, dB(t) + dI(t)\,. \quad (3.16)$$

This is a jump-diffusion model where the speed of mean reversion is common for both jumps occurring in the pure jump II process $I(t)$ and the diffusion part $B(t)$. The idea of the model is to let the Brownian motion account for the small variations in the energy price when normal trading takes place, while the arrival of information altering the supply and/or demand significantly are modelled by sudden jumps in $I(t)$. Overall, the model implies that the residuals $\sigma(t)\, dB(t) + dI(t)$ become leptokurtic. By choosing $m = n = p = 1$, and letting

$$dX(t) = -\alpha(t)X(t)\, dt + \sigma(t)\, dB(t)\,,$$

and

$$dY(t) = -\alpha(t)Y(t)\, dt + dI(t)\,,$$

we see that $d\ln S(t) = d\ln\Lambda(t) + X(t) + Y(t)$, and thus it is a model in the class (3.8). Notice that the speed of mean reversion is the same for both OU processes. In [Benth and Šaltytė-Benth (2004)] this model was analysed for natural gas and oil using a pure jump NIG model, which is obtained by assuming $\sigma = 0$ and letting $I(t)$ be a NIG Lévy process. [Eberlein and Stahl (2003)] studied a simplified version of this model without mean reversion, but with a hyperbolic Lévy process for the random innovations. For purposes of Value at Risk calculations, they found a rather good description of the stochastic characteristics of the Nord Pool spot prices with this model. We obtain the process in [Eberlein and Stahl (2003)] by choosing $\alpha(t) = 0$ in the definition of $Y(t)$ above. [Cartea and Figueroa (2005)] let $I(t)$ be a compound Poisson process, and propose this as the model for electricity spot prices. In Chapter 5 we shall perform an empirical analysis of gas prices using these models. A more sophisticated electricity spot model is proposed in [Geman and Roncoroni (2006)], where the jump component is state-dependent. They suppose that $dI(t) = h(S_t)dJ(t)$, where $J(t)$ is a time-inhomogeneous compound Poisson process and h is a function which is ± 1, depending on the level of the spot price. In this way one is able to create spikes without introducing more factors with different speeds of mean reversion. This model is not covered by our general class (3.8) due to the state-dependent function h.

The authors [Lucia and Schwartz (2002)] propose a two-factor model of the form

$$\ln S(t) = \ln \Lambda(t) + X_1(t) + X_2(t)\,, \tag{3.17}$$

with $X_1(t)$ being an OU process

$$dX_1(t) = -\alpha_1 X_1(t)\,dt + \sigma_1 dB_1(t)\,,$$

and $X_2(t)$ being a drifted Brownian motion correlated with X_1, that is,

$$dX_2(t) = \mu_2\,dt + \sigma_2\left(\rho\,dB_1(t) + \sqrt{1-\rho^2}\,dB_2(t)\right)\,.$$

The parameters of the model are constants, and we note that the second process X_2 does not revert to a mean. The correlation between the random variables B_1 and $\rho B_1 + \sqrt{1-\rho^2} B_2$ is ρ. The idea of this model is to have a non-stationary process for the long-term equilibrium price level X_2 and short-term mean-reverting component X_1, possibly correlated. Using the definition of (3.8), we have $m = 2$, $n = 0$ and $p = 2$. Moreover, the mean reversion coefficient is $\alpha_2 = 0$ in order to have a non-stationary model X_2. In [Lucia and Schwartz (2002)] the model is applied to electricity prices.

An extension of the two-factor model of [Lucia and Schwartz (2002)] is proposed by [Villaplana (2004)], where the long-term level is a mean reversion process and the short-term level is influenced by jump shocks. In the language of the model (3.8) we suppose $m = 2$, $n = 1$ and $p = 2$, where

$$\begin{aligned} dX_1(t) &= (\mu_1 - \alpha_1 X_1(t))\,dt + \sigma_1\,dB_1(t)\,,\\ dX_2(t) &= (\mu_2 - \alpha_2 X_2(t))\,dt + \sigma_2\left(\rho\,dB_1(t) + \sqrt{1-\rho^2}\,dB_2(t)\right)\,, \end{aligned}$$

and

$$dY(t) = -\alpha_1 Y(t)\,dt + dI(t)\,,$$

with $I(t)$ being a time-homogeneous compound Poisson process. The short-term variations are given by $X_1 + Y$ in this model, whereas the long-term level is X_2. We see that this model correlates the short- and long-term variations as in [Lucia and Schwartz (2002)]. [Deng (2000)] extends such models to even more factors with both diffusion and jumps. In [Hambly, Howison and Kluge (2007)], a two-factor simplification of the model of [Villaplana

(2004)] is suggested. They propose a model with $m = n = p = 1$, where the process driven by the Brownian motion is slowly mean-reverting. To account for spikes, they suggest to use a process with fast mean reversion, and driven by a time-homogeneous compound Poisson process.

We notice that the multi-factor model (3.8) can be interpreted as an OU process which reverts to a stochastic mean, also fluctuating as OU processes. Such a model was proposed and investigated empirically for electricity markets by [Barlow, Gusev and Lai (2004)]. To exemplify this, consider the special case of $m = 2$ and $n = 0$, with

$$dX_1(t) = (\mu_1 - \alpha_1 X_1(t))\, dt + \sigma_1\, dB_1(t)\,,$$
$$dX_2(t) = (\mu_2 - \alpha_2 X_2(t))\, dt + \sigma_2\, dB_2(t)\,.$$

Denote the deseasonalized log-spot price by $\widetilde{S}(t)$, that is

$$\widetilde{S}(t) = \ln S(t) - \ln \Lambda(t) = X_1(t) + X_2(t)\,.$$

Applying Itô's Formula leads to

$$d\widetilde{S}(t) = \Big(\mu_1 + \mu_2 + (\alpha_1 - \alpha_2)X_2(t) - \alpha_1 \widetilde{S}(t)\Big)\, dt + \sigma_1\, dB_1(t) + \sigma_2\, dB_2(t)\,. \tag{3.18}$$

Hence, the deseasonalized log-spot price is reverting to the mean level $\mu_1 + \mu_2 + (\alpha_1 - \alpha_2)X_2(t)$, which we can write as an OU process with speed of mean reversion α_2 to the level $\mu_1 + \mu_1\alpha_2/\alpha_1$, having volatility $\sigma_2(\alpha_1 - \alpha_2)/\alpha_1$. That is, when defining the mean level

$$M(t) \triangleq \mu_1 + \mu_2 + (\alpha_1 - \alpha_2)X_2(t)\,,$$

we have

$$dM(t) = \left(\mu_1 + \frac{\alpha_2}{\alpha_1}\mu_2 - \alpha_2 M(t)\right) dt + \frac{\sigma_2}{\alpha_1}(\alpha_1 - \alpha_2)\, dB_2(t)\,.$$

We implicitly assume that $\alpha_1 > \alpha_2$, otherwise we just change the roles of X_1 and X_2. This restriction has the interpretation that the level is mean-reverting slower than the actual price. Notice that the mean level $M(t)$ is correlated with the dynamics of the log-spot price $\widetilde{S}(t)$, which is not the case for the model considered by [Barlow, Gusev and Lai (2004)]. They assume independence between the noise in $\widetilde{S}$ and X_2. We can generalize this to more factors driving the level and introducing jumps for the level and the price process itself. [Barlow, Gusev and Lai (2004)] also propose a model where the level is not mean-reverting, but a drifted Brownian motion.

Such a model can be obtained by letting the speed of mean reversion for X_2 be equal to zero. [Crosby (2005)] argues for a general multi-factor jump-diffusion models in commodity markets. He derives forward and futures prices based on such models with shapes capturing the stylised facts of many commodity markets.

In most financial applications one wants to calculate moments of the spot price model. Typical examples involve the forward price and different options written on the spot. Also, various risk measures, like the variance, involve finding the moment of the spot price at a specific time. Thus, we need to impose integrability conditions on $S(t)$ to ensure that the expectation of some moment is well-defined. For the class of exponential models, we suppose the following exponential integrability condition on the compensator measure ℓ.

Condition G.[1] For each $j = 1, \ldots, n$, there exists a constant $c_j > 0$ such that

$$\int_0^T \int_1^\infty \{e^{c_j z} - 1\}\, \ell_j(dz, du) < \infty . \tag{3.19}$$

The constants c_j, $j = 1, \ldots, n$ will be defined more precisely later in connection with the analysis of derivatives written on the spot. Under this condition, we have, for every $j = 1, \ldots, n$,

$$\begin{aligned}
\gamma_j(t) &+ \int_0^t \int_{|z|<1} z\, \widetilde{N}_j(dz, ds) + \int_0^t \int_{|z|\geq 1} z\, N_j(dz, ds) \\
&= \gamma_j(t) + \int_0^t \int_{\mathbb{R}} z\, \widetilde{N}_j(dz, ds) + \int_0^t \int_{|z|\geq 1} z\, \ell_j(dz, ds) \\
&= \widetilde{\gamma}_j(t) + \int_0^t \int_{\mathbb{R}} z\, \widetilde{N}_j(dz, ds) ,
\end{aligned}$$

where

$$\widetilde{\gamma}_j(t) = \gamma_j(t) + \int_0^t \int_{|z|\geq 1} z\, \ell_j(dz, ds) .$$

This holds true since

$$\int_0^t \int_{|z|\geq 1} |z|\, \ell(dz, ds) \leq \int_0^t \int_{|z|\geq 1} \{e^{c_j z} - 1\} \ell_j(dz, ds) < \infty$$

[1] We refer to the condition with the letter "G" to indicate that we are dealing with the geometric model.

for every $c_j < 0$. Thus, we may represent the II process $I_j(t)$ as the sum of a deterministic drift and a pure jump martingale process. The fact that the jump part is a true martingale, and not merely a local martingale process, follows again from Condition G.

The Condition G applied on the compensator measures implies the existence of moments of the spot price process $S(t)$. This follows from the next Lemma.

Lemma 3.3 *For any $t \leq T$, it holds that $\mathbb{E}[S(t)^c] < \infty$, where $c = \min_{j \leq 1 \leq n} c_j / \sup_{u\in[0,t]} |\eta_j(u)|$.*

Proof. It is sufficient to consider $m = p = 0$ and $n = 1$, since Brownian motion has exponential moments. The proof is based on the independent increment property of I and the Lévy-Kintchine representation. Recall $Y(t)$ given in (3.14), and consider the integral term with respect to the II process $I(t)$. We need to prove that this integral term has a finite exponential moment. To do so, we introduce the short-hand notation

$$f(u) = \eta(u) \exp\left(-\int_u^t \beta(v)\,dv\right),$$

and find by dominated convergence (using that $\beta \geq 0$ and thus $|f(u)| \leq |\eta(u)|$ is bounded) that

$$\mathbb{E}\left[\exp\left(c\int_0^t f(u)\,dI(u)\right)\right] = \exp\left(c\int_0^t f(u)\,d\gamma(u)\right) \times \exp\left(\int_0^t \int \{e^{zcf(u)} - 1 - zcf(u)\}\,\ell(dz,du)\right).$$

Thus, the Lemma follows as long as the integral with respect to $\ell(dz, du)$ is finite. The assumed Condition G ensures this. □

We note that Condition G also makes the Laplace transform of Y_j well-defined. Since the moment-generating function of Y_j exists up to an order bounded by $\min_{1\leq j\leq n} c_j / \sup_{u\in[0,t]} |\eta_j(u)|$, $j = 1, \ldots, n$, the Laplace transform of $Y_j(t)$ can be defined. Later, in connection with forward and option pricing, we frequently encounter expressions involving the Laplace transform of the processes Y_j.

3.2.2 Arithmetic models

Let the stochastic process $S(t)$ be defined as

$$S(t) = \Lambda(t) + \sum_{i=1}^{m} X_i(t) + \sum_{j=1}^{n} Y_j(t), \tag{3.20}$$

where $X_i(t)$ and $Y_j(t), i = 1, \ldots, m, j = 1, \ldots, n$, are defined in (3.9) and (3.10), respectively, and the seasonality function $\Lambda(t)$ satisfies the same conditions as assumed in Subsection 3.2.1 above. The process $S(t)$ is then a semimartingale, and we suppose that the initial conditions of X_i and Y_j are such that

$$\sum_{i=1}^{m} X_i(0) + \sum_{j=1}^{n} Y_j(0) = S(0) - \Lambda(0). \tag{3.21}$$

We obtain the explicit representation of $S(t)$ by using (3.3) with $X_i(t)$ and $Y_j(t)$ given as in (3.13) and (3.14), respectively.

The following integrability conditions are imposed on the jump processes $I_j(t)$ in order to ensure that the spot price process has finite moments up to certain orders.

Condition A.[2] For each $j = 1, \ldots, n$, there exists a constant $c_j > 0$ such that

$$\int_0^{\mathcal{T}} \int_{|z|\geq 1} |z|^{c_j} \, \ell_j(du, dz) < \infty. \tag{3.22}$$

If, for instance, $c_j = 1$, then condition A ensures the existence of a finite expectation for the spot price process in view of Lemma 3.1. In connection with deriving forward prices, the spot price process must have a finite expectation. However, the expectation is with respect to a risk-neutral measure, which requires an exponential integrability condition. We will return to this question later.

As we investigate closer in Chapter 10, arithmetic processes are the natural choice when modelling the daily temperature. However, they have not gained much popularity as models for spot prices of different commodities and energies (see however, [Lucia and Schwartz (2002)] and [Benth, Cartea and Kiesel (2006)] for applications). The main reason may be that an arithmetic model apparently allows for negative prices, a phenomenon

[2]We refer to the condition with the letter "A" to indicate that we think of the arithmetic model.

which sounds odd in any normal market, since this means that the buyer of a commodity receives money rather than pays. However, in the electricity market this may have a simple explanantion, since it can be more costly for a producer to switch off the generators than to pay someone to consume electricity in the case of more supply than demand. Thus, electricity is given away along with a payment. In fact, in almost all the liberalised electricity markets, negative prices occur from time to time, although very rarely. In the UK gas market, a concrete example of negative prices for the spot was experienced on 3 October 2006, when the new Langeled pipeline from Norway to the UK was opened for testing. The market was not fully prepared for the sudden rise in gas supply combined with mild weather, and many traders had to pay off their clients to get rid of gas (see [Financial Times (2006)]).

To simplify the following discussion on negative prices, consider $S(t)$ with $n = 0$, that is, the dynamics of S does not include any terms $Y_i(t)$ with jumps. Then, $S(t)$ is nothing but a Gaussian OU process $X(t)$ which varies around a seasonal level $\Lambda(t)$. Indeed, there will always be a positive probability of having $S(t) < 0$. This probability will depend on the volatility σ and mean reversion α, but also on the mean level Λ. This is quantified in the next Lemma.

Lemma 3.4 *Suppose $n = 0$. Then*

$$P\left(S(t) < 0\right) = \Phi\left(-\frac{m(t)}{\Sigma(t)}\right),$$

where

$$m(t) = \Lambda(t) + \sum_{i=1}^{m} X_i(0) e^{-\int_0^t \alpha_i(s)\,ds},$$

$$\Sigma^2(t) = \sum_{k=1}^{p} \int_0^t \sigma_{ik}^2(s) e^{-2\int_s^t \alpha_i(u)\,du}\,ds,$$

and Φ is the cumulative standard normal distribution function.

Proof. From the explicit solution of $S(t)$ in (3.13), we see that it holds

$$S(t) \stackrel{d}{=} m(t) + \Sigma(t)\,Z,$$

where $\stackrel{d}{=}$ mean equality in distribution and Z is a standard normally distributed random variable. The result then follows from a straightforward calculation. □

Consider the case with constant coefficient functions, that is, with α_i and σ_{ik} being constants, $i = 1, \ldots, m$, $k = 1, \ldots, p$. Then we have that $m(t)$ defined in the Lemma behaves like $\Lambda(t)$, and $\Sigma^2(t)$ converges towards $\sum_{k=1}^{p} \sigma_{ik}^2 / 2\alpha_i$ for large t. Thus, when t becomes large, we find for the case of constant coefficient functions that

$$P(S(t) < 0) = \Phi\left(-\frac{\Lambda(t)}{\sum_{k=1}^{p} \frac{\sigma_{ik}^2}{2\alpha_i}}\right).$$

Introducing jump factors $Y_j(t)$ into the spot dynamics $S(t)$ will alter the probability of getting negative prices. Depending on the size and sign of the jumps, together with the speed of mean reversion, the probability can either increase or decrease.

Let us now introduce a class of arithmetic models proposed by [Benth, Kallsen and Meyer-Brandis (2007)] which has zero probability of negative prices. One may construct such a class by supposing $m = 0$, and reinterpret the seasonality function $\Lambda(t)$ as a *floor* towards which the processes Y_j revert. Consider the spot model with $m = 0$, that is,

$$S(t) = \Lambda(t) + \sum_{j=1}^{n} Y_j(t). \tag{3.23}$$

We let $Y_1(0) = S(0) - \Lambda(0)$, and $Y_j(0) = 0$, $j = 2, \ldots, n$, and suppose that the mean reversion level is equal to zero for each of the n OU processes, that is, $\delta_j = 0$ for $j = 1, \ldots, n$. Further, suppose that the pure jump II processes $I_j(t)$ are increasing, which, in particular, means that they only have positive jumps and that the compensator measures are concentrated on the positive real line. Such processes must have paths of finite variation on finite time intervals, since they will be bounded by the RCLL assumption. Thus, the Lévy-Kintchine representation will be defined by the cumulant function

$$\psi_j(t, s; \theta) = \mathrm{i}\theta(\widetilde{\gamma}_j(s) - \widetilde{\gamma}_j(t)) + \int_t^s \int_0^\infty \{\mathrm{e}^{\mathrm{i}z\theta} - 1\}\, \ell_j(dz, du), \tag{3.24}$$

for $j = 1, \ldots, n$, with

$$\widetilde{\gamma}_j(t) = \gamma_j(t) + \int_0^t \int_0^1 z\, \ell_j(dz, du). \tag{3.25}$$

This holds true since, for finite variation processes, the compensator measure satisfies the integrability condition

$$\int_0^t \int_{\mathbb{R}} \min(1, |z|)\, \ell_j(dz, du) < \infty\,. \tag{3.26}$$

The jump processes $I_j(t)$, $j = 1, \ldots, n$ can be represented as

$$I_j(t) = \widetilde{\gamma}_j(t) + \int_0^t \int_0^\infty z\, N_j(dz, du)\,,$$

where we do not have any term including the compensated random jump measure due to the finite variation.

The mean level of the spot prices is given implicitly through the deterministic function Λ and the characteristics of the jump processes Y_j (see Lemma 3.1),

$$\begin{aligned}
\Lambda_m(t) &\triangleq \Lambda(t) + \sum_{j=1}^n \mathbb{E}\left[Y_j(t)\right] \\
&= \Lambda(t) + Y_1(0) \exp\left(-\int_0^t \beta_1(v)\, dv\right) \\
&\quad + \sum_{j=1}^n \int_0^t \eta_j(u) \exp\left(-\int_u^t \beta_j(v)\, dv\right) d\widetilde{\gamma}_j(u) \\
&\quad + \sum_{j=1}^n \int_0^t \int_0^\infty z\eta_j(u) \exp\left(-\int_u^t \beta_j(v)\, dv\right) \ell_j(dz, du)\,.
\end{aligned} \tag{3.27}$$

Since the jumps are only positive, the processes Y_j will be positive. Thus, under the assumption that $\Lambda(t)$ is positive, we have that the spot model will form positive prices. The function $\Lambda(t)$ will be the level towards which all the processes Y_j will revert back, and therefore we denote this function the floor of the spot price. The floor $\Lambda(t)$ will be the lower bound of the spot price. We will have price fluctuations around the mean level $\Lambda_m(t)$ made up by jumps of the different jump processes. Note that we do not refer to $\Lambda_m(t)$ as the seasonality function for the spot price, since in energy spot models one frequently distinguishes the seasonal price level from the influence of possible price spikes. Price spikes give rise to an additional mean, which we see when observing prices. However, this additional mean one would not like to include in the so-called seasonal level.

A natural way to build up such a model is to let the first $1 \leq l < n$ mean reversion processes Y_j be responsible for the daily volatile price fluctuations, while the remaining $n - l$ processes account for the price spikes. Further, it may be a reasonable assumption that the Y_j's, $j = 1, \ldots, l$ are not seasonally varying. Thus, we may let $\eta_j(t) = \eta_j$ and $\beta_j(u) = \beta_j$ be constants, and the compensator measures be of the form $\ell_j(ds, dz) = \widetilde{\ell}_j(dz)\, ds$, which implies that $I_j(t)$ is a subordinator for $j = 1, \ldots, l$. In this case, the processes Y_j, $j = 1, \ldots, l$ are stationary, and we find that the cumulant function of $\lim_{t\to\infty} Y_j(t)$ is equal to (see Thm 17.5, p. 108 in [Sato (1999)])

$$\psi_j^{\infty}(\theta) = \int_0^{\infty} \psi(\mathrm{e}^{-\beta_j s}\theta)\, ds\,.$$

This gives us a way of identifying the seasonal function $\Lambda_s(t)$ as

$$\Lambda_s(t) = \Lambda_m(t) - \sum_{j=1}^{l} \frac{\eta_j}{\beta_j} \int_{|z|\geq 1} z\, \widetilde{\ell}_j(dz)\,. \tag{3.28}$$

Here the seasonal function is defined by the long-term mean of the spot price disregarding the influence of the price spikes.

Considering the process (3.27) on a discrete time scale, we have negative and positive price fluctuations coming from a combination of downward mean reversion and upward jumps. If the sum of jumps over the increment is stronger (that is, in sum bigger) than the total contribution from mean reversion, we observe a random price *increase*. A price *decay* is observed in the opposite case. Note that also a price decay becomes random, even though the jumps always go in the positive direction. The jumps are random, so the net effect of mean reversion and upward jump will be random.

3.3 The autocorrelation function of multi-factor Ornstein-Uhlenbeck processes

When analysing energy spot price data, we often look at the autocorrelation function (ACF). It is therefore of interest to have access to the theoretical ACF for the class of models we are proposing. We consider a stochastic process

$$Z(t) = \sum_{i=1}^{m} X_i(t) + \sum_{j=1}^{n} Y_j(t) \tag{3.29}$$

with X_i and Y_j defined in (3.9) and (3.10), respectively. We are interested in the situation where the OU processes are stationary, so we restrict our further discussion in this Section to the case where all parameter functions are constants, that is, when $\alpha_i, \beta_j, \sigma_{ik}$ and η_j are not dependent on time t.

Observe that $Z(t)$ is the deseasonalized arithmetic model in (3.20), or the deseasonalized geometric model in (3.8) on a logarithmic scale. The ACF at time t with lag $\tau > 0$ is defined as the correlation between $Z(t)$ and $Z(t+\tau)$, that is,

$$\rho(t,\tau) = \text{Corr}\left[Z(t), Z(t+\tau)\right] . \tag{3.30}$$

Note that the seasonal function $\Lambda(t)$ in the model for $S(t)$ will not make any difference for the ACF. In empirical studies it is the deseasonalized prices (or log-prices) that we want to understand the dependency structure of, and all the information is then contained in the model for $Z(t)$.

The next Proposition derives the theoretical ACF for the stochastic process $Z(t)$.

Proposition 3.4 *Assuming that all the processes Y_j are square integrable, we have that the ACF is*

$$\rho(t,\tau) = \sum_{i=1}^{m} \widehat{\omega}_i(t,\tau) e^{-\alpha_i \tau} + \sum_{j=1}^{n} \widetilde{\omega}_j(t,\tau) e^{-\beta_j \tau} .$$

Here,

$$\widehat{\omega}_i(t,\tau) = \frac{\sum_{i_1=1}^{m} \frac{\sum_{k=1}^{p} \sigma_{ik}\sigma_{i_1 k}}{\alpha_i + \alpha_{i_1}} \left(1 - e^{-(\alpha_i + \alpha_{i_1})t}\right)}{\sqrt{Var[Z(t+\tau)]\, Var[Z(t)]}}$$

$$\widetilde{\omega}_j(t,\tau) = \frac{Var[Y_j(t)]}{\sqrt{Var[Z(t+\tau)]\, Var[Z(t)]}} .$$

Proof. Let us first consider the case with $n = 0$, that is, without any jump components. Then,

$$\begin{aligned} Z(t) &= \sum_{i=1}^{m} X_i(0) \mathrm{e}^{-\alpha_i t} + \sum_{k=1}^{p} \int_0^t \sigma_{ik} \mathrm{e}^{-\alpha_i (t-s)}\, dB_k(s) \\ &= \sum_{i=1}^{m} X_i(0) \mathrm{e}^{-\alpha_i t} + \sum_{k=1}^{p} \int_0^t \sum_{i=1}^{m} \sigma_{ik} \mathrm{e}^{-\alpha_i (t-s)}\, dB_k(s) . \end{aligned}$$

Hence, we find that

$$\text{Cov}[Z(t+\tau), Z(t)]$$

$$= \mathbb{E}\left[\left(\sum_{k=1}^{p} \int_0^{t+\tau} \sum_{i=1}^{m} \sigma_{ik} \mathrm{e}^{-\alpha_i (t+\tau-s)}\, dB_k(s)\right)\right.$$

$$\times \left(\sum_{k=1}^{p} \int_0^t \sum_{i=1}^{m} \sigma_{ik} e^{-\alpha_i (t-s)} \, dB_k(s) \right) \Bigg]$$

$$= \mathbb{E} \Bigg[\left(\sum_{k=1}^{p} \int_0^t \sum_{i=1}^{m} \sigma_{ik} e^{-\alpha_i (t+\tau-s)} \, dB_k(s) + \sum_{k=1}^{p} \int_t^{t+\tau} \sum_{i=1}^{m} \sigma_{ik} e^{-\alpha_i (t+\tau-s)} \, dB_k(s) \right.$$

$$\times \left(\sum_{k=1}^{p} \int_0^t \sum_{i=1}^{m} \sigma_{ik} e^{-\alpha_i (t-s)} \, dB_k(s) \right) \Bigg] .$$

Appealing to the independent increment property of Brownian motion and the Itô isometry, we have

$$\begin{aligned} \mathrm{Cov}[Z(t+\tau), Z(t)] &= \sum_{k=1}^{p} \int_0^t \left(\sum_{i=1}^{m} \sigma_{ik} e^{-\alpha_i (t+\tau-s)} \right) \left(\sum_{i_1=1}^{m} \sigma_{i_1 k} e^{-\alpha_{i_1} (t-s)} \right) ds \\ &= \sum_{k=1}^{p} \sum_{i,i_1=1}^{m} \frac{\sigma_{ik}\sigma_{i_1 k}}{\alpha_i + \alpha_{i_1}} \left(1 - e^{-(\alpha_i + \alpha_{i_1})t} \right) e^{-\alpha_i \tau} \\ &= \sum_{i=1}^{m} \left\{ \sum_{i_1=1}^{m} \frac{\sum_{k=1}^{p} \sigma_{ik}\sigma_{i_1 k}}{\alpha_i + \alpha_{i_1}} \left(1 - e^{-(\alpha_i + \alpha_{i_1})t} \right) \right\} e^{-\alpha_i \tau} . \end{aligned}$$

Now, let us consider the jump part, and suppose that $m = p = 0$ and $n = 1$. Then,

$$Z(t) = Y(0)e^{-\beta t} + \int_0^t \eta e^{-\beta(t-s)} \, dI(s) .$$

Note that $I(t)$ is square-integrable by assumption, and therefore we may write its Lévy-Kintchine representation as

$$I(t) = \gamma(t) + \int_0^t \int_{|z|>1} z \, \ell(dz, ds) + \int_0^t \int_{\mathbb{R}} z \, \widetilde{N}(dz, ds) .$$

Hence,

$$\begin{aligned} Z(t) = Y(0)e^{-\beta t} &+ \int_0^t \eta e^{-\beta(t-s)} \, d\gamma(s) + \int_0^t \int_{|z|>1} \eta e^{-\beta(t-s)} \, \ell(dz, ds) \\ &+ \int_0^t \int_{\mathbb{R}} z\eta e^{-\beta(t-s)} \, \widetilde{N}(dz, ds) , \end{aligned}$$

which gives us a representation of $Z(t)$ in terms of a deterministic drift and a martingale term. Using the independent increment property, we are thus led to

$$\begin{aligned}\operatorname{Cov}[Z(t+\tau),Z(t)] &= \mathbb{E}\left[\int_0^t\int_{\mathbb{R}} z\eta \mathrm{e}^{-\beta(t+\tau-s)}\,\widetilde{N}(dz,ds)\right.\\ &\qquad\left.\times\int_0^t\int_{\mathbb{R}} z\eta \mathrm{e}^{-\beta(t-s)}\,\widetilde{N}(dz,ds)\right]\\ &= \mathrm{e}^{-\beta\tau}\mathbb{E}\left[\left(\int_0^t\int_{\mathbb{R}} z\eta \mathrm{e}^{-\beta(t-s)}\,\widetilde{N}(dz,ds)\right)^2\right]\\ &= \operatorname{Var}[Y(t)]\mathrm{e}^{-\beta\tau}\,.\end{aligned}$$

The general case follows by merging the two cases above. □

We note that the ACF $\rho(t,\tau)$ depends explicitly on current time t. We now exploit the stationarity of X_i and Y_j in the case when the I_j's are Lévy processes to show that the ACF becomes independent of t in the limit.

First, we have from the proof of Prop. 3.4 that the variance of $\sum_{i=1}^m X_i(t)$ is

$$\operatorname{Var}\left[\sum_{i=1}^m X_i(t)\right] = \sum_{i=1}^m\left\{\sum_{i_1=1}^m \frac{\sum_{k=1}^p \sigma_{ik}\sigma_{i_1k}}{\alpha_i+\alpha_{i_1}}\left(1-\mathrm{e}^{-(\alpha_i+\alpha_{i_1})t}\right)\right\}.$$

Hence, when $t\to\infty$, we have

$$\lim_{t\to\infty}\operatorname{Var}\left[\sum_{i=1}^m X_i(t)\right] = \sum_{i=1}^m\left\{\sum_{i_1=1}^m \frac{\sum_{k=1}^p \sigma_{ik}\sigma_{i_1k}}{\alpha_i+\alpha_{i_1}}\right\}.$$

This is also the case for the variance of $\sum_{i=1}^m X_i(t+\tau)$. Further, we recall from the discussion on stationarity at the end of Sect. 3.2 that the cumulant function of $Y_j(t)$, when $t\to\infty$, is

$$\int_0^\infty \psi_j(\theta\mathrm{e}^{-\beta_j s})\,ds\,,$$

where ψ_j is the cumulant function of the Lévy process $I_j(1)$. We can therefore express the limit of the variance of $Y_j(t)$ at infinity as

$$\begin{aligned}\lim_{t\to\infty}\operatorname{Var}[Y_j(t)] &= -\frac{d^2}{d\theta^2}\mathrm{e}^{\int_0^\infty \psi_j(\theta\eta_j\mathrm{e}^{-\beta_j s})\,ds}\Big|_{\theta=0}\\ &\quad-\left(-\mathrm{i}\frac{d}{d\theta}\mathrm{e}^{\int_0^\infty \psi_j(\theta\eta_j\mathrm{e}^{-\beta_j s})\,ds}\Big|_{\theta=0}\right)^2.\end{aligned}$$

A straightforward calculation then gives

$$\lim_{t\to\infty} \mathrm{Var}[Y_j(t)] = -\frac{\psi_j''(0)\eta_j^2}{2\beta_j}\,.$$

In conclusion, the limiting variance of $Z(t)$ (and $Z(t+\tau)$) is

$$\lim_{t\to\infty} \mathrm{Var}[Z(t)] = \sum_{i=1}^{m}\left\{\sum_{i_1=1}^{m}\frac{\sum_{k=1}^{p}\sigma_{ik}\sigma_{i_1 k}}{\alpha_i+\alpha_{i_1}}\right\} - \sum_{j=1}^{n}\frac{\psi''(0)\eta_j^2}{2\beta_j}\,.$$

We see in Prop. 3.4 that the weights will converge as well to something independent of t and τ. More specifically,

$$\widehat{\omega}_i = \lim_{t\to\infty}\widehat{\omega}_i(t,\tau) = \frac{\sum_{i_1=1}^{m}\frac{\sum_{k=1}^{p}\sigma_{ik}\sigma_{i_1 k}}{\alpha_i+\alpha_{i_1}}}{\lim_{t\to\infty}\mathrm{Var}[Z(t)]}\,,$$

$$\widetilde{\omega}_j = \lim_{t\to\infty}\widetilde{\omega}_i(t,\tau) = \frac{-\frac{\psi_j''(0)}{2\beta_j}}{\lim_{t\to\infty}\mathrm{Var}[Z(t)]}\,.$$

Hence, in stationarity the model $Z(t)$ has the following ACF only depending on the lag τ

$$\rho(\tau) = \sum_{i=1}^{m}\widehat{\omega}_i \mathrm{e}^{-\alpha_i\tau} + \sum_{j=1}^{n}\widetilde{\omega}_j \mathrm{e}^{-\beta_j\tau}\,. \tag{3.31}$$

In addition, the weights $\widehat{\omega}_i$ and $\widetilde{\omega}_j$ sum up to 1.

In Sect. 3.35 we shall utilise the connection between the empirical and theoretical ACF when analysing and estimating a multi-factor model for the gas spot price. For example, one may fit the stationary ACF in (3.31) to the empirical, and thereby identify the number of factors required with the corresponding speeds of mean reversion for modelling the energy price.

3.4 Simulation of stationary Ornstein-Uhlenbeck processes: a case study with the arithmetic spot model

In this Section we want to discuss the simulation of OU processes for which the Lévy density of the driving Lévy process is known. As an application of the simulation algorithm, we study the arithmetic model in (3.23) for the spot price dynamics including spikes, where the occurrence of these are seasonally dependent like observed in the Nord Pool market, say. The model under consideration is motivated from the studies in [Benth, Kallsen and Meyer-Brandis (2007)]. But before analysing the spot model, let us discuss some issues related to simulation.

We describe an algorithm for simulating paths of $Y(t)$, where $Y(t)$ is an OU process of the form

$$dY(t) = -\beta Y(t)\,dt + dL(t)\,, \tag{3.32}$$

where we suppose that L has a Lévy measure with density, $\widetilde{\ell}(dz) = \widetilde{\ell}_1(z)\,dz$. The algorithm was introduced by [Marcus (1987)] and [Rosinski (1991)], and explained in the context of stochastic volatility in [Barndorff-Nielsen and Shephard (2001)]. We adopt here the presentation of [Barndorff-Nielsen and Shephard (2001)]. Assume we discretize the time interval $[0, T]$ by homogeneous time intervals of length $\Delta > 0$. Then a straightforward calculation gives

$$Y(s+\Delta) = \mathrm{e}^{-\beta\Delta} Y(s) + \mathrm{e}^{-\beta\Delta}\, Z(s),$$

where

$$Z(s) = \mathrm{e}^{-\beta s} \int_s^{s+\Delta} \mathrm{e}^{\beta u}\, dL(u)\,,$$

and s is a time point in our discretization of $[0, T]$. Note that $Z(s)$ is independent of $Z(t)$ when $t \neq s$. By a change of variables, we find that

$$Z(s) \stackrel{d}{=} \int_0^{\Delta} \mathrm{e}^{\beta u}\, dL(u)\,.$$

The integral $Z(s)$ can be represented as an infinite series. Let $\{u_i\}_i$ be independent samples from a uniform probability distribution on $[0, 1]$ and $a_1 < a_2 < \cdots < a_i < \ldots$ be the arrival times of a Poisson process with intensity 1. Then

$$Z(s) \stackrel{d}{=} \sum_{i=1}^{\infty} \ell^{-1}(a_i/\Delta)\mathrm{e}^{\beta u_i \Delta}\,. \tag{3.33}$$

In the above expansion, the function $\ell^{-1}(x)$ is the inverse of $\ell^{+}(x)$, where $\ell^{+}(x)$ is the upper tail integral of the Lévy density $\widetilde{\ell}_1$, that is,

$$\ell^{+}(x) = \int_x^{\infty} \widetilde{\ell}_1(z)\,dz\,.$$

The inversion of the upper tail integral $\ell^{+}(x)$ is in general not analytically possible to calculate, thus leading to complicated (and potentially slow) simulation algorithms when the inversion must be handled numerically. We discuss a particular case where indeed the inverse is obtainable.

Suppose the stationary distribution for $Y(t)$ is in the class of Gamma distributions, that is, $Y(t) \sim \Gamma(\nu, 1/\mu_J)$, where the density is defined in (2.34). Recalling the discussion in Sect. 3.2, the Gamma distribution is self-decomposable, and the background driving Lévy process $L(t)$ is a compound Poisson process with exponentially distributed jumps with mean μ_J. Further, given that the speed of mean-reversion β of $Y(t)$ is known, we find the intensity of jumps in $L(t)$ to be $\lambda = \nu\beta$. Naturally, since the jump distribution is concentrated on the positive real line, L is a subordinator. The Lévy measure of $L(t)$ is

$$\widetilde{\ell}(dz) = \frac{\nu\beta}{\mu_J}\exp(-z/\mu_J)\,dz\,,$$

with upper tail integral

$$\ell^+(x) = \nu\beta\exp(-x/\mu_J)\,.$$

We easily find the explicit expression for the inverse of ℓ^+ as given by

$$\ell^{-1}(x) = \max\left(0, -\mu_J \ln\left(\frac{x}{\nu\beta}\right)\right)\,. \tag{3.34}$$

Introducing this function in the series expansion (3.33), [Barndorff-Nielsen and Shephard (2001)] derive a representation (in distribution) for $Z(s)$, suitable for simulation. We repeat their argument here. Let $c_1 < c_2 < c_3 < \ldots$ be the arrival times of a Poisson process with intensity $\nu\beta\Delta$, and $N(1)$ the number of jumps up to time 1. By (3.34), we find

$$\begin{aligned} Z(s) &\stackrel{d}{=} -\mu_J \sum_{i=1}^{\infty} 1_{(0,\nu\beta)}(a_i/\Delta)\ln(a_i/\nu\beta\Delta)\mathrm{e}^{\beta\Delta u_i} \\ &= \mu_J \sum_{i=1}^{\infty} 1_{(0,1)}(c_i)\ln(c_i^{-1})\mathrm{e}^{\beta\Delta u_i} \\ &= \mu_J \sum_{i=1}^{N(1)} \ln(c_i^{-1})\mathrm{e}^{\beta\Delta u_i}\,. \end{aligned}$$

We shall use the Gamma distribution in a case study to follow.

Let us consider the arithmetic spot price model (3.23) motivated by the study in [Benth, Kallsen and Meyer-Brandis (2007)]. We discuss the construction of such a process, and apply the algorithm discussed above to simulate price paths.

Suppose that the spot price is defined as

$$S(t) = \Lambda(t) + Y_1(t) + Y_2(t)\,, \tag{3.35}$$

with a seasonal floor

$$\Lambda(t) = a + bt + c\sin(2\pi(t-d)/365)\,,$$

for constants a, b, c and d. Here we assume 365 days of price quotations in the market. Furthermore, the process Y_1 models the "normal" variations in the market, while Y_2 accounts for the spikes. The innovators I_1 and I_2 are a subordinator and a time-inhomogeneous compound Poisson process, respectively. To have sample paths with spikes, Y_2 will have a fast speed of mean reversion, while Y_1 will revert to zero at a much slower rate.

The modelling idea is to first specify a seasonal floor Λ, to which the spot price is reverting. The floor can be found by fitting the deterministic function to data, and then moving the whole function downwards until the difference between the price observations and the floor is positive. The difference is referred to as the "deseasonalized" spot prices. From the empirical ACF of the deseasonalized data, one may read off the speeds of mean reversion by appealing to Prop. 3.4. Since Y_2 is modelling the spikes, it is natural to have a non-stationary jump intensity since spikes may be seasonally occurring, the Nord Pool market being a typical example where spikes are most often occurring during the winter. This complicates the connection between the empirical and theoretical ACFs. To avoid this, one may first identify the spike process Y_2 by some filtering procedure, and subtract this from the price series. The remaining series should then account for the normal variations, which is naturally assumed to be stationary. The speed of mean reversion can be found by comparing the empirical and the theoretical ACFs. Looking at the data, one can estimate the stationary distribution, and derive from it the background driving Lévy process. To perform this scheme in practice is not simple, and we shall not go into the details here but refer the reader to [Meyer-Brandis and Tankov (2007)]. However, later, in Sect. 5.3 where we consider modelling gas, we discuss briefly potential approaches for estimating such models as (3.35).

The following specification is relevant for the spot price dynamics at

Nord Pool.[3] Assume a seasonal floor

$$\Lambda(t) = 100 + 0.025 \times t + 30\sin(2\pi t/365)\,.$$

The mean reversion speeds are set to $\beta_1 = 0.085$ and $\beta_2 = 1.1$, respectively, and we assume that the stationary distribution of Y_1 is $\Gamma(\nu, 1/\mu_J)$, with $\nu = 8.06$ and $\mu_J = 7.7$. The seasonal intensity function of the inhomogeneous compound Poisson process driving the innovations of Y_2 is

$$\lambda(t) = \frac{0.14}{|\sin\left(\frac{\pi(t-90)}{365}\right)| + 1} - 1\,. \tag{3.36}$$

The idea to use this seasonal intensity function is taken from [Geman and Roncoroni (2006)], who apply it in an empirical study of several different electricity markets world-wide. They propose a parametric family of intensity functions, where (3.36) is a special case. The jump sizes are exponentially distributed with mean equal to 180.

A Monte Carlo simulation of the daily spot prices over four years is given in Fig. 3.1, where we clearly see the seasonality of the prices and occurrence of jumps. We applied the simulation algorithm above to produce paths of Y_1. The process $Y_2(t)$ was simulated directly by first simulating the (seasonal) occurrences of jumps and the corresponding jump sizes. Next, the explicit form of Y_2 is used to produce the path. The simulated processes Y_1 and Y_2 making up $S(t)$ are plotted in Figs. 3.2 and 3.3, respectively. In the latter plot the spikes induced from this model is clearly visible, along with the seasonal occurrence of these.

A full-blown empirical analysis of this model on electricity spot data from different exchanges are found in [Meyer-Brandis and Tankov (2007)]. The case study here is meant to give a flavour of the potential of an arithmetic model, and not intended as a complete study which would involve rather sophisticated methods.

[3] We are indebted to Thilo Meyer-Brandis for providing these numbers. The parameters are chosen to mimic the Nord Pool electricity spot prices, however, not based on any rigorous empirical analysis.

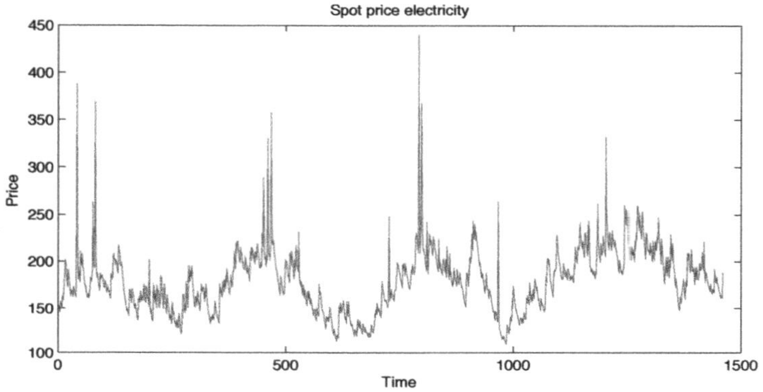

Fig. 3.1 Four years of daily spot prices simulated from the arithmetic model with seasonal spikes defined in (3.35).

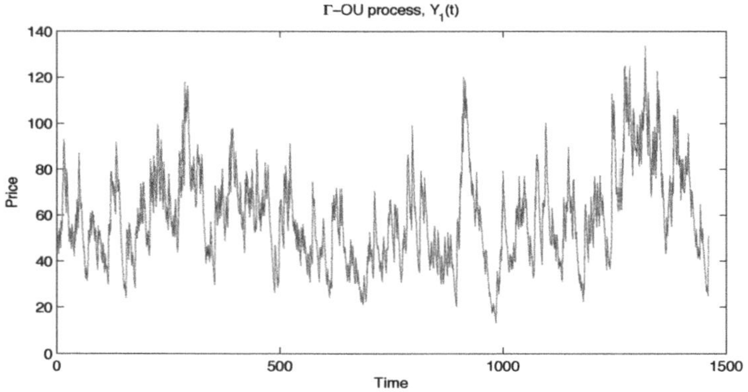

Fig. 3.2 The Gamma OU process $Y_1(t)$.

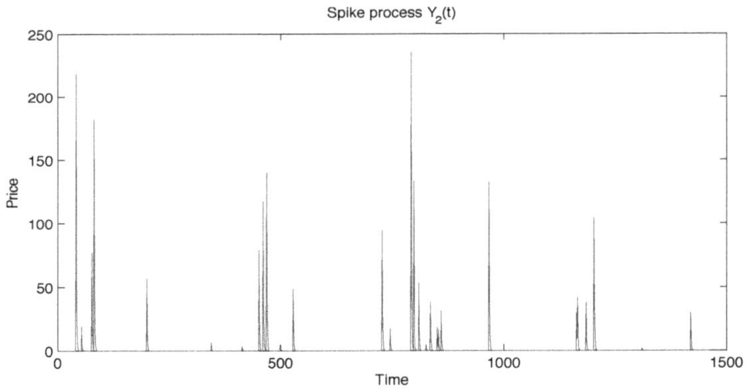

Fig. 3.3 The OU process $Y_2(t)$ with seasonal intensity for jumps given in (3.36).

Chapter 4

Pricing of Forwards and Swaps Based on the Spot Price

The purpose of this Chapter is to derive forward and swap prices from the arithmetic and geometric spot models investigated in the previous Chapter. Recall that electricity and gas futures have delivery of the underlying energy over a period, and we refer to these as swap contracts in our general treatment. Forwards, on the other hand, we understand as contracts delivering the underlying commodity at a fixed maturity time. The markets we have in mind do not allow for perfect replication in terms of a buy-and-hold strategy in the spot, and no unique price dynamics can be derived based on no-arbitrage principles. The approach will be based on the calculation of the predicted spot price under a risk-neutral probability. This probability will be constructed using the Esscher transform, which has the convenient feature of preserving the distributional properties of the driving random processes in the spot dynamics. In general, we will not be able to calculate the swap price when assuming an underlying geometric model for the spot. In view of this, we study arithmetic models for the spot where positivity is preserved, and derive analytical swap prices.

4.1 Risk-neutral forward and swap price modelling

Suppose we buy a forward contract at time t promising future delivery of some underlying spot product with price dynamics $S(t)$. Here, $S(t)$ is assumed to be a semimartingale defined by the geometric or arithmetic model, introduced in (3.3) or (3.20), respectively. Along with the spot and forward, we include in the market a risk-free asset (usually a bank account or bond) yielding a continuously compounded rate of return $r > 0$. Hence, the value of such a risk-free investment of one currency unit will be worth $\exp(rt)$ currency units at time t. When entering the forward contract, one agrees on a future delivery time and the price to be paid for receiving the underlying. Suppose that the delivery time is τ, with $0 \leq t \leq \tau < \infty$, and that the agreed price to pay upon delivery is $f(t,\tau)$. At time τ, we will effectively receive a (possibly negative) payment

$$S(\tau) - f(t,\tau)\,.$$

We may view this random payment at time τ as the payoff from a derivative on the spot. Furthermore, it is costless to enter such contracts, which, under suitable integrability assumptions on the price processes S and f, gives us a relation where we can extract the forward price

$$\mathrm{e}^{-r(\tau-t)}\mathbb{E}_Q[S(\tau) - f(t,\tau)\,|\,\mathcal{F}_t] = 0\,. \tag{4.1}$$

Here, Q is an equivalent martingale measure. We assume that $S(\tau) \in L^1(Q)$, the space of integrable random variables with respect to Q. It is reasonable to assume that we base the forward price on the available information in the market, or in other words, that $f(\cdot,\tau)$ is adapted. Hence, the relationship

$$f(t,\tau) = \mathbb{E}_Q[S(\tau)\,|\,\mathcal{F}_t]\,, \tag{4.2}$$

follows. This is the fundamental pricing relation between the spot and forward price, leading to an arbitrage-free pricing dynamics for the forward price as a process of time t. Since the energy markets we have in mind are incomplete, the choice of Q is open. Hence, in order to have one price dynamics for $f(t,\tau)$, we need some additional criterion to pin down the choice of Q.

Next, let us consider swaps, using the electricity market as the typical example. The buyer of an electricity futures receives power during a settlement period (physically or financially), against paying a fixed price per MWh. The time t value of the payoff from the continuous flow of electricity is given as

$$\int_{\tau_1}^{\tau_2} \mathrm{e}^{-r(u-t)}\left(S(u) - F(t,\tau_1,\tau_2)\right)du\,,$$

where $F(t,\tau_1,\tau_2)$ is the electricity futures price at time t for the delivery period $[\tau_1,\tau_2]$, with $t \leq \tau_1$. Recall from Chapter 1 that in the marketplace, the settlement is defined with respect to the hourly spot price, implying that we should have a summation in the expression above. However, we shall from now on stick to the slightly more mathematically convenient definition using integration. Since it is costless to enter an electricity futures contract, the risk-neutral price is defined by the equation

$$e^{-rt}\mathbb{E}_Q\left[\int_{\tau_1}^{\tau_2} e^{-r(u-t)}\left(S(u)-F(t,\tau_1,\tau_2)\right)du \,|\, \mathcal{F}_t\right]=0\,.$$

Since the electricity futures price is settled at time t based on the information available at that time, it is natural to assume $F(t,\tau_1,\tau_2)$ to be adapted. Hence, we find

$$F(t,\tau_1,\tau_2)=\mathbb{E}_Q\left[\int_{\tau_1}^{\tau_2}\frac{re^{-ru}}{e^{-r\tau_1}-e^{-r\tau_2}}S(u)\,du \,|\, \mathcal{F}_t\right]\,.$$

One may have that the settlement takes place financially at the end of the delivery period τ_2. The payoff from the contract at time τ_2 is then

$$e^{-r\tau_2}\mathbb{E}_Q\left[\int_{\tau_1}^{\tau_2}\left(S(u)-F(t,\tau_1,\tau_2)\right)du \,|\, \mathcal{F}_t\right]=0\,,$$

which yields an electricity futures price

$$F(t,\tau_1,\tau_2)=\mathbb{E}_Q\left[\int_{\tau_1}^{\tau_2}\frac{1}{\tau_2-\tau_2}S(u)\,du \,|\, \mathcal{F}_t\right]\,.$$

The same considerations could be done for gas futures contracts, and in the following we refer to $F(t,\tau_1,\tau_2)$ simply as the swap price.

Let us introduce a weight function $\widehat{w}(u)$, being equal to one if the swap is settled at the end of the delivery period, or $\widehat{w}(u)=\exp(-ru)$ if the contract is settled continuously over the delivery period. Define the function

$$w(u,s,t)=\frac{\widehat{w}(u)}{\int_s^t\widehat{w}(v)\,dv}\,, \tag{4.3}$$

where $0\leq u\leq s<t$. Observe that $w(u,s,t)=1/(t-s)$ when $\widehat{w}(u)=1$, while we have

$$w(u,s,t)=\frac{re^{-ru}}{e^{-rs}-e^{-rt}}\,,$$

for the case when $\widehat{w}(u)=\exp(-ru)$. Note that the weight function $w(u,s,t)$ integrates to one, that is,

$$\int_s^t w(u,s,t)\,du=1\,. \tag{4.4}$$

In general, we can write the link between a swap contract and the underlying spot as

$$F(t,\tau_1,\tau_2)=\mathbb{E}_Q\left[\int_{\tau_1}^{\tau_2} w(u,\tau_1,\tau_2)S(u)\,du\,|\,\mathcal{F}_t\right]. \tag{4.5}$$

Here we implicitly assume integrability conditions on the spot dynamics to make the conditional expectation and Lebesgue integration well-defined. Commuting the conditional expectation with Lebesgue integration in (4.5), yields the following relation between forwards and swaps.

Proposition 4.1 *Suppose* $\mathbb{E}_Q[\int_{\tau_1}^{\tau_2}|w(u,\tau_t,\tau_2)S(u)|\,du]<\infty$. *It holds that*

$$F(t,\tau_1,\tau_2)=\int_{\tau_1}^{\tau_2} w(u,\tau_1,\tau_2)f(t,u)\,du\,.$$

This means that holding a swap contract can be considered as holding a (weighted) continuous stream of forwards. This connection is sometimes useful in calculating swap prices.

It is known from the theory of forwards that the forward price at delivery coincides with the spot price of the underlying. This is a very natural connection in view of the fact that at delivery, there is no difference in entering the forward or buying the commodity in the spot market. We have a convergence of forward prices to the spot price when time approaches delivery. Thus, we recall the following result for contracts of forward type.

Lemma 4.1 *Suppose* $\mathbb{E}_Q[|S(\tau)|]<\infty$. *It holds that*

$$\lim_{t\uparrow\tau} f(t,\tau)=S(\tau)\,.$$

Proof. From the definition of $f(t,\tau)$, the process is a Q martingale, which is RCLL. This implies

$$\lim_{t\uparrow\tau} f(t,\tau)=f(\tau-,\tau)\,.$$

We now argue that $\Delta f(\tau,\tau)\neq 0$ with probability 0. In fact, we show that $f(t,\tau)$ does not have any fixed time discontinuities. From the time dynamics of $S(t)$ (recall definition (3.3) for the geometric model and (3.20) for the arithmetic), we see that the only terms that give rise to jumps come from the random jump measures associated to the II processes I_j. This is because the change of measure for the Brownian motions will only introduce a new drift. From Thm 3.24, page 159 in [Jacod and Shiryaev (1987)], we find that the compensator measure for I_j under an equivalent change of measure can be represented as

$$\ell_Q(dz, dt) = Y(t, z)\ell(dz, dt)$$

for a non-negative random field Y. Hence, we see that the jump terms will not have any fixed time discontinuities under Q. Further, taking the conditional expectation of $S(\tau)$ with respect to $\mathcal{F}_t$, we conclude that the dynamics of $f(t, \tau)$ cannot have any fixed time discontinuities. Therefore, it follows that $\Delta f(\tau, \tau) = 0$ *a.s.*, and

$$\lim_{t \uparrow \tau} f(t, \tau) = f(\tau, \tau) = \mathbb{E}\left[S(\tau) \mid \mathcal{F}_\tau\right] = S(\tau)\,.$$

The last equality holds since $S(\tau)$ is $\mathcal{F}_\tau$-measurable. Hence, the proof is completed. □

In the electricity and gas markets, say, where delivery takes place over a period of time rather than at a fixed point, we do not observe a convergence of swap prices to the spot at delivery. The reason for this is easily seen from the connection between forwards and swaps stated in Prop. 4.1.

Proposition 4.2 *Suppose* $\mathbb{E}_Q[\int_{\tau_1}^{\tau_2} |w(u, \tau_1, \tau_2)S(u)|\, du] < \infty$. *Then it holds that, a.s.,*

$$\lim_{t \uparrow \tau_1} F(t, \tau_1, \tau_2) = \int_{\tau_1}^{\tau_2} w(u, \tau_1, \tau_2) f(\tau_1, u)\, du\,,$$

which is different from $S(\tau_1)$ *a.s., as long as* $S(t)$ *is not a* Q *martingale.*

Proof. We apply the Fubini-Tonelli theorem (see, for example, [Folland (1984)]) to get, *a.s.*, that

$$\begin{aligned}
\lim_{t \uparrow \tau_1} F(t, \tau_1, \tau_2) &= \lim_{t \uparrow \tau_1} \int_{\tau_1}^{\tau_2} w(u, \tau_1, \tau_2) f(t, u)\, du \\
&= \int_{\tau_1}^{\tau_2} w(u, \tau_1, \tau_2) \lim_{t \uparrow \tau_1} f(t, u)\, du \\
&= \int_{\tau_1}^{\tau_2} w(u, \tau_1, \tau_2) f(\tau_1, u)\, du\,.
\end{aligned}$$

In the last equality we use that $f(t, u)$ is RCLL with no fixed times of discontinuity.

Suppose now for simplicity that $\widehat{w}(u) = 1$. If $F(\tau_1, \tau_1, \tau_2) = S(\tau_1)$, it must hold that

$$(\tau_2 - \tau_1)S(\tau_1) = \int_{\tau_1}^{\tau_2} f(\tau_1, u)\, du\,.$$

By Lebesgue's Fundamental Theorem of Calculus (see, for example, [Folland (1984)]), we find that the derivative of the right-hand side with respect to τ_2 exists *a.s.*, and equals

$$f(\tau_1, \tau_2) = \mathbb{E}_Q[S(\tau_2)\,|\mathcal{F}_{\tau_1}]\,,$$

while the left-hand side has derivative equal to $S(\tau_1)$. Thus, *a.s.*,

$$S(\tau_1) = \mathbb{E}_Q[S(\tau_2)\,|\mathcal{F}_{\tau_1}]\,.$$

In conclusion, $S(t)$ must be a martingale under Q. □

The next Proposition confirms that a swap contract delivering the commodity over a time period which collapses into a single point coincides with a forward.

Proposition 4.3 *Suppose* $\mathbb{E}_Q[\int_{\tau_1}^{\tau_2} |w(u, \tau_1, \tau_2)S(u)|\, du] < \infty$. *Then it holds that*

$$\lim_{\tau_2 \to \tau_1} F(t, \tau_1, \tau_2) = f(t, \tau_1)\,.$$

Proof. We have

$$\begin{aligned} F(t, \tau_1, \tau_2) &= \int_{\tau_1}^{\tau_2} w(u, \tau_1, \tau_2) f(t, u)\, du \\ &= \frac{\int_{\tau_1}^{\tau_2} \widehat{w}(u) f(t, u)\, du}{\int_{\tau_1}^{\tau_2} \widehat{w}(u)\, du} \\ &= \frac{\frac{1}{\tau_2 - \tau_1} \int_{\tau_1}^{\tau_2} \widehat{w}(u) f(t, u)\, du}{\frac{1}{\tau_2 - \tau_1} \int_{\tau_1}^{\tau_2} \widehat{w}(u)\, du}\,. \end{aligned}$$

Letting $\tau_2 \to \tau_1$, the desired result follows from Lebesgue's Fundamental Theorem of Calculus. □

These relations between the spot, forwards and swaps were first discussed in the context of electricity markets by [Vehviläinen (2002)].

We next discuss risk-neutral probabilities and the Esscher transform.

4.1.1 *Risk-neutral probabilities and the Esscher transform*

Recall from arbitrage theory (see, for example, [Bjørk (1998)]) that by definition, a risk-neutral probability Q is a probability being equivalent to P, such that all *tradeable* assets in the market under consideration are martingales after discounting. Electricity is a commodity that cannot be traded in the usual sense since it is not storable, except indirectly, for instance, in water reservoirs. A commodity that must be consumed once purchased, does not have a price process which necessarily must be a martingale under Q. It is only the discounted risk-free asset that must be a martingale under Q, which is trivially satisfied. Another example with similar characteristics is the temperature market, where one obviously cannot buy the underlying "commodity". The market for gas is a border case, since it can be stored. However, there are high costs and limitations connected to transportation and storage of gas, and the relationship with the convenience yield makes it difficult to derive a gas futures price based on replication. Hence, this commodity is also conveniently included in the derivations to follow.

As argued, all equivalent probabilities Q are risk-neutral in the markets we have in mind. This makes up a rather wide class of potential pricing measures, and in practice it is convenient to restrict this space. We study the class of risk-neutral probabilities given through the Esscher transform.

The Esscher transform is a generalization of the Girsanov transform of Brownian motion to jump processes. We can view the Girsanov transform as a change of measure that preserves the normality of the distribution of the Brownian motion process. In the same way the Esscher transform is preserving the distributional properties of the jump process in the sense of transforming the cumulant function by a linear change of the argument. Effectively, the Esscher transform yields an explicit change of measure, where we have access to the characteristics of the jump processes I_j also under the new risk-neutral measure. The Esscher transform introduces a set of parameters which alters the characteristics of each jump process, but preserves the independent increment property. This means that we restrict our attention to a subclass of parametrized equivalent martingale measures, where forward prices can be represented explicitly in terms of the cumulant functions of the jump processes. The parameter functions can in principle be estimated from option prices and interpreted as the market price of jump risk.

By restricting the attention to the Esscher transform (and the Girsanov transform for the Brownian motions), we obtain a flexible class of risk-neutral measures that is tractable for pricing, and at the same time can be used for estimation of the market risk premium. The transform has been used in derivatives pricing in many financial markets, starting out with the seminal paper of [Gerber and Shiu (1994)]. It was introduced by [Esscher (1932)] to study risk theory in the actuarial sciences in the following way. Suppose f is a probability density, and θ is a real number. Then, as long as the exponential moment

$$\int_{\mathbb{R}} e^{\theta y} f(y)\, dy$$

exists, we can define a new probability density

$$f(x;\theta) = \frac{e^{\theta x} f(x)}{\int_{\mathbb{R}} e^{\theta y} f(y)\, dy}.$$

This transform of the density f is the original Esscher transform, which we shall consider in the framework proposed by [Gerber and Shiu (1994)] for derivatives pricing. We are going to generalize their approach to II processes including time-dependent parameters $\theta(t)$. The theory of [Esscher (1932)] was developed to approximate the aggregated claim amount distribution around a point of interest, x_0, and the parameter θ is chosen so that the new mean is x_0.

Let now $\theta(t)$ be a $(p+n)$-dimensional vector of real-valued continuous functions on $[0, \mathcal{T}]$

$$\theta(t) = \left(\widehat{\theta}_1(t), \ldots, \widehat{\theta}_p(t), \widetilde{\theta}_1(t), \ldots, \widetilde{\theta}_n(t)\right). \tag{4.6}$$

Define for $0 \leq t \leq \tau$ the stochastic exponential

$$Z^\theta(t) = \prod_{k=1}^{p} \widehat{Z}_k^\theta(t) \times \prod_{j=1}^{n} \widetilde{Z}_j^\theta(t), \tag{4.7}$$

where, for $k = 1, \ldots, p$,

$$\widehat{Z}_k^\theta(t) = \exp\left(\int_0^t \widehat{\theta}_k(s)\, dB_k(s) - \frac{1}{2}\int_0^t \widehat{\theta}_k^2(s)\, ds\right), \tag{4.8}$$

and, for $j = 1, \ldots, n$,

$$\widetilde{Z}_j^\theta(t) = \exp\left(\int_0^t \widetilde{\theta}_j(s)\, dI_j(s) - \phi_j(0, t; \widetilde{\theta}_j(\cdot))\right). \tag{4.9}$$

As long as we have $\sup_{0\leq t\leq T}|\widetilde{\theta}_j(t)|\leq c_j$, where c_j is a constant ensuring that Condition G is satisfied, it follows from Itô's Formula that $\widetilde{Z}_j^\theta(t)$ is a positive local martingale with expectation equal to one. Hence, it is a martingale process. Similarly, since $\widehat{\theta}_k(s)$ is a continuous function, the Novikov condition (see [Karatzas and Shreve (1991)]) obviously holds, which therefore implies that $\widehat{Z}_k^\theta(t)$ is a martingale as well. Hence, we can define an equivalent probability measure Q^θ such that $Z^\theta(t)$ is the density process of the Radon-Nikodym derivative dQ^θ/dP, that is,

$$\left.\frac{dQ^\theta}{dP}\right|_{\mathcal{F}_t}=Z^\theta(t)\,, \tag{4.10}$$

for $0\leq t\leq T$. The expectation operator with respect to the probability Q^θ is denoted by $\mathbb{E}_\theta[\cdot]$. We observe that the Radon-Nikodym derivative dQ^θ/dP can be factorised as

$$\left.\frac{dQ^\theta}{dP}\right|_{\mathcal{F}_t}=\prod_{k=1}^p\widehat{Z}_k^\theta(t)\times\prod_{j=1}^n\widetilde{Z}_j^\theta(t)\,.$$

Hence, we associate a price of risk to each random source given by the Brownian motions B_k and the jump factors I_j, $k=1,\ldots,p, j=1,\ldots,n$ in the model of the spot price.

The idea is that the market players charge a price for the risk of not being able to hedge. These are given by the $\widehat{\theta}_k$'s. Next they assign prices $\widetilde{\theta}_j$ to the jump risk given by Y_j. The factors are conveniently assumed to be independent, and therefore we do not have any price for the correlation risk between factors. However, since the X_i's will be correlated in general, there will be an implicit correlation risk between the factors.

Let us study how the characteristics of B and I are changing when we apply the Esscher transform. The details are stated in the next Proposition.

Proposition 4.4 *With respect to the probability measure Q^θ, the processes*

$$B_k^\theta(t)=B_k(t)-\int_0^t\widehat{\theta}_k(u)\,du$$

are Brownian motions for $k=1,\ldots,p$ and $0\leq t\leq T$. Furthermore, for each $j=1,\ldots,n$, $I_j(t)$ is an II process on $0\leq t\leq T$ with drift

$$\gamma_j(t)+\int_0^t\int_{|z|<1}z\{e^{\widetilde{\theta}_j(u)z}-1\}\,\ell_j(dz,du)\,,$$

and compensator measure $e^{\widetilde{\theta}_j(t)z}\,\ell_j(dz,dt)$*. Under* Q^θ*, we denote the random jump measure associated with* I_j *by* N_j^θ*, and its compensator with* $\widetilde{N}_j^\theta$*.*

Proof. The result for the Brownian motion part comes from a simple application of the Girsanov transform (see, for example, [Karatzas and Shreve (1991)]). Let $p = 0$ and $n = 1$. We calculate the characteristic function of $I(t)$ under the Esscher transformed measure Q^θ. The calculation is based on the independent increment property of the process I together with Bayes' Formula. For $0 \le s < t$,

$$
\begin{aligned}
&\mathbb{E}_\theta\left[\exp\left(\mathrm{i}\lambda(I(t)-I(s))\right) \mid \mathcal{F}_s\right] \\
&\quad = \mathbb{E}\left[\exp\left(\mathrm{i}\lambda(I(t)-I(s))\right)\frac{\widetilde{Z}^\theta(t)}{\widetilde{Z}^\theta(s)} \mid \mathcal{F}_s\right] \\
&\quad = \mathbb{E}\left[\exp\left(\mathrm{i}\lambda(I(t)-I(s)) + \mathrm{i}\int_s^t (-\mathrm{i}\widetilde{\theta}(u))\,dI(u)\right)\right]\exp(-\psi(s,t;-\mathrm{i}\widetilde{\theta}(\cdot))) \\
&\quad = \exp\left(\psi(s,t;\lambda-\mathrm{i}\widetilde{\theta}(\cdot)) - \psi(s,t;-\mathrm{i}\widetilde{\theta}(\cdot))\right) \\
&\quad = \exp\Big(\mathrm{i}\lambda(\gamma(t)-\gamma(s)) + \mathrm{i}\lambda\int_s^t\int_{|z|<1} z\left(\mathrm{e}^{\widetilde{\theta}(u)z}-1\right)\ell(du,dz) \\
&\qquad + \int_s^t\int_{\mathbb{R}}\left\{\mathrm{e}^{\mathrm{i}\lambda z}-1-\mathrm{i}\lambda z 1_{|z|<1}\right\}\mathrm{e}^{\widetilde{\theta}(u)z}\ell(dz,du)\Big).
\end{aligned}
$$

Hence, the result follows. □

Note that if we do not have any jumps, that is, $n = 0$, the Esscher transform is nothing but a traditional Girsanov transform of the Brownian motions driving the dynamics. Observe that if the price of jump risk $\widetilde{\theta}_j$ is positive, the positive jumps of I_j are more emphasised under Q^θ than under P, while the negative jumps are less emphasised under the risk-neutral measure.

We denote by $N_j^\theta(dz,dt)$ the random jump measure associated to I_j. This random jump measure is counting the number of jumps falling in the interval $[z, z+dz)$ over the time $[t, t+dt)$. Since this depends on the path of I_j, we must have that $N_j^\theta = N_j$ by definition. However, the frequencies of occurrence of the jumps, and the probability distribution of the jump sizes are changed when moving from P to Q^θ, which we see reflected in the change of the compensator measure. Hence, $\widetilde{N}_j^\theta$ is not coinciding with $\widetilde{N}_j$. However, we can perform the following calculation

$$
\widetilde{N}_j^\theta(dz,dt) = N_j^\theta - \mathrm{e}^{\widetilde{\theta}_j(t)z}\,\ell_j(dz,dt)
$$

$$= N_j(dz, dt) - \ell_j(dz, dt) - \left(e^{\widetilde{\theta}_j(t)z} - 1\right) \ell_j(dz, dt)$$
$$= \widetilde{N}(dz, dt) - \left(e^{\widetilde{\theta}_j(t)z} - 1\right) \ell_j(dz, dt) .$$

Therefore, $\widetilde{N}_j$ translates to $\widetilde{N}_j^\theta$ by subtraction of a drift, exactly as the Girsanov transform of B_k to B_k^θ.

4.1.2 *The Esscher transform for some specific models*

In this Subsection we study how the Esscher transform works on different choices of II processes $I(t)$. We restrict our discussion to the case of one II process ($m = 1$), and consider only constant choices of $\widetilde{\theta}$, the parameter function of the Esscher transform (or the market price of jump risk).

Consider first the case of a time-inhomogeneous compound Poisson process. From Subsect. 2.6.1, recall the compensator measure to be

$$\ell(dz, dt) = \lambda(t) F_X(dz)\, dt ,$$

where F_X is the distribution of the jump size random variable X and $\lambda(t)$ is the time-dependent jump intensity. From Prop. 4.4, we find the compensator measure under Q^θ to be

$$\ell^\theta(dt, dz) = \lambda(t) e^{\widetilde{\theta} z} F_X(dz)\, dt .$$

A common choice of jump size is the exponential distribution with expectation μ_J. Then, we find the compensator measure under Q^θ to be

$$\ell^\theta(dt, dz) = \frac{\lambda(t)}{\mu_J} \exp\left(-\left(\frac{1}{\mu_J} - \widetilde{\theta}\right) z\right) dz\, dt .$$

Hence, with $\widetilde{\theta} < 1/\mu_J$, $I(t)$ will remain a compound Poisson process under Q^θ, with jump size being exponentially distributed with expectation $1/(1/\mu_J - \widetilde{\theta})$ and intensity $\lambda(t)/(1 - \mu_J \widetilde{\theta})$.

Another popular choice of X is the normal distribution. In this case it can be shown that the Esscher transform is altering the mean of the jump size, but not the distributional properties.

We next turn our attention to the popular class of NIG and hyperbolic Lévy processes. From Subsect. 2.6.2 we know that these two Lévy processes are special cases of the GH Lévy process. Inspecting the expression for the Lévy measure $\widetilde{\ell}_{\mathrm{GH}}(dz)$ in (2.25) of the GH Lévy process, we see that the structure of the Lévy measure is preserved completely, and the only change is that the skewness parameter β in $\widetilde{\ell}_{\mathrm{GH}}(dz)$ is transformed to $\beta + \widetilde{\theta}$. Hence, we still have a GH Lévy process with all parameters unchanged except skewness, which is $\beta + \widetilde{\theta}$ under Q^{θ}. From this we see that a positive price of jump risk leads to a more right-skewed distribution, and therefore also more emphasis on the bigger jumps after Esscher transformation.

We end our discussion with the CGMY Lévy process. From the Lévy measure $\widetilde{\ell}_{\mathrm{CGMY}}(dz)$ defined in (2.6.3), it follows that the G parameter is transformed into $G+\widetilde{\theta}$ and the M parameter into $M-\widetilde{\theta}$. Thus, the resulting compensator measure will assign less emphasis on negative jumps, and more on positive ones in the case of $\widetilde{\theta} > 0$. Also, in this case the Esscher transform preserves the distribution.

If we choose a time-dependent price of jump risk, $\widetilde{\theta}(t)$, we get a time-dependent change in the compensator measures in the examples above. Since for each time step t we preserve the distributional properties, we may say that the resulting II process under Q^{θ} is GH or CGMY distributed for each increment, however, the parameters of the distribution will now depend on time.

4.2 Currency conversion for forward and swap prices

In this section, we will discuss currency conversion for financial derivatives contracts. Sometimes it is convenient to change the denomination of a financial contract from one currency to another. This is a relevant problem for foreign traders in a market. In the Nord Pool electricity market we have seen a transition from NOK to EUR denominated contracts, and, for instance, Swedish, Danish and Norwegian particpants in this market are exposed to currency risk since the contracts are not denominated in their respective kroner. A similar problem is encountered in Chapters 7 and 8, where we calibrate different models to historical Nord Pool prices including both NOK and EUR denominated contracts. There it is most convenient to denominate all contracts in a common currency.

Our currency model is a simple one, where we assume that domestic interest rates and foreign interest rates are both constant (see, for instance, Chapter 17 in [Bjørk (1998)] for a nice textbook treatment of modelling currency markets). Below we derive the forward exchange rate and the forward commodity price conversion rate. For a commodity contract with a delivery period, we demonstrate how this parity formula must be modified. Note that our focus is on currency conversion so that contracts with prices in different currencies can be consistently converted to a common currency.

Currency risk in trading and hedging cross-border electricity contracts is not treated in this book. We refer to [Hansen and Jensen (2004)] for an explicit treatment of currency risk in electricity derivatives.

Let Q and Q^* denote the domestic and foreign risk-neutral probability measures, respectively. Domestic and foreign interest rates are assumed to be constants and denoted by r and r^*, respectively. The price at time t of a domestic zero coupon bond with maturity $\tau > t$, denoted $P(t,\tau)$, is defined by

$$P(t,\tau) = \mathbb{E}_Q\left[\mathrm{e}^{-\int_t^\tau r\,ds} \,|\, \mathcal{F}_t\right] = \mathrm{e}^{-r(\tau-t)}\,. \tag{4.11}$$

Similarly, a foreign zero coupon bond, $P^*(t,\tau)$, is given by

$$P^*(t,\tau) = \mathbb{E}_{Q^*}\left[\mathrm{e}^{-\int_t^\tau r^*\,ds} \,|\, \mathcal{F}_t\right] = \mathrm{e}^{-r^*(\tau-t)}\,. \tag{4.12}$$

These relationships follow from standard no-arbitrage arguments.

Let now $X(t)$ be the spot exchange rate prevailing at time t and measured in the ratio

$$\frac{\text{units of domestic currency}}{\text{units of foreign currency}}\,.$$

We suppose that $X(t)$ is a positive semimartingale process. Denote by $f_{\mathrm{FRA}}(t,\tau)$ the agreed price at time t for delivery of one unit foreign currency at time τ. This forward contract is simply called a *forward exchange rate* or *forward exchange rate agreement* (FRA). The payoff of a long position at time τ is $X(\tau) - f_{\mathrm{FRA}}(t,\tau)$. Under the domestic risk-neutral measure we have (assuming $\mathbb{E}_Q[X(\tau)] < \infty$) that

$$f_{\mathrm{FRA}}(t,\tau) = \mathbb{E}_Q\left[X(\tau) \,|\, \mathcal{F}_t\right]\,, \tag{4.13}$$

since it is costless to enter the FRA. Now we use the foreign risk-neutral measure to derive the spot-forward exchange rate relationship. From a foreign point of view the exchange rate should be replaced by the rate

$$X^*(t) = \frac{1}{X(t)}\,,$$

which is quoted in

$$\frac{\text{units of foreign currency}}{\text{units of domestic currency}}\,.$$

Let $f^*_{\mathrm{FRA}}(t,\tau)$ be the forward exchange rate agreement for delivery of $X^*(t)$ at time τ. Using the same line of reasoning as above (and assuming

$\mathbb{E}_{Q^*}[X^*(\tau)] < \infty$), we have

$$f^*_{\mathrm{FRA}}(t,\tau) = \mathbb{E}_{Q^*}\left[X^*(\tau)\,|\,\mathcal{F}_t\right] .$$

The covered interest rate parity gives us the forward exchange rate, $f_{\mathrm{FRA}}(t,\tau)$, defined as

$$f_{\mathrm{FRA}}(t,\tau) = X(t)\mathrm{e}^{(r-r^*)(\tau-t)} , \tag{4.14}$$

or, similarly,

$$f^*_{\mathrm{FRA}}(t,\tau) = X^*(t)\mathrm{e}^{(r^*-r)(\tau-t)} . \tag{4.15}$$

This relation is well known, and was derived by [Oldfield and Messian (1977)] in a continuous-time framework using the risk-neutral hedging arguments of [Black and Scholes (1973)]. As an alternative, it can be proved by setting up a synthetic FRA, and discounting it under the risk-neutral measure. Consider the following portfolio. You buy a foreign zero coupon bond today (the price is $\mathrm{e}^{-r^*(\tau-t)}$), an investment which delivers one unit of foreign currency at time τ ($X(\tau)$). Finance this by borrowing the necessary amount domestically ($X(t)\mathrm{e}^{-r^*(\tau-t)}$). Pay off your debt at time τ (total of $X(t)\mathrm{e}^{-r^*(\tau-t)}\mathrm{e}^{r(\tau-t)}$). This is an investment with zero outlay today, and so, after discounting the payoff under the risk-neutral domestic measure, we get the equation

$$\mathrm{e}^{-r(\tau-t)}\mathbb{E}_Q\left[\left(X(\tau) - X(t)\mathrm{e}^{-r^*(\tau-t)}\mathrm{e}^{r(\tau-t)}\right)\,|\,\mathcal{F}_t\right] = 0 .$$

This equation together with (4.13) give the covered interest rate parity in (4.14). The covered interest rate parity in terms of $f^*_{\mathrm{FRA}}(t,\tau)$ can be derived similarly.

We now consider spot and forward commodity prices in foreign and domestic currency. Let the spot rate $S(t)$ denote one unit of the commodity in domestic currency. The price of the same commodity quoted in foreign currency is $S^*(t)$. You can buy such a commodity using either domestic or foreign currency. No-arbitrage arguments, and assuming no market frictions, give the following spot price relationship

$$S^*(t)X(t) = S(t) .$$

Denote by $f(t,\tau)$ the domestic price at time t of a forward contract with delivery at $\tau > t$. We recall the definition of the forward price $f(t,\tau)$ in (4.2), and consider a forward contract for the same commodity, but with the commodity (and the forward contract) denominated in foreign currency.

Let $f^*(t,\tau)$ denote the forward price in foreign currency. Naturally we have the following relationship (when assuming that $\mathbb{E}_{Q^*}[|S^*(\tau)|] < \infty$)

$$f^*(t,\tau) = \mathbb{E}_{Q^*}\left[S^*(\tau) \,|\, \mathcal{F}_t\right]. \tag{4.16}$$

The forward price for the commodity denoted in domestic and foreign currency is linked through the exchange rate in the following way

$$f(t,\tau) = f_{\mathrm{FRA}}(t,\tau) f^*(t,\tau) = X(t) \mathrm{e}^{(r-r^*)(\tau-t)} f^*(t,\tau). \tag{4.17}$$

The domestic forward price is the forward price converted at the forward exchange rate. The proof follows along the lines of the covered interest rate parity. Consider a long foreign commodity forward position. The payoff at time τ, in domestic currency, is

$$X(\tau)\left(S^*(\tau) - f^*(t,\tau)\right).$$

This is a zero cost investment, and therefore we have the equation

$$\mathrm{e}^{-r(\tau-t)} \mathbb{E}_Q\left[\left(f(t,\tau) - X(\tau) f^*(t,\tau)\right) \,|\, \mathcal{F}_t\right] = 0.$$

Then (4.17) follows from (4.13) and (4.14).

Let us put the same arguments on commodity swap prices in foreign and domestic currency. Recall from (4.5) and Prop. 4.1 the relationship between the spot, forward and swap prices. Denote by $F^*(t,\tau_1,\tau_2)$ the swap price on the same commodity in foreign currency. Under natural integrability conditions on $S^*(u)$, we have that

$$\begin{aligned} F^*(t,\tau_1,\tau_2) &= \mathbb{E}_{Q^*}\left[\int_{\tau_1}^{\tau_2} w^*(u,\tau_1,\tau_2) S^*(u)\, du \,\Big|\, \mathcal{F}_t\right] \\ &= \int_{\tau_1}^{\tau_2} w^*(u,\tau_1,\tau_2) f^*(t,u)\, du \end{aligned}$$

with

$$w^*(u,\tau_1,\tau_2) = \frac{\widehat{w}^*(u)}{\int_{\tau_1}^{\tau_2} \widehat{w}^*(v)\, dv}.$$

Again, $\widehat{w}^*(u) = 1$ if the settlement takes place at the end of the delivery period, while settlement during the delivery period gives $\widehat{w}^*(u) = \mathrm{e}^{-r^* u}$. The proposition below shows how foreign and domestic swap prices are related through a change of currency.

Proposition 4.5 *Suppose that* $\mathbb{E}_Q[\int_{\tau_1}^{\tau_2} |w(u,\tau_1,\tau_2)X(u)|\,du] < \infty$ *and* $\mathbb{E}_Q[\int_{\tau_1}^{\tau_2} |w(u,\tau_1,\tau_2)S(u)|\,du] < \infty$. *Foreign and domestic swap prices are related in the following way*

$$F(t,\tau_1,\tau_2) = \int_{\tau_1}^{\tau_2} w(u,\tau_1,\tau_2) f_{FRA}(t,u)\,du \times F^*(t,\tau_1,\tau_2)\,.$$

Proof. Consider a short position in a foreign denominated swap contract which settles continuously during delivery. Measured in domestic currency, the market value at date t is

$$\begin{aligned}
\mathbb{E}_Q&\left[\int_{\tau_1}^{\tau_2} \mathrm{e}^{-r(u-t)} X(u)\left(F^*(t,\tau_1,\tau_2) - S^*(u)\right) du \,|\, \mathcal{F}_t\right] \\
&= \mathbb{E}_Q\left[\int_{\tau_1}^{\tau_2} \mathrm{e}^{-r(u-t)} X(u) F^*(t,\tau_1,\tau_2)\,du \,|\, \mathcal{F}_t\right] \\
&\quad - \mathbb{E}_Q\left[\int_{\tau_1}^{\tau_2} \mathrm{e}^{-r(u-t)} S(u)\,du \,|\, \mathcal{F}_t\right] = 0\,.
\end{aligned}$$

Applying the definitions of $f(t,\tau)$ and $f_{\mathrm{FRA}}(t,\tau)$, it follows that

$$F^*(t,\tau_1,\tau_2) = \frac{\int_{\tau_1}^{\tau_2} \mathrm{e}^{-r(u-t)} f(t,u)\,du}{\int_{\tau_1}^{\tau_2} \mathrm{e}^{-r(u-t)} f_{\mathrm{FRA}}(t,u)\,du}\,.$$

Hence, the Proposition is proved. □

Therefore, a swap price can be denoted in foreign currency by converting it with an appropriate weighting of the forward exchange rate over the settlement period of the contract.

4.3 Pricing of forwards

We consider the forward price $f(t,\tau)$ at time t for a contract with settlement at time $\tau \geq t$. The forward price will be derived from a geometric or arithmetic spot price model $S(t)$, as defined in Subsects. 3.2.1 and 3.2.2, respectively. Our analysis will use the risk-neutral probabilities defined by the Esscher transform.

4.3.1 *The geometric case*

Let us assume a geometric spot price model as introduced in Subsect. 3.2.1. The forward price at time $t \geq 0$ for contracts with settlement at $\tau \geq t$ is explicitly given in the following Proposition.

Proposition 4.6 *Let $0 \leq t \leq \tau$ and suppose $S(t)$ is the geometric spot price model (3.8). Suppose that Condition G in Subsect. 3.2.1 holds for $j = 1, \ldots, n$ with*

$$\sup_{0 \leq s \leq \tau} |\eta_j(s) e^{-\int_s^\tau \beta_j(u)\,du} + \widetilde{\theta}_j(s)| \leq c_j\,.$$

Then we have that the forward price $f(t,\tau)$ is

$$\begin{aligned} f(t,\tau) = \Lambda(\tau)\Theta(t,\tau;\theta(\cdot)) \\ &\times \exp\left(\sum_{i=1}^m \int_t^\tau \mu_i(u) e^{-\int_u^\tau \alpha_i(v)\,dv}\,du\right) \\ &\times \exp\left(\sum_{j=1}^n \int_t^\tau \delta_j(u) e^{-\int_u^\tau \beta_j(v)\,dv}\,du\right) \\ &\times \exp\left(\sum_{i=1}^m e^{-\int_t^\tau \alpha_i(v)\,dv} X_i(t) + \sum_{j=1}^n e^{-\int_t^\tau \beta_j(v)\,dv} Y_j(t)\right)\,, \end{aligned}$$

where $\Theta(t,\tau;\theta(\cdot))$ is given as

$$\begin{aligned} \ln\Theta(t,\tau;\theta(\cdot)) &= \sum_{j=1}^n \psi_j(t,\tau; -i(\eta_j(\cdot) e^{-\int_\cdot^\tau \beta_j(v)\,dv} + \widetilde{\theta}_j(\cdot))) - \psi_j(t,\tau; -i\widetilde{\theta}_j(\cdot)) \\ &\quad + \frac{1}{2}\sum_{k=1}^p \int_t^\tau \left(\sum_{i=1}^m \sigma_{ik}(u) e^{-\int_u^\tau \alpha_i(v)\,dv}\right)^2 du \\ &\quad + \sum_{i=1}^m \sum_{k=1}^p \int_t^\tau \sigma_{ik}(u)\widehat{\theta}_k(u) e^{-\int_u^\tau \alpha_i(v)\,dv}\,du\,. \end{aligned}$$

Proof. Without loss of generality, we assume $n = 1$. From (3.3), we have

$$\begin{aligned} X_i(\tau) &= \mathrm{e}^{-\int_t^\tau \alpha_i(v)\,dv} X_i(t) + \int_t^\tau \mu_i(u) \mathrm{e}^{-\int_u^\tau \alpha_i(v)\,dv}\,du \\ &\quad + \sum_{k=1}^p \int_t^\tau \sigma_{ik}(u) \mathrm{e}^{-\int_u^\tau \alpha_i(v)\,dv}\,dB_k(u) \\ Y(\tau) &= \mathrm{e}^{-\int_t^\tau \beta(v)\,dv} Y(t) + \int_t^\tau \delta(u) \mathrm{e}^{-\int_u^\tau \beta(v)\,dv}\,du \\ &\quad + \int_t^\tau \eta(u) \mathrm{e}^{-\int_u^\tau \beta(v)\,dv}\,dI(u)\,, \end{aligned}$$

for $i = 1, \dots, m$. Since $X_i(t)$ and $Y(t)$ are $\mathcal{F}_t$-measurable, we find that

$$
\begin{aligned}
f(t,\tau) &= \mathbb{E}_\theta\left[S(\tau)\,|\,\mathcal{F}_t\right] \\
&= \Lambda(\tau)\mathbb{E}_\theta\Big[\mathrm{e}^{\sum_{i=1}^m\sum_{k=1}^p\int_t^\tau \sigma_{ik}(u)\exp(-\int_u^\tau \alpha_i(v)\,dv)\,dB_k(u)} \\
&\qquad \times \mathrm{e}^{\int_t^\tau \eta(u)\exp(-\int_u^\tau \beta(v)\,dv)\,dI(u)}\,|\mathcal{F}_t\Big] \\
&\qquad \times \exp\left(\sum_{i=1}^m\int_t^\tau \mu_i(u)\mathrm{e}^{-\int_u^\tau \alpha_i(v)\,dv}\,du + \int_t^\tau \delta(u)\mathrm{e}^{-\int_u^\tau \beta(v)\,dv}\,du\right) \\
&\qquad \times \exp\left(\sum_{i=1}^m \mathrm{e}^{-\int_t^\tau \alpha_i(v)\,dv}X_i(t) + \mathrm{e}^{-\int_t^\tau \beta(v)\,dv}Y(t)\right).
\end{aligned}
$$

To prove the Proposition, we calculate the expectation in the expression above and show that it coincides with $\Theta(t,\tau;\theta(\cdot))$. From the characteristics of B_k and I under the Esscher transform (see Prop. 4.4), we have that they are independent under Q^θ. We can therefore argue in the following way.

Due to the integrability Condition G imposed in the Proposition, we can use the same line of arguments as in the proof of Prop. 4.4 to get

$$
\begin{aligned}
&\mathbb{E}_\theta\left[\mathrm{e}^{\int_t^\tau \eta(u)\exp(-\int_u^\tau \beta(v)\,dv)\,dI(u)}\,|\mathcal{F}_t\right] \\
&\quad = \exp\left(\psi(t,\tau;-\mathrm{i}(\eta(\cdot)\mathrm{e}^{-\int_\cdot^\tau \beta(v)\,dv} + \widetilde{\theta}(\cdot))) - \psi(t,\tau;-\mathrm{i}\widetilde{\theta}(\cdot))\right).
\end{aligned}
$$

This proves the part of $\Theta(t,\tau;\theta(\cdot))$ accounting for the jump risk. To calculate the Brownian motion part, we first change the Brownian motion B_k defined under P to the risk-neutral one by appealing to Prop. 4.4. Hence, using the independence of Brownian increments, we find

$$
\begin{aligned}
&\mathbb{E}_\theta\left[\mathrm{e}^{\sum_{i=1}^m\sum_{k=1}^p\int_t^\tau \sigma_{ik}(u)\exp(-\int_u^\tau \alpha_i(v)\,dv)\,dB_k(u)}\,|\,\mathcal{F}_t\right] \\
&\quad = \mathbb{E}_\theta\left[\mathrm{e}^{\sum_{k=1}^p\int_t^\tau\left\{\sum_{i=1}^m \sigma_{ik}(u)\exp(-\int_u^\tau \alpha_i(v)\,dv)\right\}dB_k(u)}\,|\,\mathcal{F}_t\right] \\
&\quad = \prod_{k=1}^p \mathbb{E}_\theta\left[\mathrm{e}^{\int_t^\tau\left\{\sum_{i=1}^m \sigma_{ik}(u)\exp(-\int_t^\tau \alpha_i(v)\,dv)\right\}dB_k^\theta(u)}\right] \\
&\qquad\qquad \times \exp\left(\int_t^\tau\sum_{i=1}^m \sigma_{ik}(u)\widehat{\theta}_k(u)\exp(-\int_u^\tau \alpha_i(v)\,dv)\,du\right).
\end{aligned}
$$

But,

$$
\mathbb{E}_\theta\left[\mathrm{e}^{\int_t^\tau\left\{\sum_{i=1}^m \sigma_{ik}(u)\exp(-\int_t^\tau \alpha_i(v)\,dv)\right\}dB_k^\theta(u)}\right]
$$

$$= \exp\left(\frac{1}{2}\int_t^\tau \left\{\sum_{i=1}^m \sigma_{ik}(u)\exp(-\int_u^\tau \alpha_i(v)\,dv)\right\}^2 du\right),$$

which proves the Proposition. □

The forward price $f(t,\tau)$ is dependent on the level of all factors $X_i(t)$ and $Y_j(t)$ in such a way that we cannot reduce the connection to the current level of the spot. However, in the case of one factor driving the spot price dynamics $S(t)$, the forward price will be a function of the spot only. This is stated in the next Corollary.

Corollary 4.1 *If $m = 1, n = 0$ or $m = 0, n = 1$, it holds that*

$$f(t,\tau) = \Lambda(\tau)\Theta(t,\tau;\theta(\cdot))\exp\left(\int_t^\tau \zeta(u)e^{-\int_u^\tau \xi(v)\,dv}\,du\right)\left(\frac{S(t)}{\Lambda(t)}\right)^{e^{-\int_t^\tau \xi(v)\,dv}},$$

with $\Theta(t,\tau,\theta(\cdot))$ defined in Prop. 4.6 above, and $\xi = \alpha, \zeta = \mu$ for $m = 1$, or $\xi = \beta, \zeta = \delta$ for $n = 1$.

Obviously, we can think of other models where the forward price may be represented explicitly as a function of the spot, such as if we have two factors with the same speed of mean reversion, say.

The next result describes the forward price for contracts far from maturity, that is, the contracts in the long end of the forward curve. For constant coefficients, we prove that the forward price simply becomes the risk-adjusted seasonal function Λ when maturity time τ becomes large. We have assumed that the seasonal function is continuously differentiable on the finite interval $[0,T]$. In the Proposition below, we extend our considerations to all $\tau > t$, and therefore some asymptotic properties on $\Lambda(\tau)$ are needed. In most situations, the seasonality of an energy may be decomposed into something varying cyclically over, for example, the year, and a trend due to inflation, say. The cyclical component is bounded, while the trend is typically linear. We consider such a case, and further assume that the coefficients in the spot model are constants and the II processes I_j have compensator measures being bounded by Lévy measures.

Proposition 4.7 *Suppose that $\mu_i, \alpha_i, \sigma_{ik}, \delta_j, \beta_j$ and η_j are all constant, for $i = 1,\ldots,m, j = 1,\ldots,n$ and $k = 1,\ldots,p$. Next, suppose that the market price for risk $\theta(\cdot)$ is bounded. Further, assume that*

$$\Lambda(\tau) = \Lambda_{trend}(\tau) + \Lambda_{season}(\tau),$$

where Λ_{season} is bounded and Λ_{trend} is at most of linear growth. Let the compensator measures $\ell_j(dz, ds)$ be bounded by $\widetilde{\ell}_j(dz)\, ds$, where $\widetilde{\ell}_j$ are Lévy measures satisfying the exponential integrability conditions

$$\int_{|z|\geq 1} e^{(|\eta_j|+c_j+\epsilon)|z|}\, \widetilde{\ell}_j(dz) < \infty\,, \quad j = 1, \ldots, m\,,$$

with $c_j = \sup_s |\widetilde{\theta}_j(s)|$. Finally, suppose that the drifts of the II processes are absolutely continuous, $\gamma_j(dt) = \gamma_j(t)\, dt$, with $\gamma_j(t)$ being bounded. Then,

$$f(t,\tau) - \Lambda(\tau)\Theta(t,\tau;\theta(\cdot))e^{\sum_{i=1}^m \frac{\mu_i}{\alpha_i}(1-e^{-\alpha_i(\tau-t)})+\sum_{j=1}^n \frac{\delta_j}{\beta_j}(1-e^{-\beta_j(\tau-t)})} \rightarrow_{\tau\rightarrow\infty} 0\,,$$

where $\Theta(t,\tau,\theta(\cdot))$ is defined in Prop. 4.6.

Proof. Consider the explicit forward price $f(t,\tau)$ in Prop. 4.6, and assume without loss of generality that $\mu_i = \delta_j = 0$ and $m = 1$. Then, for constant coefficients, we have

$$\begin{aligned} &|f(t,\tau) - \Lambda(\tau)\Theta(t,\tau;\theta(\cdot))| \\ &\qquad = |\Lambda(\tau)\Theta(t,\tau;\theta(\cdot))||\mathrm{e}^{\sum_{i=1}^m X_i(t)\mathrm{e}^{-\alpha_i(\tau-t)}+Y(t)\mathrm{e}^{-\beta(\tau-t)}} - 1|\,. \end{aligned}$$

Thus, letting $\tau \to \infty$, we obtain the result of the Proposition as long as $|\Lambda(\tau)\Theta(t,\tau;\theta(\cdot))|$ is at most linearly growing. Now, $\Lambda(\tau)$ is by assumption at most linearly growing. We demonstrate that $|\Theta(t,\tau;\theta(\cdot))|$ is uniformly bounded in τ.

First, we see that

$$|\int_t^\tau \sigma_{ik}\widehat{\theta}_k(u)\mathrm{e}^{-\alpha_i(\tau-t)}\, du| \leq \frac{\sigma_{ik}}{\alpha_i}\sup_{u\geq t}|\widehat{\theta}_k(u)| < \infty\,,$$

due to boundedness of $\widehat{\theta}_k$. Further, we have

$$\int_t^\tau \left(\sum_{i=1}^m \sigma_{ik}\mathrm{e}^{-\alpha_i(\tau-t)}\right)^2 du \leq \sum_{i,j=1}^m \frac{\sigma_{ik}\sigma_{jk}}{\alpha_i+\alpha_j} < \infty\,.$$

We now consider the jump part of $\Theta(t,\tau;\theta(\cdot))$. It is sufficient to let $t = 0$. We know that the jump part contribution to $\Theta(0,\tau;\theta(\cdot))$ follows from the expectation

$$\mathbb{E}_\theta\left[\mathrm{e}^{\eta\int_0^\tau \mathrm{e}^{-\beta(\tau-s)}\, dI(s)}\right]\,.$$

It implies (see Prop. 4.4) that we need to show that

$$\int_0^\tau \int_{\mathbb{R}} \{e^{z\eta e^{-\beta(\tau-s)}} - 1 - \eta e^{-\beta(\tau-s)} z 1_{|z|<1}\} e^{z\widetilde{\theta}(s)} \ell(dz, ds) \tag{4.18}$$

and

$$\int_0^\tau e^{-\beta(\tau-s)}\gamma(s)\, ds + \int_0^\tau \int_{|z|<1} z\eta e^{-\beta(\tau-s)} \{e^{z\widetilde{\theta}(s)} - 1\} \ell(dz, ds) \tag{4.19}$$

are bounded uniformly in $\tau \geq 0$. Let us first consider the drift part.

Since $\gamma(t)$ is assumed to be bounded, it holds that

$$|\int_0^\tau e^{-\beta(\tau-s)}\gamma(s)\, ds| \leq \frac{\sup_t |\gamma(t)|}{\beta}.$$

Furthermore, we estimate the integral part using the boundedness of $\widetilde{\theta}$ and the restriction on $\ell(dz, ds)$. Since $|e^z - 1| \leq| z|e^{|z|}$, we have

$$\begin{aligned}
|\int_0^\tau \int_{|z|<1} & z e^{-\beta(\tau-s)} \left(e^{\widetilde{\theta}(s)z} - 1\right) \ell(dz, ds)| \\
&\leq \int_0^\tau e^{-\beta(\tau-s)} \int_{|z|<1} |z||e^{\widetilde{\theta}(s)z} - 1| \widetilde{\ell}(dz)\, ds \\
&\leq \frac{ce^c}{\beta}(1 - e^{-\beta\tau}) \int_{|z|<1} |z|^2 \widetilde{\ell}(dz) \\
&\leq \frac{ce^c}{\beta} \int_{|z|<1} |z|^2 \widetilde{\ell}(dz),
\end{aligned}$$

which is finite whenever $\widetilde{\ell}$ is a Lévy measure. In the estimations above, $c = \sup_{s\geq 0} |\widetilde{\theta}(s)|$. Hence, (4.19) is bounded uniformly in τ.

We estimate (4.18). First, we find from the boundedness of $\widetilde{\theta}$ and condition of $\ell(dz, ds)$, that

$$\begin{aligned}
|\int_0^\tau \int_{\mathbb{R}} & \left(e^{z\eta e^{-\beta(\tau-s)}} - 1 - \eta e^{-\beta(\tau-s)} z 1_{|z|<1}\right) e^{\widetilde{\theta}(s)z} \ell(dz, ds)| \\
&\leq \int_0^\tau \int_{\mathbb{R}} |e^{z\eta e^{-\beta s}} - 1 - \eta e^{-\beta s} z 1_{|z|<1}| e^{c|z|} \widetilde{\ell}(dz)\, ds.
\end{aligned}$$

Consider the case where $|z| < 1$. We find

$$\int_0^\tau \int_{|z|<1} |e^{z\eta e^{-\beta s}} - 1 - \eta e^{-\beta s} z| e^{c|z|} \widetilde{\ell}(dz)\, ds$$

$$
\begin{aligned}
&\le \int_0^\tau \int_{|z|<1} \sum_{k=2}^{\infty} \frac{(|z||\eta| e^{-\beta s})^k}{k!} e^{c|z|}\,\widetilde{\ell}(dz)\,ds \\
&\le \int_0^\tau \int_{|z|<1} z^2\eta^2 e^{-2\beta s} e^{|z|(|\eta| e^{-\beta s}+c)}\,\widetilde{\ell}(dz)\,ds \\
&\le e^{|\eta|+c}\frac{\eta^2}{2\beta}\int_{|z|<1} z^2\,\widetilde{\ell}(dz)\,,
\end{aligned}
$$

which is finite since $\widetilde{\ell}(dz)$ is a Lévy measure.

Next we treat the case where $|z| \ge 1$. We find

$$
\begin{aligned}
\int_0^\tau \int_{|z|\ge 1} |e^{z\eta e^{-\beta s}} - 1| e^{c|z|}\,\widetilde{\ell}(dz)\,ds &\le \int_0^\tau \int_{|z|\ge 1} \sum_{k=1}^{\infty} \frac{(|z||\eta| e^{-\beta s})^k}{k!} e^{c|z|}\,\widetilde{\ell}(dz)\,ds \\
&\le |\eta| \int_0^\tau \int_{|z|\ge 1} |z| e^{-\beta s} e^{|z|(|\eta|+c)}\,\widetilde{\ell}(dz)\,ds \\
&\le \frac{|\eta|}{\beta}\int_{|z|\ge 1} |z| e^{|z|(|\eta|+c)}\,\widetilde{\ell}(dz) \\
&\le C\int_{|z|\ge 1} e^{|z|(|\eta|+c+\epsilon)}\,\widetilde{\ell}(dz)\,,
\end{aligned}
$$

for some $\epsilon > 0$ and a constant C independent of τ. Hence, from Condition G, we obtain a uniform bound in τ. This completes the proof. □

It is possible to weaken the conditions in the Proposition above to include time-dependent coefficients, and also more general II processes. We will not go into more details regarding this.

From Prop. 4.7, we find that in the long end of the forward curve,

$$
f(t,\tau) \approx \Theta(t,\tau;\theta(\cdot))\Lambda(\tau)\,,
$$

when $\mu_i = \delta_j = 0$, $i = 1,\ldots,m, j = 1,\ldots,n$.[1] The forward prices tend to the risk-adjusted seasonal mean level when time to delivery becomes large. This result is rather intuitive, since all spot price variations should go back to their seasonal mean due to the mean reversion effect. However, from the introduction of the market price of risk via θ, we have a risk-adjustment of this mean.

In the next Proposition we state the risk neutral dynamics of f.

[1] This is the natural case if we interpret Λ as the seasonal mean.

Proposition 4.8 *Suppose that Condition G holds with*

$$\sup_{0\leq t\leq\tau} |\eta_j(t)e^{-\int_t^\tau \beta_j(v)\,dv} + \widetilde{\theta}_j(t)| \leq c_j$$

for $j = 1,\ldots,n$. *The dynamics of* $t \mapsto f(t,\tau)$ *with respect to* Q^θ *is*

$$\frac{df(t,\tau)}{f(t-,\tau)} = \sum_{k=1}^{p}\left\{\sum_{i=1}^{m}\sigma_{ik}(t)\exp\left(-\int_t^\tau \alpha_i(u)\,du\right)\right\} dB_k^\theta(t)$$
$$+\sum_{j=1}^{n}\left\{\int_{\mathbb{R}}\exp\left(z\eta_j(t)e^{-\int_t^\tau \beta_j(u)\,du}\right)-1\right\}\widetilde{N}_j^\theta(dt,dz)\,.$$

Proof. Consider $f(t,\tau)$ in Prop. 4.6. We have that f is a martingale with respect to Q^θ. Hence, by Itô's Formula the only terms that will be included in the dynamics of $t \mapsto f(t,\tau)$ are those involving dB_k^θ and $\widetilde{N}_j^\theta$. To obtain the dynamics from Itô's Formula, it is convenient to first rewrite the dynamics of $Y_j(t)$. Using the integrability hypothesis in Condition G, we have,

$$dY_j(t) = (\delta_j(t) - \beta_j(t)Y_j(t))\,dt + \eta_j(t)\,d\gamma_j(t)$$
$$+\int_{|z|<1} z\left\{e^{\widetilde{\theta}_j(t)z}-1\right\}\ell_j(dz,dt) + \int_{|z|\geq 1} z\eta_j(t)\,\ell_j(dz,dt)$$
$$+\int_{\mathbb{R}} z\eta_j(t)\,\widetilde{N}_j^\theta(dz,dt)\,.$$

Hence, from Itô's Formula, the dynamics follows. □

We see that the forward price dynamics becomes a geometric model, and in the case when we do not have any jump terms Y_i, we are back to a geometric Brownian motion with time-dependent volatility

$$\frac{df(t,\tau)}{f(t,\tau)} = \sum_{k=1}^{p}\left\{\sum_{i=1}^{m}\sigma_{ik}(t)\exp\left(-\int_t^\tau \alpha_i(u)\,du\right)\right\} dB_k^\theta(t)\,. \qquad (4.20)$$

Hence, we find that the volatilities of the forward are decreasing with time to delivery, being smaller than the spot volatility. When time to delivery approaches zero, however, the forward volatility converges to the volatilities of the underlying spot $\sigma_{ik}(t)$. This is known as the Samuelson effect (see [Samuelson (1965b)]), and is a direct result of the mean-reverting spot price dynamics. In economical terms, it may be explained as follows. The arrival of information to the market has a much bigger influence when there is short time left to maturity than for the long-term contracts. For long-term contracts, the market has time to adjust before delivery takes place, making them less sensitive to changes in the spot.

We observe a similar Samuelson effect when including jumps in the spot dynamics, where so-called "jump volatility" is expressed through the integrands

$$\exp\left(z\eta_j(t)e^{-\int_t^\tau \beta_j(u)\,du}\right) - 1\,.$$

Suppose that $\eta_j(t) > 0$. When $t \to \tau$, the integrands are converging to $\{\exp(z\eta_j(t)) - 1\}$, which is identical to the corresponding terms of the spot price dynamics (see Prop. 3.3). However, when $t < \tau$, we find for $z \geq 0$, that

$$0 \leq \left\{\exp\left(z\eta_j(t)e^{-\int_t^\tau \beta_j(u)\,du}\right) - 1\right\} \leq \{\exp(z\eta_j(t)) - 1\}\,,$$

meaning that the positive jumps in the spot price dynamics are scaled down in the forward price dynamics, and the downscaling is exponential with respect to the mean reversion. For the negative jumps ($z < 0$) we find

$$0 \geq \left\{\exp\left(z\eta_j(t)e^{-\int_t^\tau \beta_j(u)\,du}\right) - 1\right\} \geq \{\exp(z\eta_j(t)) - 1\}\,.$$

Thus, also the negative jumps in the spot are scaled down in the forward. All in all, the "jump volatility" of the forward dynamics is a downscaling of the jump volatility of the spot, in line with the observations we made for the Brownian motion terms. The downscaling of the jump volatility is dependent on time to maturity. The farther away from maturity, the less influence the jump volatility gets from the spot. The influence is "discounted" by the speed of mean reversion. The stronger the speed of mean reversion, the faster jumps in the spot price are whipped out along the term structure. This means that if we have a spike in the spot price, then the influence of this spike on the forward price is only in the very short end of the forward curve, whereas the long end of the curve is not affected at all. Changes in the spot having slow mean reversion will translate over longer ranges of the forward curve.

Observe that the dynamics of $f(t,\tau)$ is only dependent on the volatilities σ_{ik} and η_j, together with the mean reversion coefficients α_i and β_j. The initial state of $f(0,\tau)$, that is, today's forward curve, is dependent on the seasonality function Λ of the spot, along with the levels μ_i and δ_j and the market price of risk θ. We see this by letting $t = 0$ in the explicit dynamics for $f(t,\tau)$ in Prop. 4.6.

We end our discussion on the forward price implied by the geometric spot model with the market dynamics of $t \mapsto f(t,\tau)$.

Proposition 4.9 *Suppose that Condition G holds with*

$$\sup_{0\leq t\leq \tau} |\eta_j(t)e^{-\int_t^\tau \beta_j(u)\,du} + \widetilde{\theta}_j(t)| \leq c_j$$

for every $j = 1,\ldots,n$. The P dynamics of $t \mapsto f(t,\tau)$ is given by

$$\begin{aligned}
\frac{df(t,\tau)}{f(t-,\tau)} = &-\sum_{k=1}^{p}\left\{\sum_{i=1}^{m}\sigma_{ik}(t)\widehat{\theta}_k(t)\exp\left(-\int_t^\tau \alpha_i(v)\,dv\right)\right\}dt \\
&-\sum_{j=1}^{n}\int_{\mathbb{R}}\left\{\exp\left(z\eta_j(t)e^{-\int_t^\tau \beta_j(v)\,dv}\right)-1\right\} \\
&\quad\times\left\{\exp\left(\widetilde{\theta}_j(t)z\right)-1\right\}\ell_j(dz,dt) \\
&+\sum_{k=1}^{p}\left\{\sum_{i=1}^{m}\sigma_{ik}(t)\exp\left(-\int_t^\tau \alpha_i(u)\,du\right)\right\}dB_k(t) \\
&+\sum_{j=1}^{n}\left\{\int_{\mathbb{R}}\exp\left(z\eta_j(t)e^{-\int_t^\tau \beta_j(u)\,du}\right)-1\right\}\widetilde{N}_j(dt,dz)\,.
\end{aligned}$$

Proof. Consider the dynamics of $t \mapsto f(t,\tau)$ under Q^θ in Prop. 4.8. From Prop. 4.4, we have

$$dB_k^\theta(t) = dB_k(t) - \widehat{\theta}_k(t)\,dt\,.$$

Furthermore, following the discussion after the proof of Prop 4.4, it holds that

$$\widetilde{N}_j^\theta(dz,dt) = \widetilde{N}_j(dz,dt) - \left(e^{\widetilde{\theta}_j(t)z}-1\right)\ell_j(dz,dt)\,.$$

The result follows by direct insertion after using Condition G to validate the finiteness of the integrals with respect to $\ell_j(dz,dt)$ in the drift. □

Letting $\tau \to \infty$, the drift part of $f(t,\tau)$ is eliminated, at least for constant coefficients. Hence, the long end of the forward curve behaves like a martingale with similar diffusional volatility as for the Q^θ dynamics. The jump part has the same scaling as in the Q^θ dynamics, however, $\widetilde{N}_j^\theta$ will have different distributional properties under Q^θ as $\widetilde{N}_j$ under P. For instance, in the P dynamics we do not see any influence of $\widetilde{\theta}_j$ on the jump terms, which is the case under Q^θ.

Note that we can weaken the integrability Condition G assumed in the Proposition above. Going to the proof, we can split the integral with respect to $\widetilde{N}_j^\theta$ into two parts; integration over $|z|<1$ and $|z|\geq 1$. This would leave us with one integral with respect to $\widetilde{N}_j^\theta$ that we could treat as in the proof, and the remaining part yielding integral terms with respect to N_j^θ and ℓ_j. We leave the details to the interested reader.

We proceed further and analyse the forward price dynamics under the arithmetic model for the spot price.

4.3.2 *The arithmetic case*

Suppose that $S(t)$ is modelled by the arithmetic spot price model defined in (3.20). The results we derive here are analogous to the geometric case, and in some sense just simplifications of the latter. We start by deriving the forward price dynamics $f(t,\tau)$ of a contract maturing at time $\tau\leq\mathcal{T}$.

Proposition 4.10 *Suppose that there exists an $\epsilon>0$ such that Condition G holds with*

$$\sup_{0\leq t\leq\mathcal{T}}|\widetilde{\theta}_j(t)|+\epsilon\leq c_j\,,$$

for $j=1,\ldots,n$. The forward price $f(t,\tau)$ is given by

$$\begin{aligned}f(t,\tau)&=\Lambda(\tau)+\Theta(t,\tau;\theta)\\&\quad+\sum_{i=1}^m\int_t^\tau\mu_i(u)e^{-\int_u^\tau\alpha_i(v)\,dv}\,du+\sum_{j=1}^n\int_t^\tau\delta_j(u)e^{-\int_t^\tau\beta_j(v)\,dv}\,du\\&\quad+\sum_{i=1}^m X_i(t)e^{-\int_t^\tau\alpha_i(s)\,ds}+\sum_{j=1}^n Y_j(t)e^{-\int_t^\tau\beta_j(s)\,ds}\,,\end{aligned}$$

for $0\leq t\leq\tau$, where

$$\begin{aligned}\Theta(t,\tau;\theta)&=\sum_{k=1}^p\sum_{i=1}^m\int_t^\tau\sigma_{ik}(u)\widehat{\theta}_k(u)e^{-\int_u^\tau\alpha_i(v)\,dv}\,du\\&\quad+\sum_{j=1}^n\int_t^\tau\eta_j(u)e^{-\int_u^\tau\beta_j(v)\,dv}\,d\gamma_j(u)\\&\quad+\sum_{j=1}^n\int_t^\tau\int_{\mathbb{R}}\eta_j(u)e^{-\int_u^\tau\beta_j(v)\,dv}z\left(e^{\widetilde{\theta}_j(u)z}-1_{|z|<1}\right)\ell_j(dz,du)\,.\end{aligned}$$

Proof. For simplicity, consider $n = m = 1$. We find for $t \leq \tau$,

$$\mathbb{E}_\theta\left[S(\tau)\,|\,\mathcal{F}_t\right] = \Lambda(\tau) + \mathbb{E}_\theta\left[X(\tau)\,|\,\mathcal{F}_t\right] + \mathbb{E}_\theta\left[Y(\tau)\,|\,\mathcal{F}_t\right].$$

By (3.3) and Prop. 4.4, we have

$$\begin{aligned}X(\tau) = X(t)\mathrm{e}^{-\int_t^\tau \alpha(v)\,dv} &+ \int_t^\tau \mu(u)\mathrm{e}^{-\int_u^\tau \alpha(v)\,dv}\,du \\ &+ \sum_{k=1}^p \int_t^\tau \sigma_k(u)\widehat{\theta}_k(u)\mathrm{e}^{-\int_u^\tau \alpha(v)\,dv}\,du \\ &+ \sum_{k=1}^p \int_t^\tau \sigma_k(u)\mathrm{e}^{-\int_u^\tau \alpha(v)\,dv}\,dB_k^\theta(u).\end{aligned}$$

The process $X(t)$ is $\mathcal{F}_t$-measurable. Furthermore, since B_k^θ is a Brownian motion under Q^θ, we find that

$$\mathbb{E}_\theta\left[\int_t^\tau \sigma_k(u)\mathrm{e}^{-\int_u^\tau \alpha(v)\,dv}\,dB_k^\theta\,|\,\mathcal{F}_t\right] = 0.$$

This completes the derivation of the contribution from the $X_i(\tau)$'s to the forward price.

Again, by (3.3), we have

$$\begin{aligned}Y(\tau) = Y(t)\mathrm{e}^{-\int_t^\tau \beta(v)\,dv} &+ \int_t^\tau \delta(u)\mathrm{e}^{-\int_u^\tau \beta(v)\,dv}\,du \\ &+ \int_t^\tau \eta(u)\mathrm{e}^{-\int_u^\tau \beta(v)\,dv}\,dI(u).\end{aligned}$$

Invoking the characteristics of $I(u)$ under Q^θ presented in Prop 4.4 together with the integrability assumptions in the Proposition, we have

$$\begin{aligned}dI(u) = d\gamma(u) &+ \int_{\mathbb{R}} z\left\{\mathrm{e}^{\widetilde{\theta}(u)z} - 1_{|z|<1}\right\}\ell(dz, du) \\ &+ \int_{\mathbb{R}} z\,\widetilde{N}^\theta(dz, du).\end{aligned}$$

Hence,

$$\begin{aligned}Y(\tau) = Y(t)\mathrm{e}^{-\int_t^\tau \beta(v)\,dv} &+ \int_t^\tau \delta(u)\mathrm{e}^{-\int_u^\tau \beta(v)\,dv}\,du \\ &+ \int_t^\tau \eta(u)\mathrm{e}^{-\int_u^\tau \beta(v)\,dv}\,d\gamma(u)\end{aligned}$$

$$+\int_t^\tau \int_{\mathbb{R}} \eta(u) \mathrm{e}^{-\int_u^\tau \beta(v)\,dv} z\left\{\mathrm{e}^{\widetilde{\theta}(u)z} - 1_{|z|<1}\right\} \ell_j(dz, du)$$
$$+\int_t^\tau \int_{\mathbb{R}} z\eta(u) \mathrm{e}^{-\int_u^\tau \beta(v)\,dv} \widetilde{N}^\theta(dz, du)\,.$$

The compensated random jump measure $\widetilde{N}^\theta$ is a martingale with respect to Q^θ, which implies that

$$\mathbb{E}_\theta\left[\int_t^\tau \int_{\mathbb{R}} z\eta(u) \mathrm{e}^{-\int_u^\tau \beta(v)\,dv} \widetilde{N}^\theta(dz, du) \,|\, \mathcal{F}_t\right] = 0\,.$$

Hence, the Proposition follows from applying the adaptedness of $Y(t)$. □

As a special case of an arithmetic model, we consider the dynamics proposed by [Benth, Kallsen and Meyer-Brandis (2007)] ensuring positive spot prices. Recall from Subsect. 3.2.2 that in this case $m = 0$ since we do not assume any diffusional variations in the dynamics, while the compensator measures ℓ_j are all supported on the positive real line since I_j have only positive jumps. Further, $\delta_j = 0$ since we assume the mean reversion levels to be zero. We see that the forward price reduces to

$$f(t, \tau) = \Lambda(\tau) + \Theta(t, \tau; \theta(\cdot)) + \sum_{i=1}^n Y_j(t) \mathrm{e}^{-\int_t^\tau \beta_j(s)\,ds}\,,$$

where

$$\Theta(t, \tau; \theta(\cdot)) = \sum_{j=1}^n \int_t^\tau \eta_j(u) \mathrm{e}^{-\int_u^\tau \beta_j(v)\,dv}\, d\gamma_j(u)$$
$$+\sum_{j=1}^n \int_t^\tau \int_0^\infty \eta_j(u) \mathrm{e}^{-\int_u^\tau \beta_j(v)\,dv} z\{\mathrm{e}^{\widetilde{\theta}_j(u)z} - 1_{|z|<1}\} \ell_j(dz, du)\,.$$

Observe that we can move the compensating term in the integrals with respect to ℓ_j to the drift integrals $d\gamma_j$, and redefine $\Theta(t, \tau; \theta(\cdot))$ in terms of $d\widetilde{\gamma}_j$ (recall the definition of $\widetilde{\gamma}_j$ in (3.25)) instead

$$\Theta(t, \tau; \theta(\cdot)) = \sum_{j=1}^n \int_t^\tau \eta_j(u) \mathrm{e}^{-\int_u^\tau \beta_j(v)\,dv}\, d\widetilde{\gamma}_j(u)$$
$$+\sum_{j=1}^n \int_t^\tau \int_0^\infty \eta_j(u) \mathrm{e}^{-\int_u^\tau \beta_j(v)\,dv} z\mathrm{e}^{\widetilde{\theta}_j(u)z} \ell_j(dz, du)\,.$$

Note that since Q^θ is equivalent to P, the two measures have the same zero sets. Hence, since the spot model is positive under P, it must be so under

Q^θ as well. The forward price is calculated as the conditional expectation of the spot with respect to the risk-neutral probability, and therefore it follows that the forward price dynamics becomes positive in the arithmetic class of spot models defined by [Benth, Kallsen and Meyer-Brandis (2007)].

Let us return to the general case. The risk-neutral dynamics of the forward price process in Prop. 4.10 is now straightforward to derive.

Proposition 4.11 *The dynamics of the stochastic process $t \mapsto f(t,\tau)$, $t \leq \tau$ with respect to the risk-neutral measure Q^θ is given by*

$$\begin{aligned} df(t,\tau) &= \sum_{k=1}^{p}\sum_{i=1}^{m} \sigma_{ik}(t) e^{-\int_t^\tau \alpha_i(s)\,ds}\, dB_k^\theta(t) \\ &\quad + \sum_{j=1}^{n} \eta_j(t) e^{-\int_t^\tau \beta_j(s)\,ds} \int_{\mathbb{R}} z\, \widetilde{N}^\theta(dz,dt)\,. \end{aligned} \tag{4.21}$$

Proof. The argument is analogous to the proof of Prop. 4.8. □

We continue our exposition of the arithmetic case with the implied P dynamics of the forward price.

Proposition 4.12 *Suppose that there exists an $\epsilon > 0$ so that Condition G holds with*

$$\sup_{0\leq t\leq T} |\widetilde{\theta}_j(t)| + \epsilon \leq c_j\,,$$

for $j = 1,\ldots,n$. The P dynamics of the stochastic process $t \mapsto f(t,\tau)$ for $t \leq \tau$, is

$$\begin{aligned} df(t,\tau) &= -\sum_{k=1}^{p}\sum_{i=1}^{m} \sigma_{ik}(t)\widehat{\theta}_k(t) e^{-\int_t^\tau \alpha_i(v)\,dv}\, dt \\ &\quad - \sum_{j=1}^{n} \eta_j(t) e^{-\int_t^\tau \beta_j(v)\,dv} \int_{\mathbb{R}} z\left(e^{\widetilde{\theta}_j(t)z} - 1\right) \ell_j(dz,dt) \\ &\quad + \sum_{k=1}^{p}\sum_{i=1}^{m} \sigma_{ik}(t) e^{-\int_t^\tau \alpha_i(v)\,dv}\, dB_k(t) \\ &\quad + \sum_{j=1}^{n} \int_{\mathbb{R}} z\eta_j(t) e^{-\int_t^\tau \beta_j(v)\,dv}\, \widetilde{N}_j(dz,dt)\,. \end{aligned}$$

Proof. Observe that the integral terms with respect to $\ell_j(dz,du)$ are all well defined by the integrability assumption in the Proposition. The argument follows closely the proof of Prop. 4.9. □

From the results above, we see that the arithmetic form of the spot price dynamics is transferred to the forward price dynamics. Furthermore, the volatility is dampened exponentially by the speeds of mean reversion, exactly as in the geometric case. We also see a convergence towards a martingale dynamics in the long end of the forward curve, at least when we have constant coefficients in the dynamics. Let us discuss the asymptotics of the forward curve when time to maturity becomes large. To reduce the number of terms, we restrict our attention to the simpler case $\mu_i = \delta_j = 0$, for $i = 1, \ldots, m$ and $j = 1, \ldots, n$. This means that all the factor processes X_i and Y_j revert to zero, and $\Lambda(\tau)$ is the mean price level of the spot. Further, assume that α_i and β_j, are constants for $i = 1, \ldots, m$ and $j = 1, \ldots, n$. In this case, we have from Prop. 4.10

$$|f(t,\tau) - \Lambda(\tau) - \Theta(t,\tau;\theta(\cdot))| \leq \sum_{i=1}^{m} |X_i(t)| \mathrm{e}^{-\alpha_i(\tau-t)} + \sum_{j=1}^{n} |Y_j(t)| \mathrm{e}^{-\beta_j(\tau-t)} .$$

When $\tau \to \infty$, the right-hand side tends to zero. Hence, it follows that

$$f(t,\tau) - \Lambda(\tau) - \Theta(t,\tau;\theta(\cdot)) \to_{\tau\to\infty} 0 .$$

As in the geometric case, we conclude that the forward prices behave asymptotically as the seasonal mean level and a risk-adjustment factor. Note that we do not need any growth conditions on $\Lambda(\tau)$ to establish the result.

In the chapter on option pricing, Sect. 9.2, we will consider pricing of spread and Asian options based on the positive arithmetic model. For both option types, the arithmetic model is particularly suited for pricing using the Fourier approach.

4.4 Pricing of swaps

In the electricity and gas markets we recall that the basic forward contracts deliver over a period rather than at a fixed maturity time. We therefore move our attention to the pricing of swaps when the spot price process is either a geometric or an arithmetic model. We show that the explicit pricing formula and dynamics are lost in the former case. On the other hand, the arithmetic model still permits the derivation of an explicit forward price dynamics. It makes this class of models particularly interesting in the electricity market. However, we start our discussion with the case of a geometric spot price dynamics.

4.4.1 *The geometric case*

It is in general not possible to state an explicit formula for the swap price $F(t, \tau_1, \tau_2)$ of contracts with settlement over the period $[\tau_1, \tau_2]$ when we choose to work with a geometric model of the spot price dynamics as in (3.8). We now elaborate on the approximation of the swap price.

Recall the relation in Prop. 4.1 between forwards and swaps

$$F(t, \tau_1, \tau_2) = \int_{\tau_1}^{\tau_2} w(u, \tau_1, \tau_2) f(t, u) \, du \, .$$

For a geometric spot price model, Prop. 4.6 gives the explicit formula for the forward prices $f(t, \tau)$, which we for the moment write as

$$\begin{aligned} f(t,\tau) &= \widetilde{f}_\theta(t,\tau) \\ &\quad \times \exp\left(\sum_{i=1}^m e^{-\int_t^\tau \alpha_i(v)\,dv} X_i(t) + \sum_{j=1}^n e^{-\int_t^\tau \beta_j(v)\,dv} Y_j(t)\right) , \end{aligned}$$

with

$$\begin{aligned} \widetilde{f}_\theta(t,\tau) &= \Lambda(\tau)\Theta(t,\tau;\theta(\cdot)) \\ &\quad \times \exp\left(\sum_{i=1}^m \mu_i(u) e^{-\int_u^\tau \alpha_i(v)\,dv}\,du + \sum_{j=1}^n \delta_j(u) e^{-\int_u^\tau \beta_j(v)\,dv}\,du\right) . \end{aligned}$$

Hence, we find that

$$\begin{aligned} F(t,\tau_1,\tau_2) &= \int_{\tau_1}^{\tau_2} w(u,\tau_1,\tau_2)\widetilde{f}_\theta(t,u) \\ &\quad \times \exp\left(\sum_{i=1}^m e^{-\int_t^u \alpha_i(v)\,dv} X_i(t) + \sum_{j=1}^n e^{-\int_t^u \beta_j(v)\,dv} Y_j(t)\right) du \, . \end{aligned} \tag{4.22}$$

This integral does in general not have any analytic solution, and numerical integration is required for its valuation. In other words, if we want to derive the swap price dynamics $F(t, \tau_1, \tau_2)$, we must do this by numerical integration. Let us discuss the issue of time discretization of the integral in (4.22).

We have chosen to represent the swap price as an integral of forwards (or equivalently, as an integral of the predicted spot prices). In the marketplace, contracts are not settled in continuous time, but rather at discrete time points over the delivery period. Typically, financial contracts are settled each hour or day in the delivery period. Henceforth, the integral representation in (4.22) is simply a convenient mathematical approximation of the actual contract traded in the market. However, this connection gives us an obvious choice of the time discretzation when we want to estimate numerically the integral yielding the price $F(t,\tau_1,\tau_2)$. We must choose the discretization coinciding with the actual settlement times for the contract. This, in practical terms, means that a numerical evaluation of (4.22) should be based on the summation

$$F(t,\tau_1,\tau_2) \approx \sum_{u=1}^{N} w(u,\tau_1,\tau_2) f(t,u),$$

where $u = 1,\ldots,N$ are the days, or hours, in the delivery period where the settlement against the spot price takes place according to the contractual specifications. Note that we have implicitly assumed that time is measured on the same scale as the settlements (for example, days or hours).

We see that the swap price $F(t,\tau_1,\tau_2)$ in (4.22) depends explicitly on $X_i(t)$ and $Y_j(t)$. The swap price dynamics is therefore Markovian in the $m+n$ variables $X_i(t)$ and $Y_j(t)$, which is to our advantage when we price options, say, written on $F(t,\tau_1,\tau_2)$. Note that a European call or put option on the swap will not be an Asian type option, because we do not integrate $X_i(t)$ and $Y_j(t)$ over time, but scale instead these random variables with respect to an integral over the delivery period. A simple Monte Carlo algorithm can be implemented which only requires simulation of the factor at one time instance, and not over the whole path. Thus, the valuation of options written on $F(t,\tau_1,\tau_2)$ is simpler numerically than "classical" Asian options (see [Weron (2005)] for pricing of Asian options on electricity spot).

There exist a few cases where we can derive explicit forward prices for contracts with delivery over a period in the geometric case. Let us restrict our attention to a non-stationary dynamics of the spot price represented by setting the mean reversion coefficients α_i and β_j equal to zero. Hence, the spot price becomes a geometric Brownian motion in the case of no jumps. We find the following risk-neutral dynamics of $F(t,\tau_1,\tau_2)$.

Proposition 4.13 *Suppose that Condition G holds with*

$$2\sup_{0\le t\le T}|\eta_j(t)| + |\widetilde{\theta}_j(t)| \le c_j\,,$$

for $j = 1,\ldots,n$. Suppose further that the mean reversion coefficients α_i and β_j are set equal to zero for $i = 1,\ldots,m$ and $j = 1,\ldots,n$. Then the *risk-neutral dynamics of the forward price $F(t,\tau_1,\tau_2)$ is given by*

$$\frac{dF(t,\tau_1,\tau_2)}{F(t-,\tau_1,\tau_2)}=\sum_{k=1}^{p}\sum_{i=1}^{m}\sigma_{ik}(t)\,dB_k^{\theta}(t)+\sum_{j=1}^{n}\int_{\mathbb{R}}\{e^{\eta_j(t)z}-1\}\,\widetilde{N}_j^{\theta}(dz,dt)\,.$$

Proof. From the integrability condition we have that the risk-neutral dynamics of $f(t,\tau)$ stated in Prop. 4.8 is well-defined when $\alpha_i=\beta_j=0$. Then we have

$$\begin{aligned}df(t,u)&=\sum_{k=1}^{p}\sum_{i=1}^{m}\sigma_{ik}(t)f(t,u)\,dB_k^{\theta}(t)\\&\quad+\sum_{j=1}^{n}f(t-,u)\int_{\mathbb{R}}\{e^{\eta_j(t)z}-1\}\,\widetilde{N}_j^{\theta}(dz,dt)\,.\end{aligned}$$

Hence, integrating this over the delivery period, we find

$$\begin{aligned}&\int_{\tau_1}^{\tau_2}w(u,\tau_1,\tau_2)f(t,u)\,du\\&\quad=\int_{\tau_1}^{\tau_2}w(u,\tau_1,\tau_2)f(0,u)\,du\\&\qquad+\sum_{k=1}^{p}\sum_{i=1}^{m}\int_0^t\sigma_{ik}(s)\int_{\tau_1}^{\tau_2}w(u,\tau_1,\tau_2)f(s,u)\,du\,dB_k^{\theta}(s)\\&\qquad+\sum_{j=1}^{n}\int_0^t\int_{\mathbb{R}}\{e^{\eta_j(s)z}-1\}\int_{\tau_1}^{\tau_2}w(u,\tau_1,\tau_2)f(s,u)\,du\,\widetilde{N}_j^{\theta}(dz,ds)\,,\end{aligned}$$

where we have used the stochastic Fubini Theorem to commute stochastic integration with Lebesgue integration.

To conclude the proof, we verify that we can use the stochastic Fubini Theorem. First, we see that due to the integrability conditions on the compensator measure, we have that

$$\sigma_{ik}(s)\left(\int_{\tau_1}^{\tau_2}w^2(u,\tau_1,\tau_2)f^2(s,u)\,du\right)^{1/2}$$

is Itô integrable. This is so because when writing out $f^2(s,u)$ we obtain terms like

$$\exp\left(2\int_0^s\eta_j(t)e^{-\int_s^u\beta_j(v)\,dv}\,dI_j(t)\right)$$

and

$$\exp\left(2\sum_{k=1}^{p}\int_0^s \sigma_{ik}(t)\mathrm{e}^{-\int_s^u \alpha_i(v)\,dv}\,dB_k^\theta(t)\right).$$

The latter has a finite expectation since it is a lognormal variable. The former integral has a finite expectation as a result of the integrability condition in the Proposition. We can conclude that the Itô integral is well-defined. Thus, from the stochastic Fubini Theorem we are allowed to commute the integration with respect to du with that of dB^θ. Similar arguments show that (2.17) holds with $X(s,z)$ defined as

$$X(s,z)=\left(\mathrm{e}^{\eta_j(s)z}-1\right)\left(\int_{\tau_1}^{\tau_2} w^2(u,\tau_1,\tau_2)f^2(s,u)\,du\right)^{1/2}.$$

In the verification of (2.17), we use expectation with respect to Q^θ and the compensator measure of N^θ. Hence, $X(s,z)$ is integrable with respect to the compensated random jump measure $\widetilde{N}^\theta(dz,ds)$. Thus, the stochastic Fubini Theorem yields that we can commute the integrations with respect to du and $\widetilde{N}^\theta(dz,ds)$. Hence, the Proposition follows. □

Letting the mean reversion coefficients be equal to zero corresponds to a volatility of forward being independent on delivery time. This is an unrealistic feature if we want to model commodity markets accurately, since the forward contracts in these markets typically possess the Samuelson effect, as discussed earlier. This effect says that the volatility is decreasing with time to maturity of the contract, and therefore volatility must be dependent on delivery time. Hence, having zero speed of mean reversion creates a market which does not have the right properties for modelling the riskiness of forwards, and henceforth swaps.

4.4.2 *The arithmetic case*

Suppose now that the spot price process is modelled as the arithmetic dynamics in (3.20). In this case, following Prop. 4.10, the arithmetic property is transferred to the forward price dynamics. In the next Proposition, we show that this property also holds for swap contracts.

Proposition 4.14 *Suppose that there exists an $\epsilon > 0$ such that Condition G is satisfied with*

$$\sup_{0\le t\le T}|\widetilde{\theta}_j(t)|+\epsilon\le c_j\,,$$

for $j = 1, \ldots, n$. The forward price is given by

$$
\begin{aligned}
F(t, \tau_1, \tau_2) = &\int_{\tau_1}^{\tau_2} w(u, \tau_1, \tau_2)\Lambda(u)\, du + \Theta(t, \tau_1, \tau_2; \theta(\cdot)) \\
&+ \sum_{i=1}^{m} \int_t^{\tau_2} \int_{\max(v,\tau_1)}^{\tau_2} w(u, \tau_1, \tau_2)\mu_i(v) e^{-\int_v^u \alpha_i(s)\, ds}\, du\, dv \\
&+ \sum_{j=1}^{n} \int_t^{\tau_2} \int_{\max(v,\tau_1)}^{\tau_2} w(u, \tau_1, \tau_2)\delta_j(v) e^{-\int_v^u \beta_j(s)\, ds}\, du\, dv \\
&+ \sum_{i=1}^{m} X_i(t) \int_{\tau_1}^{\tau_2} w(u, \tau_1, \tau_2) e^{-\int_t^u \alpha_i(s)\, ds}\, du \\
&+ \sum_{j=1}^{n} Y_j(t) \int_{\tau_1}^{\tau_2} w(u, \tau_1, \tau_2) e^{-\int_t^u \beta_j(s)\, ds}\, du\,,
\end{aligned}
$$

for $0 \leq t \leq \tau_1 < \tau_2$, where

$$
\begin{aligned}
&\Theta(t, \tau_1, \tau_2; \theta(\cdot)) \\
&= \sum_{k=1}^{p} \sum_{i=1}^{m} \int_t^{\tau_2} \int_{\max(v,\tau_1)}^{\tau_2} w(u, \tau_1, \tau_2)\sigma_{ik}(v)\widehat{\theta}_i(v) e^{-\int_v^u \alpha_i(s)\, ds}\, du\, dv \\
&\quad + \sum_{j=1}^{n} \int_t^{\tau_2} \int_{\max(v,\tau_1)}^{\tau_2} w(u, \tau_1, \tau_2)\eta_j(v) e^{-\int_v^u \beta_j(s)\, ds}\, du\, d\gamma_j(v) \\
&\quad + \sum_{j=1}^{n} \int_t^{\tau_2} \int_{\mathbb{R}} \int_{\max(v,\tau_1)}^{\tau_2} w(u, \tau_1, \tau_2)\eta_j(v) e^{-\int_v^u \beta_j(s)\, ds} \\
&\qquad\quad \times z \left(e^{\widetilde{\theta}_j(v) z} - 1_{|z|<1} \right) du\, \ell_j(dz, dv)\,.
\end{aligned}
$$

Proof. To prove this, we appeal to the identity in Prop. 4.1

$$
F(t, \tau_1, \tau_2) = \int_{\tau_1}^{\tau_2} w(u, \tau_1, \tau_2) f(t, u)\, du\,,
$$

which holds since $\mathbb{E}_\theta[\int_{\tau_1}^{\tau_2} |w(u, \tau_1, \tau_2)S(u)|\, du] < \infty$ due to the integrability condition and the boundedness of $w(u, \tau_1, \tau_2)$. Using the explicit formula for $f(t, u)$ in Prop. 4.10, the Proposition follows from the Fubini-Tonelli Theorem (see [Folland (1984)]). □

Notice the arithmetic structure of the swap price, inherited from the spot dynamics and the forward price. The attractiveness of using an arithmetic spot price model in the context of electricity and gas markets is clear from the explicitness of the swap dynamics.

Let us discuss the asymptotics of the swap prices in the long end of the curve. For simplicity, we restrict our attention to the case of constant coefficients, and where in particular $\mu_i = \delta_j = 0$ for $i = 1, \ldots, m$ and $j = 1, \ldots, n$. From Prop. 4.14, we find

$$\begin{aligned}
|F(t,\tau_1,\tau_2) - &\int_{\tau_1}^{\tau_2} w(u,\tau_1,\tau_2)\Lambda(u)\,du - \Theta(t,\tau_1,\tau_2;\theta(\cdot))| \\
&\leq \sum_{i=1}^{m} |X_i(t)| \int_{\tau_1}^{\tau_2} |w(u,\tau_1,\tau_2)| \mathrm{e}^{-\alpha_i(u-t)}\,du \\
&\quad + \sum_{j=1}^{n} |Y_j(t)| \int_{\tau_1}^{\tau_2} |w(u,\tau_1,\tau_2)| \mathrm{e}^{-\beta_j(u-t)}\,du \\
&\leq C \sum_{i=1}^{m} |X_i(t)| (1 - \mathrm{e}^{-\alpha_i(\tau_2-\tau_1)}) \mathrm{e}^{-\alpha_i(\tau_1-t)} \\
&\quad + \sum_{j=1}^{n} |Y_j(t)| (1 - \mathrm{e}^{-\beta_j(\tau_2-\tau_1)}) \mathrm{e}^{-\beta_j(\tau_1-t)}\,,
\end{aligned}$$

where we have used the boundedness of $w(u,\tau_1,\tau_2)$. Letting $\tau_1 \to \infty$ and the length of the delivery period $\tau_2 - \tau_1$ be fixed, we obtain that

$$F(t,\tau_1,\tau_2) - \int_{\tau_1}^{\tau_2} w(u,\tau_1,\tau_2)\Lambda(u)\,du - \Theta(t,\tau_1,\tau_2;\theta(\cdot)) \to_{\tau_1\to\infty} 0\,.$$

Hence, the swap price behaves asymptotically as the weighted average seasonal function $\Lambda(u)$ and a risk-adjustment function Θ. This is in line with the asymptotics of forwards.

The dynamics of the swap price in Prop. 4.14 is derived by appealing to the dynamics of $f(t,\tau)$ in Prop. 4.11 together with the stochastic Fubini Theorem.

Proposition 4.15 *Suppose that there exists an $\epsilon > 0$ such that Condition G is satisfied with*

$$\sup_{0\leq t\leq T} |\widetilde{\theta}_j(t)| + \epsilon \leq c_j\,,$$

for $j = 1, \ldots, n$. The risk-neutral dynamics of the stochastic process $t \mapsto F(t, \tau_1, \tau_2)$, $0 \le t \le \tau_1 < \tau_2$, is given by

$$dF(t,\tau_1,\tau_2) = \sum_{k=1}^{p}\sum_{i=1}^{m} \sigma_{ik}(t) \int_{\tau_1}^{\tau_2} w(u,\tau_1,\tau_2) e^{-\int_t^u \alpha_i(s)\,ds}\,du\,dB_k^\theta(t)$$
$$+ \sum_{j=1}^{n} \int_{\mathbb{R}} z\eta_j(t) \int_{\tau_1}^{\tau_2} w(u,\tau_1,\tau_2) e^{-\int_t^u \beta(s)\,ds}\,du\,\widetilde{N}_j^\theta(dt,dz)\,.$$

Proof. From Prop. 4.11 and the identity

$$F(t,\tau_1,\tau_2) = \int_{\tau_1}^{\tau_2} w(u,\tau_1,\tau_2) f(t,u)\,du$$

we find (assuming that $m = n = p = 1$)

$$dF(t,\tau_1,\tau_2) = \int_{\tau_1}^{\tau_2} w(u,\tau_1,\tau_2)\sigma(t)\mathrm{e}^{-\int_t^u \alpha(s)\,ds}\,dB^\theta(t)\,du$$
$$+ \int_{\tau_1}^{\tau_2} w(u,\tau_1,\tau_2)\eta(t)\mathrm{e}^{-\int_t^u \beta(s)\,ds} \int_{\mathbb{R}} z\,\widetilde{N}^\theta(dz,dt)\,du\,.$$

From the stochastic Fubini theorem, we may commute integration with respect to du, and B^θ and $\widetilde{N}^\theta$, which implies the desired dynamics for $F(t,\tau_1,\tau_2)$.

We verify that we can use the stochastic Fubini Theorem. Calculating $\int_{\tau_1}^{\tau_2} w(u,\tau_1,\tau_2) f^2(s,u)\,du$, it follows from the boundedness of $w(u,\tau_1,\tau_2)$ and the triangle inequality that we can bound it by terms like

$$\int_{\tau_1}^{\tau_2} \left(\int_0^s \eta_j(t)\mathrm{e}^{-\int_s^u \beta_j(v)\,,dv} \int_{\mathbb{R}} z\,\widetilde{N}^\theta(dz,dt)\right)^2 du$$

and

$$\int_{\tau_1}^{\tau_2} \left(\sum_{k=1}^{p} \int_0^s \sigma_{ik}(t)\mathrm{e}^{-\int_s^u \alpha_i(v)\,dv}\,dB_k^\theta(t)\right)^2 du\,.$$

The latter has a finite expectation. The former integral has a finite expectation as a result of the integrability condition in the Proposition. We conclude that the Itô integral is well-defined. Thus, from the stochastic Fubini Theorem we are allowed to commute integration with respect to du and dB^θ. Similar considerations hold for the commutation of integration with respect to du and the compensated random jump measure, in light of (2.17) and the integrability conditions in the Proposition. □

The volatility in the dynamics of $F(t,\tau_1,\tau_2)$ has an average Samuelson effect. To see this, let $w(u,\tau_1,\tau_2) = 1/(\tau_2-\tau_1)$, and assume for simplicity that the coefficients of mean reversion α_i and β_j, are constant for $i = 1,\dots,m$ and $j = 1,\dots,n$. We find that

$$\sigma_{ik}(t)\int_{\tau_1}^{\tau_2} w(u,\tau_1,\tau_2)\mathrm{e}^{-\alpha_i(u-t)}\,du = \frac{\sigma_{ik}(t)}{\alpha_i(\tau_2-\tau_1)}\left(\mathrm{e}^{-\alpha_i(\tau_1-t)} - \mathrm{e}^{-\alpha_i(\tau_2-t)}\right),$$

for $k = 1,\dots,p$ and $i = 1,\dots,m$. The same argument holds true for the jump volatility. The weight $w(u,\tau_1,\tau_2)$ defined via $\widehat{w}(u) = \exp(-ru)$ simply scales this averaging. Hence, we may think of the arithmetic model as yielding an average Samuelson effect on the swap volatility in the sense of smoothing the "classical" effect over the delivery period. Observe, however, that we do not have a convergence to the spot volatility as the time to delivery approaches zero.

We also see that the volatility is eliminated in the long end of the swap curve, that is, for contracts with long time before delivery starts. This effect holds at least when we have constant parameters in the spot model. This means that the model predicts vanishing risk-neutral price variability in the long end, which should be reflected in a nearly constant swap price over time under the risk-neutral probability. Going to the market probability P, the model predicts close to a deterministic price movement for contracts which are far from maturity. The interpretation of "far from delivery" is highly dependent on the speed of mean reversion. The slower mean reversion, the longer a shock in a factor lasts along the swap curve.

We end this Chapter with a few words on the positive arithmetic model proposed by [Benth, Kallsen and Meyer-Brandis (2007)] and presented in Subsect. 3.2.2. The model has a zero mean reversion level, and no Brownian motion terms. Thus, the explicit swap price dynamics in this case becomes

$$\begin{aligned} F(t,\tau_1,\tau_2) &= \int_{\tau_1}^{\tau_2} w(u,\tau_1,\tau_2)\Lambda(u)\,du + \Theta(t,\tau_1,\tau_2;\theta(\cdot)) \\ &\quad + \sum_{i=1}^{n} Y_j(t)\int_{\tau_1}^{\tau_2} w(u,\tau_1,\tau_2)\mathrm{e}^{-\int_t^u \beta_j(v)\,dv}\,du\,. \end{aligned}$$

Here, the risk-adjusting function (after moving the compensating term to the drift γ_j) is,

$$\Theta(t,\tau_1,\tau_2;\theta(\cdot)) = \sum_{j=1}^{n}\int_t^{\tau_2}\int_{\max(v,\tau_1)}^{\tau_2} w(u,\tau_1,\tau_2)\eta_j(v)\mathrm{e}^{-\int_v^u \beta_j(s)\,ds}\,du\,d\widetilde{\gamma}_j(v)$$

$$
\begin{aligned}
&+\sum_{j=1}^{n}\int_t^{\tau_2}\int_0^{\infty}\int_{\max(v,\tau_1)}^{\tau_2} w(u,\tau_1,\tau_2)z\eta_j(v) \\
&\qquad \times \mathrm{e}^{-\int_v^u \beta_j(s)\,ds}\mathrm{e}^{\widetilde{\theta}_j(v)z}\,du\,\ell_j(dz,dv)\,,
\end{aligned}
$$

where we recall the definition of $\widetilde{\gamma}_j$ in (3.25). Obviously, the swap prices are positive due to the positivity of $f(t,u)$. In conclusion, we have explicit swap prices derived from a spot model which ensures positivity.

Chapter 5

Applications to the Gas Markets

In this Chapter we investigate the natural gas markets. We analyse empirically the spot price of gas in the UK and show that its time evolution is captured reasonably well by a geometric model with NIG distributed noise. A jump-diffusion model is also investigated, and we look at the implications of the different spot model choices to the predicition of the gas futures term structure. In the final section we discuss the extension to multi-factor spot models and possible ways to estimate such on data.

5.1 Modelling the gas spot price

In this Section we analyse a data set of spot prices from the UK market for natural gas. We focus our attention on a geometric model for the spot price of gas of the form

$$d\ln S(t) = d\ln \Lambda(t) + dX(t) + dY(t)\,,$$

with

$$dX(t) = -\alpha X(t)\,dt + \sigma\,dB(t)$$

and

$$dY(t) = -\alpha Y(t)\,dt + dI(t)\,.$$

We recall this as a special case of the geometric models analysed in Subsect. 3.2.1, where we assume $m = p = n = 1$, and let the coefficients be constant. Since the speed of mean reversion is α for both processes $X(t)$ and $Y(t)$, we recognise $S(t)$ as the Schwartz model with jumps (see the examples discussed in Subsect. 3.2.1). The dynamics of $\ln S(t)$ may be written as

$$d\ln S(t) = d\ln\Lambda(t) - \alpha\left(\ln S(t) - \ln\Lambda(t)\right)dt + \sigma\, dB(t) + dI(t)\,. \qquad (5.1)$$

We assume $I(t)$ to be a Lévy process, and we show later in Subsect. 5.1.2 that a case of particular interest is when $I(t)$ can be represented as a sum of two compound Poisson processes. Another interesting class of models we have in mind is defined by letting $I(t)$ be a NIG Lévy process and ignoring the Brownian motion term (that is, assuming $m = 0$ so that process $X(t)$ is absent from the dynamics of $S(t)$). This stochastic process was studied in relation to gas and oil prices in [Benth and Šaltytė-Benth (2004)], and shown to explain the dynamics of the prices well. In discrete time, we are proposing an AR(1) dynamics for the deseasonalized logarithmic spot price of gas with different models of the residuals. This structure will be exploited in the estimation procedure, which we now describe.

5.1.1 *Empirical analysis of UK gas spot prices*

We have available daily gas spot prices (the Heren day-ahead index, see Sect. 1.2) from the UK measured over a period ranging from 2 February 2001 till 24 October 2006, resulting in 1,454 price quotes. Only prices for working days are included, and the notations are collected at the NBP (see Sect. 1.2). In the data series we have 25 missing observations (constituting 1.7% of the total amount of data), which were substituted by the mean of the two closest nearby recordings. The logarithm of the data series is plotted in Fig. 5.1. As we see from the plot there may be some outliers present in the data in the form of large positive and negative price spikes.

The presence of outliers may strongly influence the analysis of trend and seasonality in the considered time series. It may be reasonable to remove these outliers, or price spikes, before proceeding to the estimation of parameters in trend and seasonal function. To detect the possible outliers, we analyse daily changes in the logarithmically transformed gas spot prices. Obviously, these price changes are not normally distributed, as can be seen from the histogram in Fig. 5.2. To check for outliers in data that are not normally distributed, the following simple descriptive statistics can be used. Given the lower and upper quartiles, Q_1 and Q_3, respectively, and the interquartile range IQR, defined as the difference between the upper and the lower quartile, an observation is called an outlier if it is smaller than $Q_1 - 3 \times IQR$, or larger than $Q_3 + 3 \times IQR$. Following this rule, we detected 43 outliers in the logarithmic gas spot prices.

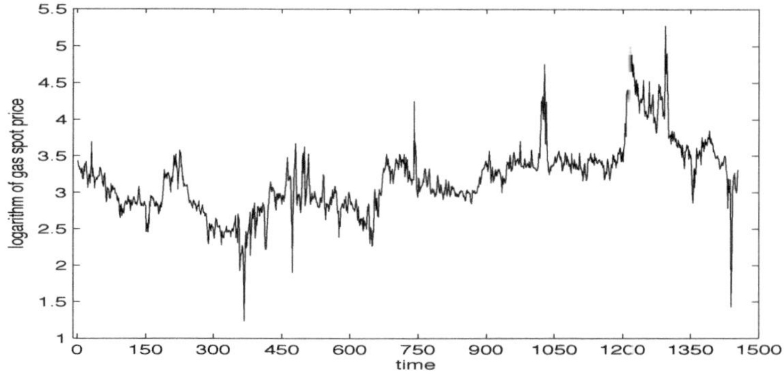

Fig. 5.1 The logarithm of gas spot prices from the UK, 6 February 2001 until 24 October 2006.

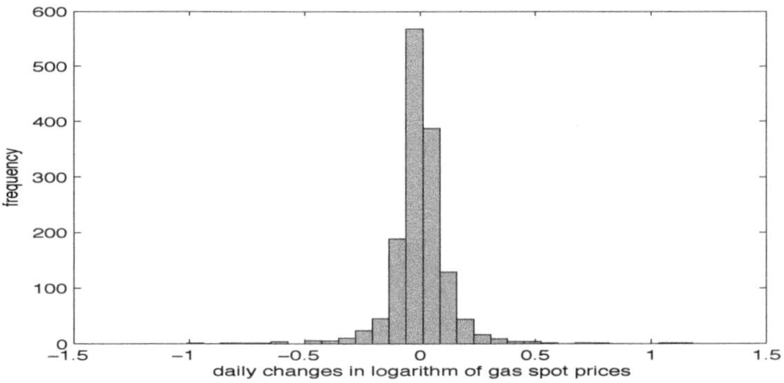

Fig. 5.2 Histogram of daily changes in the logarithmic gas spot prices.

We model the trend and seasonal component of the logarithmic spot prices with the mean level function

$$\ln \Lambda(t) = a_0 + a_1 t + a_2 \cos(2\pi(t - a_3)/250)\,. \tag{5.2}$$

We assume 250 trading days in a year, hence the periodicity of 250 in $\Lambda(t)$. This function represents the average level which the gas prices revert back to. It consists of a linear trend describing the inflation in price level, and a seasonal term explaining possible variations over the year. Before fitting this function, we substituted the detected outliers in the time series of logarithmic prices with the average of the two closest observations. The parameters of the function (5.2) were estimated using the least squares approach, [1] and the results are reported in Table 5.1. All four parameters are significant at the 5% level, indicating that there are both significant seasonal variations and increase in gas spot prices over the considered period. We checked for weekly, monthly and quarterly effects in the logarithmic gas spot prices, however they were not significant and thus ignored.

Table 5.1 Fitted parameters of $\ln \Lambda(t)$

a_0	a_1	a_2	a_3
2.69	0.0007	−0.234	118.1

After estimating the mean function, we insert back the outliers to the data set, and then remove the effect of $\ln \Lambda(t)$ by subtracting it from the logarithmic prices. Now we analyse the detrended and deseasonalized logarithmic prices presented in Fig. 5.3.

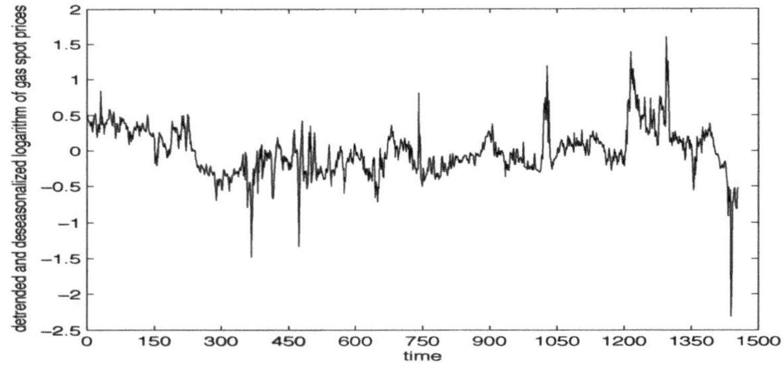

Fig. 5.3 Detrended and deseasonalized logarithm of gas spot prices.

The empirical ACF of the obtained residuals is plotted in Fig. 5.4. It shows that there is a strong memory effect in the data. The partial ACF (PACF) plot in Fig. 5.5 confirms this, and indicates that we need an autoregressive (AR) model to explain the dynamics. Based on the PACF, we propose to use an AR(1) process

$$z(t+1) = \gamma z(t) + \epsilon(t),$$

[1] We applied the `nlinfit` procedure in Matlab.

where $\epsilon(t)$ is an i.i.d. process, and time is measured on a discrete daily scale. The autoregression coefficient γ is assumed to be a constant. We note the connection to our continuous-time model. Letting $Z(t) \triangleq X(t) + Y(t) = \ln S(t) - \ln \Lambda(t)$, we have

$$dZ(t) = -\alpha Z(t)\,dt + \sigma\,dB(t) + dI(t)\,.$$

Discretizing this using daily increments (an Euler discretization, see [Kloeden and Platen (1992)]), we find

$$Z(t+1) \approx (1-\alpha)Z(t) + \sigma\Delta B(t) + \Delta I(t)\,,$$

with $\Delta B(t) = B(t+1) - B(t)$ and $\Delta I(t) = I(t+1) - I(t)$. Hence, $\sigma\Delta B(t) + \Delta I(t)$ is an i.i.d. sequence of random variables for $t = 1, \ldots$. Thus, we see that the discrete-time version of $Z(t)$ coincides with the time series $z(t)$, where $\gamma = 1 - \alpha$ and $\epsilon(t) = \sigma\Delta B(t) + \Delta I(t)$.

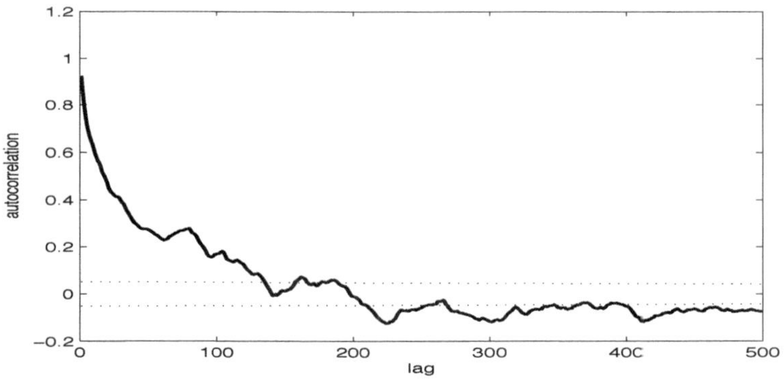

Fig. 5.4 ACF of detrended and deseasonalized logarithm of gas spot prices with 95% confidence interval.

Regressing today's data on the previous day, we estimate the mean reversion constant γ to be

$$\widehat{\gamma} = 0.927\,.$$

This estimate is significant at the 1% level. Admittedly, the p-value must be treated with care since the assumptions of the regression model may be violated. However, the successive analysis clearly indicates that the AR(1) process fits data rather well, and therefore we believe that the obtained p-value is reasonable.

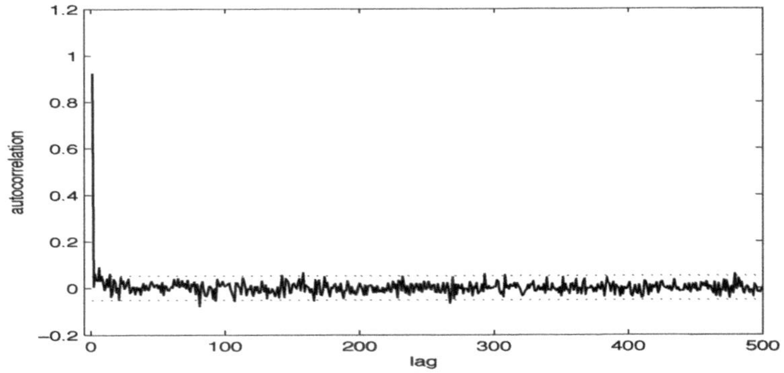

Fig. 5.5 PACF of detrended and deseasonalized logarithm of gas spot prices with 95% confidence interval.

The estimate of γ is close to 1, meaning that the speed of mean reversion is rather slow. This implies an estimate for the mean reversion speed α being

$$\widehat{\alpha} = 0.073\,. \tag{5.3}$$

We did not detect any signs of seasonality in mean reversion. Hence, a constant γ, and thus α, seems to be a reasonable choice.

The residuals obtained from the AR(1) process are presented as a time series plot in Fig. 5.6. It is clear from the ACF plot (see Fig. 5.7) that the

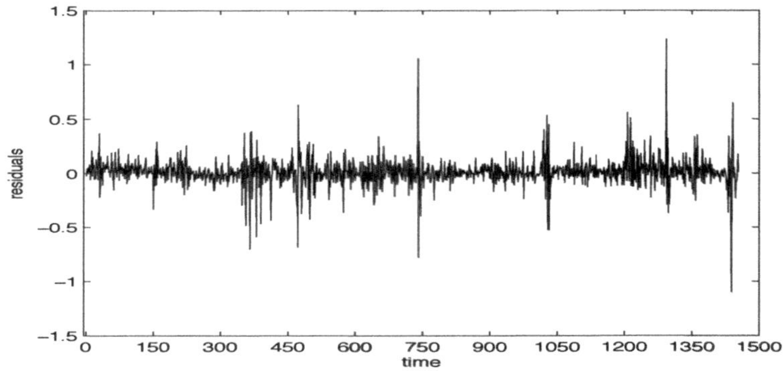

Fig. 5.6 Residuals after trend, seasonal component and AR(1) process were eliminated.

obtained residuals are uncorrelated. The histogram of the residuals with

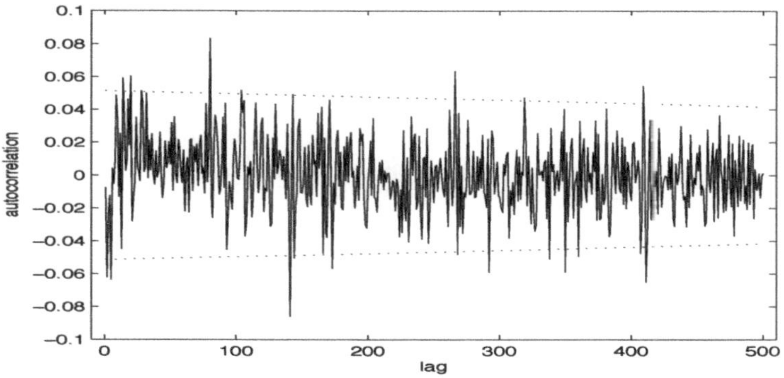

Fig. 5.7 ACF of the residuals (with 95% confidence interval) after trend, seasonal component and AR(1) process were eliminated.

the fitted normal density curve is presented in Fig. 5.8. Observe that the residuals are far from being normally distributed. The mean is equal 0.01 with standard deviation 0.12. They have an extremely high peak (kurtosis is equal 19.52) and very heavy tails. The skewness is equal to 0.08, indicating that the distribution of residuals is quite symmetric.

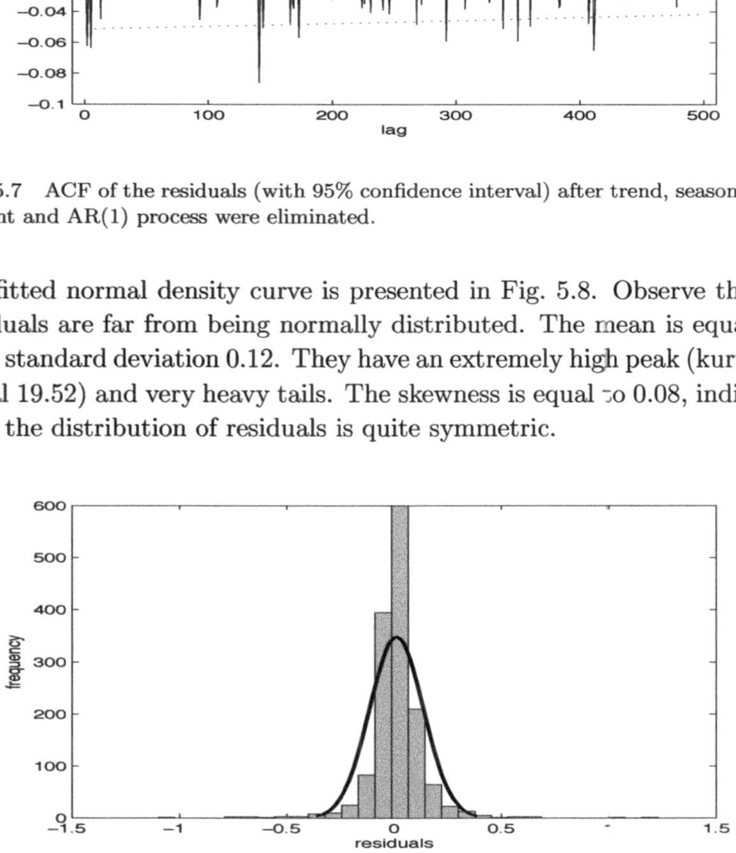

Fig. 5.8 Histogram of the residuals with normal density curve after trend, seasonal component and AR(1) process were eliminated.

We propose two ways of modelling the residuals. First, we consider a model which is a mix of a Brownian motion and a compound Poisson process, where we distinguish the jumps from "normal variations". As a second model for the residuals we choose the NIG distribution, which stems from a NIG Lévy process.

5.1.2 *Residuals modelled as a mixed jump-diffusion process*

Observe that the residuals in Fig. 5.6 have mostly small fluctuations around zero, but from time to time rather extreme jumps appear. This motivates the use of a mix of a Brownian motion and a jump process as in the exponential Schwartz model. The next step in our analysis is thus to identify the jumps in the data and model these by a Lévy process. After the jumps are identified, we let the remaining residuals be modelled by a Brownian motion. We apply recursive filtering to implement this procedure.

The recursive filtering procedure identifies as a jump all data which deviate in absolute value more than a preset level from the mean. The most commonly used levels are two or three standard deviations. The filtering is performed recursively in the sense that after identifying jumps, these are removed and the level is recalculated for a new round of identification of jumps. We iterate until the level is unchanged, and no new jumps are found by the procedure. The recursive filtering algorithm is described in more detail in [Clewlow and Strickland (2000)].

The algorithm converged after eight iterations using a level of two standard deviations, and the results from each step are documented in Table 5.2. From this table, we can read off the standard deviation (or volatility) of

Table 5.2 Summary of the recursive filtering procedure

Iteration	Std.dev.	Cumul. # jumps	Daily jump frequency
1	0.092	27	0.0186
2	0.080	56	0.0385
3	0.077	68	0.0468
4	0.075	77	0.0530
5	0.073	84	0.0578
6	0.072	88	0.0605
7	0.072	91	0.0626
8	0.072	91	0.0626

the filtered residuals to be 0.072. This means that we have estimated σ in the process $X(t)$ as

$$\widehat{\sigma} = 0.072\,. \tag{5.4}$$

A histogram of the residuals after the 91 identified jumps were filtered out is presented in Fig. 5.9. It has a rather clear bell shape, with kurtosis and skewness equal 0.412 and 0.053, respectively. The Kolmogorov-Smirnov statistics is not significant at the 5% level, therefore the hypothesis of normally distributed filtered residuals is not rejected.

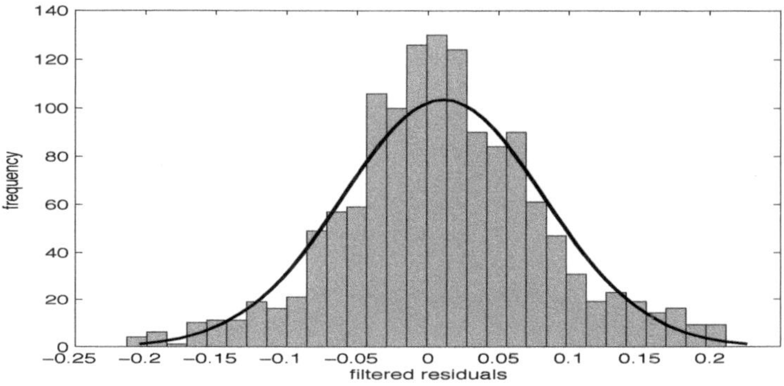

Fig. 5.9 Histogram of the filtered residuals.

Let us turn the attention to the filtered jumps. From the recursive filtering, we know that the jumps arrive with an intensity of 0.0626, which means between six and seven jumps on average a year.[2] It seems to be standard in the literature to assume that the jump sizes are normally distributed (see, for example, [Clewlow and Strickland (2000)]). However, as we observe from the histogram of the jump sizes in Fig. 5.10, this does not seem to be a reasonable choice in our case. Indeed, the normality hypothesis is rejected at the 5% level. Instead of searching for alternative jump distributions, we propose to model $I(t)$ as the sum of two compound Poisson processes, describing the positive and negative jumps separately.

Suppose the Lévy process $I(t)$ is

$$I(t) = I^+(t) + I^-(t)\,, \tag{5.5}$$

where

$$I^{\pm}(t) = \sum_{i=1}^{N^{\pm}(t)} J_i^{\pm}\,, \tag{5.6}$$

and the Poisson processes $N^{\pm}(t)$ have intensities $\lambda^{\pm}$. Here, $J_i^{\pm}$ are two sequences of i.i.d. random variables modelling the jump sizes. Note in

[2]We did not detect any seasonality effects of the jump intensity

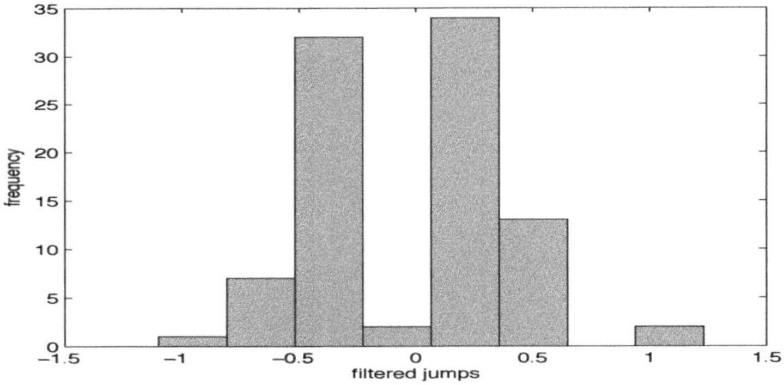

Fig. 5.10 Histogram of the filtered jumps.

passing that with this definition, $I(t)$ is a Lévy process with paths of finite variation.

Inspecting the results of the recursive filtering algorithm, we estimate the jump intensities for positive and negative jumps to be

$$\widehat{\lambda}^{+} = 0.0337\,, \tag{5.7}$$

and

$$\widehat{\lambda}^{-} = 0.0289\,, \tag{5.8}$$

respectively. As we see, the positive jumps (sudden increase in price) are a bit more likely than the negative ones (sudden price drop). In Fig. 5.11 and Fig. 5.12, respectively, we plot the histogram of the positive and negative jumps.

These histograms suggest that we may use an exponential or lognormal distribution as a model for the jump sizes (obviously using the corresponding mirrored distributions for the negative jumps). The lognormal distribution with parameters m and s^2 is defined as

$$f_{\mathrm{Log}}(x; m, s^2) = \frac{1}{x\sqrt{2\pi s^2}} \exp\left(-\frac{(\ln x - m)^2}{2s^2}\right)\,, \tag{5.9}$$

whereas the exponential distribution with parameter μ_J is given in (3.7). Hence, we suppose that J_i^+ and $-J_i^-$ are either exponentially or lognormally distributed. The estimation of the parameters for the distribution of positive jumps is done using the maximum likelihood (ML) approach. The same approach is applied for the negative jumps after multiplying them

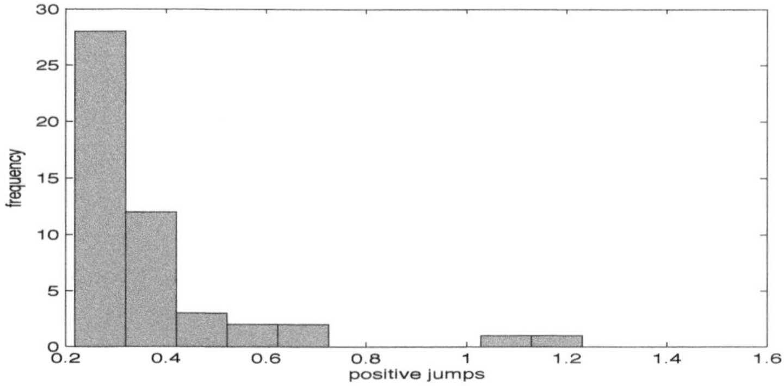

Fig. 5.11 Histogram of the filtered positive jumps.

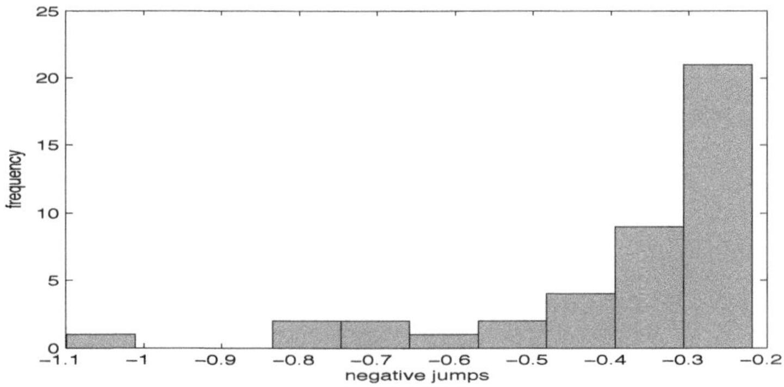

Fig. 5.12 Histogram of the filtered negative jumps.

with −1. The estimated parameters are reported in Table 5.3. Hence, we have a full specification of the stochastic process governing the dynamics of the spot price of gas. This can now be used for pricing of gas futures and other derivatives.

We now investigate the alternative way for representing the time dynamics of the residuals, using a NIG Lévy process.

5.1.3 *NIG distributed residuals*

In [Benth and Šaltytė-Benth (2004)] Brent oil and UK gas spot prices were investigated using a NIG-Lévy type Schwartz model. It was shown the su-

Table 5.3 Estimated parameters of the lognormal and exponential distributions

Parameter	Positive jumps	Negative jumps
$\widehat{m}$	−1.1137	−1.0644
$\widehat{s}$	0.3857	0.4137
$\widehat{\mu}_J$	0.3590	0.3791

perior fit of the NIG distribution to the residuals of the logarithmic spot price model. The model in [Benth and Šaltytė-Benth (2004)] is given in (5.1), with $\sigma = 0$ and $I(t)$ being a Lévy process where $I(1)$ is NIG distributed.

In this subsection we redo the analysis in [Benth and Šaltytė-Benth (2004)] with the data set at hand. Given the residuals from the regression analysis of the deseasonalized and detrended logarithmic spot prices, we estimate the NIG distribution using ML. The residuals are sampled on a daily time scale, which implies that we fit a NIG distribution to the daily increments of $I(t)$. Suppose therefore that $I(1) \sim \mathrm{NIG}(\zeta, \beta, \delta, \mu)$.[3] The ML estimates of the parameters are reported in Table 5.4 below.

Table 5.4 Estimated parameters of the NIG distribution

$\widehat{\zeta}$	$\widehat{\beta}$	$\widehat{\delta}$	$\widehat{\mu}$
4.869	−0.125	0.0874	0.00280

In Fig. 5.13 (top graph) we plot the fitted NIG density together with the empirical density of the residuals. We also include the fitted normal distribution. In the bottom graph of Fig. 5.13 a logarithmic scale is used in order to emphasise the heavy-tailed nature of the data. Clearly, the normal distribution fits poorly the gas residuals. The NIG distribution captures the heavy tails very well, however, as seen best in the top graph, the empirical distribution assigns less probability in the center than predicted by the fitted NIG. The same was observed in [Benth and Šaltytė-Benth (2004)] for a much smaller data set. This may suggest another distribution, for example, the GH distribution. This distribution is more flexible since it has an additional parameter compared to the NIG distribution. We tried

[3]Since we have reserved the notation α for the mean reversion coefficient, we use the notation ζ instead of the more common α for the tail-heaviness parameter of the NIG distribution.

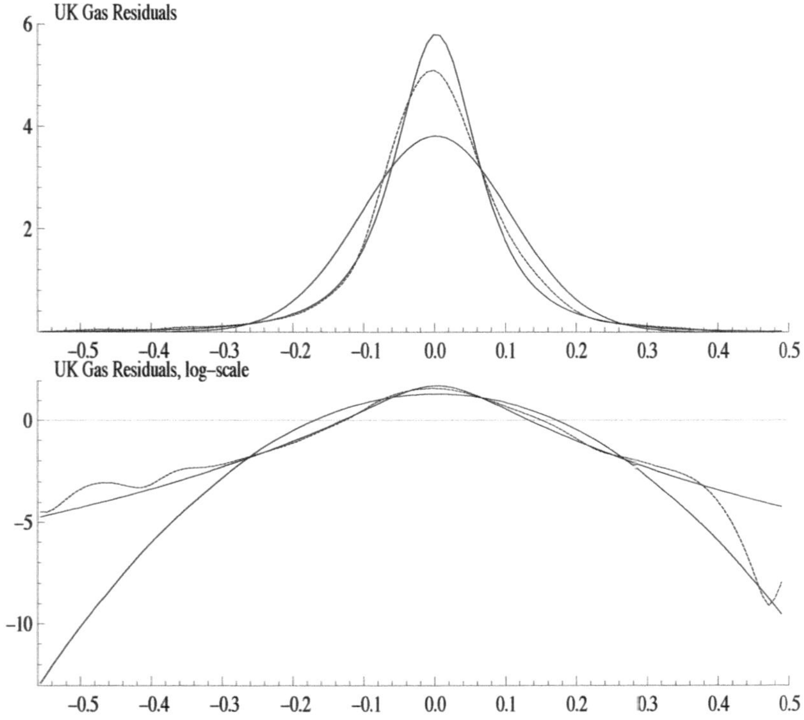

Fig. 5.13 The empirical density together with fitted normal and NIG distributions of the gas residuals (in log-frequency scale below). The curve in the bottom figure with wavy tails is the empirical density, with the normal density being the curve which tails off the quickest.

to fit the GH distribution instead, however the result was basically the same. The NIG distribution is preferable due to its simpler analytical properties.

To understand the heavy tails observed for the gas residuals better, we map the NIG parameters into coordinates (ξ, χ) of the shape triangle (recall (2.32)). The gas residuals have estimated shape triangle values $\widehat{\xi} = 0.84$ and $\widehat{\chi} = -0.02$. Although being slightly negatively skewed, the NIG distribution is reasonably symmetric. The estimate of ξ is close to 1, telling us that the residuals are far from normally distributed. This confirms the conclusions drawn by examining the density plots above.

In conclusion, we see that the NIG is superior to the normal distribution when fitting to the residuals. However, such an approach may miss important information concerning the path properties of the spot dynamics, like, for instance, seasonality in jump frequency and size. In some option pricing problems, it is important to have a model which gives a good distributional description of the residuals, which the NIG distribution is providing.

5.2 Pricing of gas futures

As we know from the survey of the gas markets in Sect. 1.2, the futures contracts deliver gas over a specified period, usually a month. We are going to analyse the forward curve implied from the spot model which we fitted to the UK data above. This analysis will provide us with a smooth curve of forward prices. To convert this to prices for gas futures traded in the market, we need to find the average forward price over the delivery period. We do this numerically for contracts with monthly delivery.

Our first question is how the distributional choice for the residuals affects the gas futures prices. We take the parameters from the subsection above to set up two spot price models; one where the residuals are normally distributed, and one where they are NIG distributed. Using the estimates above, we can calculate the gas futures prices for contracts with monthly delivery, starting January 1, and ranging over the next 24 months. Hence, we consider the gas futures curve of monthly prices for the next two years. Recall that we only take into account the trading days (ignoring weekends and holidays). This means that we normalise all months to have the length of 21 days, which makes up a year consisting of 252 days. This number was approximately what we assumed for the spot price dynamics (there, 250 was chosen). Further, when converting forward prices $f(t,\tau)$ to monthly gas futures prices, we assume that the following relation holds true,

$$F(t,\tau_1,\tau_2) = \frac{1}{\tau_2-\tau_1}\int_{\tau_1}^{\tau_2} f(t,u)\,du\,.$$

This integral is numerically evaluated through an approximation given by daily sampling of $f(t,u)$. The forward price $f(t,\tau)$ is calculated in Prop. 4.6 as,

$$f(t,\tau) = \frac{\Lambda(\tau)}{\Lambda(t)^{\mathrm{e}^{-\alpha(\tau-t)}}}\Theta(t,\tau;0)S(t)^{\mathrm{e}^{-\alpha(\tau-t)}}\,,$$

where

$$\ln\Theta(t,\tau;0) = \int_t^\tau \psi(-\mathrm{i}\mathrm{e}^{-\alpha(\tau-u)})\,du\,.$$

For the moment, we choose the market price of risk for simplicity to be zero. However, later we return to the question of assessing the market price of risk. Recall that ψ is the cumulant function, whereas $\psi(-\mathrm{i}c)$ becomes the logarithm of the moment generating function. When assuming a Brownian motion model, we find for $c \in \mathbb{R}$

$$\psi(-\mathrm{i}c) = \frac{1}{2}c^2\sigma^2 ,$$

while the NIG case implies (recall (2.31))

$$\psi(-\mathrm{i}c) = \mu c + \delta\left(\sqrt{\zeta^2 - \beta^2} - \sqrt{\zeta^2 - (c+\theta)^2}\right) ,$$

as the logarithm of the moment generating function. These two functions will be inserted into the expression for Θ, and integrated (numerically for NIG). We remark that c must be restricted so that $|\beta + c| < \zeta$, when the NIG case is considered.

We recall the estimates from the previous Section (in particular, Subsect. 5.1.3). Remark that $\widehat{\zeta} > |\widehat{\beta}| + 1$, and therefore $\psi(-\mathrm{i}e^{-\alpha(\tau-u)})$ is well-defined. We use the initial spot price of $S(0) = 22.25$, the closing spot price of 31 December 2003. The resulting curves of gas futures prices are plotted in Fig. 5.14, where the NIG case is depicted with '+', and the normal with '$*$'. The NIG model has heavier tails and is more peaky in the

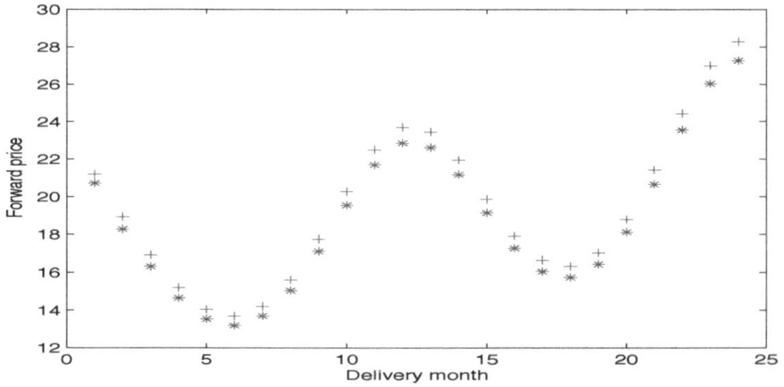

Fig. 5.14 Monthly gas futures prices based on NIG (+) and Brownian motion ($*$) models.

center, and in our case it assigns slightly higher futures prices.

The relative difference between the Brownian motion and the NIG case is given in Fig. 5.15, where we see that the difference converges to about

3.7% in the long end, meaning that the NIG assigns for about 3.7% higher gas futures prices relative to Brownian motion. In the very short end the

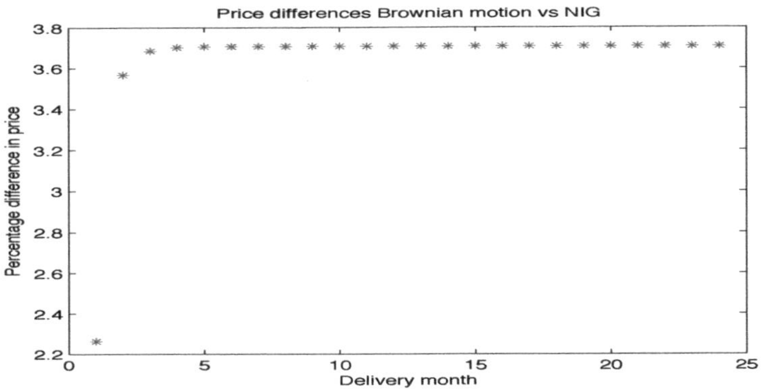

Fig. 5.15 The relative difference of monthly gas futures prices based on Brownian motion and NIG Lévy models.

difference is slightly above 2.2%, fastly increasing (due to mean reversion) up to the asymptotic difference.

We proceed to a comparison of the theoretically derived gas futures prices with the prices quoted in the market. The gas futures curve defined by the closing prices on 31 December 2003 is tabulated in Table 5.5 and

Table 5.5 Gas futures prices on 31 December 2003

Delivery period	Closing price
Week 2	32.50
Week 3	32.00
February	27.90
March	23.80
April	20.25
May	18.88
Q2	18.95
Q3	18.80
Q4	23.88

plotted in Fig. 5.16. In this Figure, we draw the closing future prices as horizontal lines ranging over the delivery period, where time is measured in days. In Fig. 5.17 the corresponding theoretical gas futures prices are plotted for the NIG Lévy model with market price of risk equal to zero.

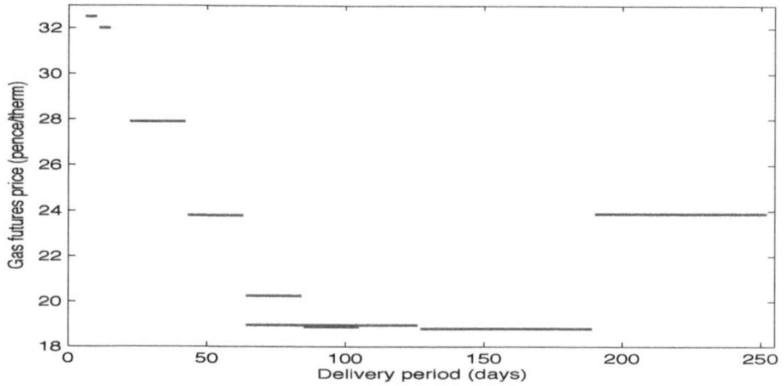

Fig. 5.16 The gas futures curve on 31 December 2003.

The market prices are all above the theoretical prices, yielding a positive

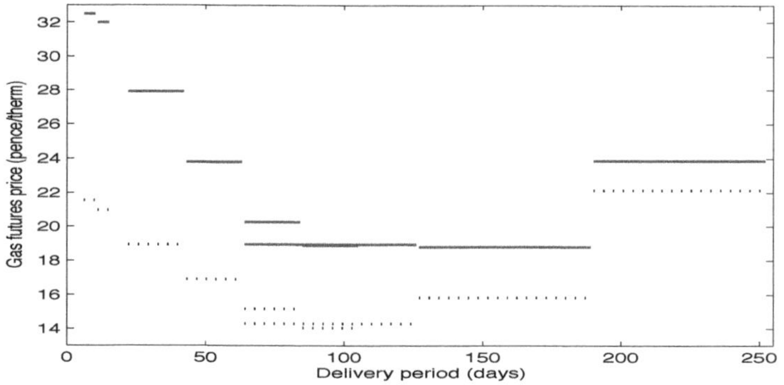

Fig. 5.17 The theoretical gas futures curve (dotted lines) on 31 December 2003, with market price of risk equal to zero. The complete lines are the observed gas futures prices.

risk premium for this particular day. Let us investigate the value of the market price of risk $\widetilde{\theta}$ required to match the observed gas futures prices.

We calculate the implied values of $\widetilde{\theta}$ for each contract by using Newton's search algorithm for zero-points.[4] The results are reported in Table 5.6. Worth noticing is the decay of the market price of risk with respect to the time to delivery. In the short end, we have the largest value of $\widetilde{\theta}$, which then decreases with time to delivery. An explanation for this may be that although the NIG distribution fits the residuals in the spot dynamics, it

[4] We applied the `fsolve`-routine in Matlab.

Table 5.6 The implied market prices of risk

Delivery period	Implied $\widetilde{\theta}$
Week 2	2.928
Week 3	2.358
February	1.587
March	1.302
April	1.090
May	1.114
Q2	1.050
Q3	0.656
Q4	0.292

does not capture the jump risk appropriately. The distributional properties of the spot dynamics are reasonably well captured, but not the pathwise properties. Another explanation may be that the market assigns different risk premia along the curve due to different actors operating in the different segments, having different preferences, and seasonality. According to the findings of [Cartea and Williams (2006)], the market price of risk may change sign along the gas futures curve. They find a positive market price of risk in the long end, while the sign is negative in the short end. This indicates a complicated structure of the market price of risk, but it also tells us that it may vary quite significantly. This may be attributed to spot models which do not capture all the relevant stylised facts of the price dynamics, or to make connections between the spot and gas futures market more complex.

In order to gain further understanding of the stylised facts of the market price of risk, one needs to do the same analysis for a series of gas futures curves. We will not pursue such an investigation here. An alternative approach is a direct modelling of the gas futures as we discuss in Chapter 6. See [Kjaer and Ronn (2006)] for an application of this approach to NYMEX gas futures, where the correlation structure of the returns is studied.

5.3 Inference for multi-factor processes

In this Section we discuss some possible techniques for estimating the general multi-factor spot models presented in Chapter 3 with and without jumps. All the examples we have looked at in this Chapter have been one-factor models. One may suspect that there are several factors driving the spot price, for example, slow mean reversion in normal market conditions and fast mean reversion in periods with spikes. We now look at possible ways to do inference on such models, where we have unobservable factors in the process.

5.3.1 *Kalman filtering*

The empirical analysis of gas spot prices has shown that the process consists of two parts, daily variations and jumps occurring relatively rare. It is natural to believe that the mean reversion and the noise are very different for these two subprocesses. In the empirical analysis above, we applied recursive filtering to separate jumps from "normal" daily variations *after* having estimated the speed of mean reversion. Hence, we fail to separate the different speeds of mean reversion associated with jumps and normal variations.

An alternative approach is the Kalman filter (see, for example, [Harvey (2003)] for a complete account), where all parameters can be estimated simultaneously. The technique was invented by [Kalman (1960)] to solve engineering problems where the state process has to be extracted from noisy measurements. The Kalman filter is a recursive procedure for computing the optimal estimator of the state vector at time t, based on the available past information. It is based on the assumption of normality of the noise and the process. However, when the normality assumption is violated, the Kalman filter is still an optimal estimator within the class of all linear estimators.

We give an example on the use of the Kalman filter on our gas data. The detrended and deseasonalized logarithmic spot price dynamics is assumed to follow the process $X_1(t) + X_2(t)$, where

$$dX_i(t) = -\alpha_i X_i(t)\,dt + \sigma_i\,dB_i(t)\,,$$

for $i = 1, 2$, and B_1 and B_2 are two independent Brownian motions. [Barlow, Gusev and Lai (2004)] applied the Kalman filter on a version of this model in the context of electricity markets. A detailed presentation of the filter for this model is found there, together with estimates of the parameters for spot price data collected at the Alberta Power Pool, California Power Exchange and the Australian NEMMCO (two other, more sophisticated, spot models were also proposed and analysed empirically based on Kalman filtering).

One of the findings in [Barlow, Gusev and Lai (2004)] was the wide confidence bands on the mean reversion estimates, which seems to be a common problem (see also [Clewlow and Strickland (2000)]). We therefore propose to use the empirical ACF for estimating the mean reversion parameters, based on the assumption of stationarity. From (3.31) we find the stationary ACF of $X_1(t) + X_2(t)$ to be

$$\rho(\tau) = \widehat{\omega}_1 \mathrm{e}^{-\alpha_1 \tau} + \widehat{\omega}_2 \mathrm{e}^{-\alpha_2 \tau}, \tag{5.10}$$

where τ is the lag. The coefficients $\widehat{\omega}_i$, $i = 1, 2$ are defined as

$$\widehat{\omega}_i = \frac{\frac{\sigma_i^2}{2\alpha_i}}{\frac{\sigma_1^2}{2\alpha_1} + \frac{\sigma_2^2}{2\alpha_2}}. \tag{5.11}$$

Using the least squares approach,[5] we fitted $\rho(\tau)$ to the empirical ACF in order to derive the two speeds of mean reversion α_1 and α_2. In this fitting we boldly assume that the data have reached stationarity. The estimated values are presented in Table 5.7. Both empirical and fitted ACF are plotted in Fig. 5.18. Here, we clearly see the separation into a slow (α_1) and fast (α_2) speed of mean reversion. In the ACF we recognise this as a steeper slope for the first lags than for the higher lags. The steep slope is attributed to the fast mean reversion, whereas the slow mean reversion is recognised for higher lags. Recall that we estimated the mean reversion to be 0.073 in the one-factor models analysed in Subsect. 5.1.1. Thus, this estimate seems to be a weighted average of the two mean reversion parameters α_1 and α_2, where the small variations are given higher weight than the more spiky ones since those seem to appear less frequently. Hence, a one-factor model yields a speed of mean reversion which averages out the fast and slow mean reversions.

Table 5.7 Fitted parameters of $\rho(\tau)$

$\widehat{\omega}_1$	$\widehat{\omega}_2$	α_1	α_2
0.73	0.27	0.02	0.28

The Kalman filter was next used to estimate the volatilities σ_1 and σ_2 of the two processes $X_1(t)$ and $X_2(t)$, respectively. To apply the Kalman filter, we need a starting point to search for the optimal estimates of σ_1 and σ_2, based on the available history of data for $X_1(t) + X_2(t)$. From the

[5]The `nlinfit` function of Matlab was used.

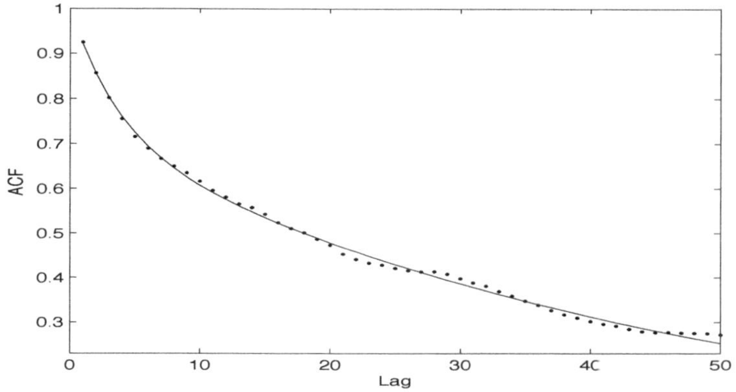

Fig. 5.18 Empirical and fitted ACF for detrended and deseasonalized logarithm of gas spot price.

definition of $\widehat{\omega}_i$, $i = 1, 2$, we find that

$$\sigma_i^2 = 2\alpha_i \times \mathrm{Var}[X_1(t) + X_2(t)]\,,$$

where the last term is the stationary variance of the sum of two processes X_1 and X_2. This is estimated to be 0.11, yielding initial values of σ_1 and σ_2 to be 0.057 and 0.129, respectively. The Kalman filter estimated the two volatilities to be $\widehat{\sigma}_1 = 0.065$ and $\widehat{\sigma}_2 = 0.573$. We observe the small volatility estimated for the process $X_1(t)$ which is slowly mean-reverting, whereas the fast mean-reverting process $X_2(t)$ has a much bigger volatility. We observe that the estimated volatility in the recursive filtering procedure from Subsect. 5.1.2 was 0.072, somewhat higher than the Kalman filter estimate for the volatility of X_1. The Kalman filter procedure, given the speeds of mean reversions, is able to identify two processes, one modelling the "normal" variations and another one more spiky process. However, since we use a Brownian motion model for the random innovations of $X_2(t)$, we do not see spikes resembling exactly the ones we observe in the data, and to mend this we should substitute $B_2(t)$ with a compound Poisson process. The problem is to use the filtering to estimate the jump size and intensity parameters. The particle filter is a possible approach to follow in this case (see [Doucet, de Freitas and Gordon (2000)]). A nonparametric alternative approach is proposed in [Meyer-Brandis and Tankov (2007)], who present an estimation procedure for the arithmetic model in [Benth, Kallsen and Meyer-Brandis (2007)]. They apply their estimation procedure on EEX spot price data. [Green and Nossman (2007)] propose a hybrid approach based on Markov Chain Monte Carlo methods and traditional statistical techniques to fit a two-factor jump diffusion model with stochastic volatility to electricity spot prices observed on Nord Pool.

We remark in passing that we implemented the Kalman filter in Matlab, where in particular the built-in `fminsearch` function was used in the optimisation. Stability tests showed that the estimates are robust with respect to small changes in the initial values of the volatilities.

We next proceed with discussing different approaches to estimate a multi-factor model (with jumps) using additional data from forwards and swaps.

5.3.2 *Inference using forward and swap data*

When having a multi-factor model with jump processes in one or more of the factors, it may be an idea to exploit several data sources. One may, for instance, combine spot and gas futures prices when estimating. We discuss this idea briefly for a two-factor model of the detrended and deseasonalized logarithmic spot price dynamics of the form $X(t) + Y(t)$ with

$$dX(t) = -\alpha X(t)\,dt + \sigma\,dB(t)$$

and

$$dY(t) = -\beta Y(t)\,dt + dI(t)\,,$$

where $I(t)$ is an II process independent of the Brownian motion $B(t)$. We think of the process $Y(t)$ as the model for the price spikes frequently observed in the gas markets due to limited storage and supply. There are reasons to believe that the frequencies of these spikes are seasonally varying (admittedly, we did not detect any such seasonality in the empirical study of UK gas prices, but this could be due to a limited amount of data). In the winter, the demand for gas for household heating is larger than in the summer, say, and the limitations of storage makes it difficult to cover up for this additional demand. Hence, there are chances of price spike occurrences due to sudden unexpected drops in temperature leading to an increased demand for gas. This leads us to II processes which, in their simplest form, is time-inhomogeneous compound Poisson processes, with time-dependent jump frequency (and even the jump sizes).

Since $Y(t)$ models the spikes in the gas price, the speed of mean reversion β is typically much higher than α, which signifies the speed of mean reversion of the price variations in the market in "normal circumstances". We can exploit this difference when using gas futures prices with different times to maturity for estimation.

For the sake of simplicity in the further argumentation, we assume that the gas futures have fixed delivery times and the market charges a constant price of risk $\theta = (\widehat{\theta}, \widetilde{\theta})$. Then, from Prop. 4.6, we see that the forward price for a contract with delivery at time τ is

$$f(t,\tau) = \Lambda(\tau)\Theta(t,\tau;\theta)\exp\left(\mathrm{e}^{-\alpha(\tau-t)}X(t) + \mathrm{e}^{-\beta(\tau-t)}Y(t)\right)$$

with

$$\begin{aligned}\ln\Theta(t,\tau;\theta) &= \psi(t,\tau;-\mathrm{i}(\mathrm{e}^{-\beta(\tau-\cdot)} + \widetilde{\theta})) - \psi(t,\tau;-\mathrm{i}\widetilde{\theta}) \\ &\quad + \frac{\sigma^2}{2\alpha}(1-\mathrm{e}^{-2\alpha(\tau-t)}) + \frac{\sigma\widehat{\theta}}{\alpha}(1-\mathrm{e}^{-\alpha(\tau-t)}).\end{aligned}$$

We recall that $\Lambda(\tau)$ is the trend and seasonality function, and ψ is the cumulant function of $I(t)$.

Let us now consider contracts in the long end of the forward curve. Since β is big, it follows that $\exp(-\beta(\tau-t)) \approx 0$, and the contribution from $Y(t)$ in the expression of the forward price vanishes. Hence, in the long end of the forward curve, we can apply the approximation

$$f(t,\tau) \approx \Lambda(\tau)\exp\left(\frac{\sigma^2}{2\alpha}(1-\mathrm{e}^{-\alpha(\tau-t)}) + \frac{\sigma\widehat{\theta}}{\alpha}(1-\mathrm{e}^{-\alpha(\tau-t)}) + \mathrm{e}^{-\alpha(\tau-t)}X(t)\right).$$

Long term contracts depend on parameters of $X(t)$ only.

In the short end of the curve, we have that $\tau - t$ is close to zero, which implies that both $\exp(-\alpha(\tau-t))$ and $\exp(-\beta(\tau-t))$ is close to one. However, since β is much bigger than α, the convergence towards one is much faster for the contribution from $X(t)$. Therefore, the term $\exp(-\beta(\tau-t))$ will still be less than one when $\exp(-\alpha(\tau-t))$ is essentially one, implying that we can approximate the forward price dynamics in the short end of the curve as

$$\begin{aligned}f(t,\tau) \approx \Lambda(\tau)\exp\Big(&\psi(t,\tau;-\mathrm{i}(\mathrm{e}^{-\beta(\tau-\cdot)} + \widetilde{\theta}) - \psi(t,\tau;-\mathrm{i}\widetilde{\theta}) \\ &+X(t) + \mathrm{e}^{-\beta(\tau-t)}Y(t)\Big).\end{aligned}$$

Hence, we have an expression which depends on the parameters of $Y(t)$. Of course, forward price data in the middle parts of the curve will depend on both sets of parameters. Also, by including forward price data, we get two additional parameters to estimate, namely $\widehat{\theta}$ and $\widetilde{\theta}$. On the other hand, we have available much longer price data series in addition to the spot prices.

The estimation procedure is as follows. Find the trend and seasonality function Λ from the spot price data. Next, let $Y(0) = 0$ and $X(0) = \ln(S(0)) - \ln(\Lambda(0))$, and use the forward price and its short- and long-term approximations together with spot prices to estimate the remaining parameters on time series of spot and forward data. This last task can technically be rather complicated and cumbersome. Additional complexity is added by the fact that gas futures deliver over a contracted period rather than at a fixed time point. Hence, the forward prices above must be averaged over the delivery periods to be matched with observations. We can also try to derive forward price data by smoothing the observed gas futures price observations. This is the topic of Chapter 7.

In the paper by [Singleton (2001)], an estimation approach based on the characteristic functions is proposed. We outline this approach in the context of gas markets, but note that it is obviously applicable in other energy markets as well following the same lines.

Suppose that the spot price dynamics $S(t)$ is modelled by $N = m + n$ OU processes $X_i(t)$, $i = 1, \ldots, m$ and $Y_j(t)$, $j = 1, \ldots, n$. The OU processes are defined as in Chapter 3, and we suppose that the seasonality $\Lambda(t)$ has been removed from the dynamics. From Prop. 3.2 we have the conditional characteristic function (CCF)[6] available for each OU process in the spot dynamics. To estimate the unknown parameters in the OU processes, we have observations on spot $\{S(t)\}_{t=1}^{T}$ and gas futures prices $\{F(t, \tau_1, \tau_2)\}_{t=1}^{T}$. By ranging over different times to delivery, we get several series of gas futures prices.

Consider first the case of an arithmetic spot price model as discussed in Subsect. 3.2.2. Then, from Prop. 4.14, we have an affine structure on the gas futures prices, in the sense that the futures price can be expressed as

$$
\begin{aligned}
F(t, \tau_1, \tau_2) = a(t, \tau_1, \tau_2) &+ \sum_{i=1}^{m} X_i(t) \int_{\tau_1}^{\tau_2} w(u, \tau_1, \tau_2) \mathrm{e}^{-\int_t^u \alpha_i(s)\, ds}\, du \\
&+ \sum_{j=1}^{n} Y_j(t) \int_{\tau_1}^{\tau_2} w(u, \tau_1, \tau_2) \mathrm{e}^{-\int_t^u \beta_j(s)\, ds}\, du ,
\end{aligned}
$$

for a given function a. We recall that $w(u, \tau_1, \tau_2)$ is the weight function and α_i and β_j are the speeds of mean reversions for X_i and Y_j, respectively.

[6]Here we use the notion CCF following [Singleton (2001)].

Thus, the CCF of $F(t, \tau_1, \tau_2)$ is easily calculated from the CCF of $X_i(t)$ and $Y_j(t)$. Now, by defintion, the density function of a random variable can be expressed as the inverse Fourier transform of the characteristic function. Hence, we can write a conditional ML function for the samples of spot and gas futures prices, which we can maximise to obtain estimates of the unknown parameters. By modelling with N factors, we can either base our ML estimation on the observation of the spot and $N-1$ different gas futures price series, or on N different gas futures price series only. Note that the market prices for risk are not known, but will be estimated as an integral part of the set of unknown parameters.

If we choose a geometric spot model, we do not get an affine structure of the gas futures prices for the most relevant models. One solution is to smoothen the gas futures curve using the methodology we will present in Chapter 7, and thereby preparing forward data which can be used directly as for the arithmetic case.

Estimation procedures in the frequency domain are also proposed and discussed in [Singleton (2001)]. The advantage with these is that we avoid to perform a Fourier inversion in higher dimensions, which may be computationally time consuming. Instead, we must transform the data into the frequency domain to construct the empirical CCF. Estimators based on the empirical CCF are reviewed in [Yu (2004)]. We refer to [Singleton (2001)] and [Yu (2004)] for more details on the different estimation procedures based on the CCF.

Chapter 6

Modelling Forwards and Swaps Using the Heath-Jarrow-Morton Approach

In the fixed income markets, instead of modelling the prices via one- or multi-factor spot rate models, the dynamics of the forward rates are directly specified. This approach leads to simple evaluations of bond prices through integration in time, and is known as the HJM approach (see [Heath, Jarrow and Morton (1992)]). The HJM approach has later been adopted to modelling forward and futures prices in commodity markets, and this will be the topic of the current chapter.

Most commodity markets trade in forward contracts with settlement at a fixed time. In this case the adoption of the HJM approach is rather straightforward, and we start this chapter with a thorough discussion of such contracts. However, in electricity, gas or weather markets, the commodity is delivered over a period, and it is no longer obvious how to apply the HJM approach. We describe the approach for electricity and gas contracts, which we commonly denote as swaps.

As we will see, the straightforward implementation of HJM for swaps leads to intractable models. The alternative is to model only those contracts which are traded. This resembles closely what is known as swap and LIBOR models in the interest rate markets (see, for example, [Brigo and Mercurio (2001)] for an introduction of such models). These models are also known as market models.

One may also generate models for swap prices by integrating the forward price over the delivery period. However, forward prices are not directly observed in the electricity or gas markets, which introduces some problems with estimation. One may derive data by smoothing the swap curve, however, this may be a dubious path to follow since the data may depend on the algorithm chosen. In Chapter 7 we discuss smoothing of the swap curve, while in this chapter we concentrate on the theoretical foundation for direct modelling of swaps and forwards.

6.1 The HJM modelling idea for forward contracts

We discuss the HJM modelling approach applied to energy markets. In order to convey the idea, we restrict in this Section the discussion to forwards only.

In fixed income markets, the HJM approach models the forward rates directly, and frequently a geometric Brownian motion model is used. The dynamics of the forward rates are normally stated under the risk-neutral measure, being the convenient measure if the purpose is to price options. The direct analogy to energy markets would be to let the forward price dynamics in the risk-neutral setting be given as[1]

$$\frac{df(t,\tau)}{f(t,\tau)} = \sigma(t,\tau)\,dW(t)\,,$$

where W is a standard Brownian motion. The function σ, modelling the volatility term structure in the market, is assumed to be positive. Usually, this term structure is supposed to be continuous in both current time t and time of delivery τ, $0 \leq t \leq \tau \leq \mathcal{T}$. The market has a maximal time of delivery given by $\mathcal{T}$. The explicit dynamics of the forward is

$$f(t,\tau) = f(0,\tau)\exp\left(-\frac{1}{2}\int_0^t \sigma^2(u,\tau)\,du + \int_0^t \sigma(u,\tau)\,dW(u)\right)\,,$$

with $f(0,\tau)$ being the initial forward curve observed today in the market. Thus, the forward price will have independent and normally distributed logreturns under the risk-neutral measure. Note that since it is costless to enter a forward contract, it has zero expected return. Thus, the dynamics is without drift.

When we model forward prices the main questions which arise are: how to specify an empirically sound forward curve dynamics, and how to price options written on forwards? Of course, there are several other issues related to the forward price evolution, but the two mentioned above are our main concern. When modelling the forward price dynamics with a geometric Brownian motion, the customary approach is to use the differential representation of it. Estimation is then easily performed using the normal distribution hypothesis for the logreturns together with the fact that the

[1] We consider for simplicity only one source of randomness. Later, we make the models more general.

volatility does not change under a Girsanov transform. This implies that the volatility can be estimated directly from observations. The price of plain vanilla call and put options will follow from the Black-76 formula, which is a modification of the well-known Black-Scholes option pricing formula. We refer to Chapter 9 for more details on this.

Commodities like gas and electricity have frequently large jumps in the spot price, which in theory should be reflected in the forward price, at least for contracts close to maturity and where the delivery period is short. Hence, when creating forward price models directly, it is natural to include jump processes like we did for the spot price dynamics. However, it is not so clear what is the natural generalization of a geometric Brownian motion with included jump processes. One possibility is to state the dynamics in differential form

$$\frac{df(t,\tau)}{f(t-,\tau)} = \sigma(t,\tau)\,dW(t) + \eta(t,\tau)\,dJ(t)$$

under the risk-neutral measure. Here J is an II process. However, with this dynamics we may obtain negative forward prices, and moreover, the explicit representation becomes highly complicated. By the Itô Formula, we can prove that the explicit dynamics becomes

$$f(t,\tau) = f(0,\tau)\exp\left(\int_0^t a(s,\tau) - \frac{1}{2}\sigma^2(s,\tau)\,ds + \int_0^t \sigma(s,\tau)\,dW(s)\right)$$
$$\prod_{s\leq t}(1+\eta(s,\tau)\Delta J(s))\exp\left(-\eta(s,\tau)\Delta J(s)\right).$$

Observe that we get negative forward prices whenever jumps of magnitude smaller than -1 is allowed in the II process. Thus, to ensure positive forward prices, we need to assume that $\Delta J(s) > -1$, which is equivalent to saying that the compensator measure $\ell(dz, ds)$ is supported on the interval $z \in (-1, \infty)$. This issue together with the rather complicated explicit form of $f(t,\tau)$ are serious ones when we want to fit the model to data, since it becomes a delicate task to derive the distributional properties of the log-returns. This could, however, be overcome by considering a discretized version of the dynamics giving a representation of the returns instead. Another, simpler way, is to state the explicit dynamics directly rather than the dynamics in differential form. This will allow for a richer class of jump processes, and a much simpler way to find the distributional properties of the logreturns implied by the modelling, even after a change of measure. This is the approach suggested by [Barndorff-Nielsen (1998)] where asset prices are modelled by an exponential NIG Lévy process. In our context, the analogous modelling perspective would be to define the forward curve dynamics in exponential form under the risk-neutral measure Q as

$$f(t,\tau) = f(0,\tau)\exp\left(\int_0^t a(u,\tau)\,du + \int_0^t \sigma(u,\tau)\,dW(u) + \int_0^t \eta(u,\tau)\,dJ(u)\right).$$

Note that the characteristics of the II process are dependent on the measure change, and since we observe the logreturns under the market probability P, we need to know the characteristics of J under P, and not under the risk-neutral probability. Stating the forward dynamics directly under the risk-neutral measure, forces us to describe the measure change in order to recover the model under P. With respect to the risk-neutral probability Q, we know that the forward price must be a martingale process, which puts on restrictions on the drift $a(u,\tau)$. From the Esscher transform it is simple to characterise the jump process under P, and thus to have accessible the distributional properties of the logreturns.

In many circumstances option prices are used to estimate the forward dynamics, and thus the risk-neutral dynamics is the right starting point. Note that this is difficult in many power markets since the options are rather thinly traded. Typically, the forward market is liquid, which means good access to reliable data under P, whereas the option market is less liquid and thereby providing rather questionable information for estimating the forward curve dynamics.

When pricing call and put options, one must have access to the risk-neutral forward price at the exercise time. It is advantageous to have specified the dynamics of the forward in explicit form. The risk-neutral expression for the forward price allows for both Monte Carlo pricing techniques and the use of fast Fourier transform (FFT) solvers. In some cases, when the dynamics is only driven by a Brownian motion, we can also derive explicit pricing and hedging formulas. This is the topic of Subsect. 9.1.1 in Chapter 9, where we derive the Black-76 option pricing formula for our forward models.

When modelling swap contracts, certain conditions need to be satisfied in order to ensure an arbitrage-free dynamics. This raises new and challenging questions for the possibility of creating dynamical pricing models for such contracts. The problem relates to the decomposition of a delivery period into smaller periods, and thereby (at least artificially) creating portfolios of swaps yielding the same as the original swap. Indeed, this happens in an organized market like the Nord Pool, where you may invest in three monthly contracts exactly covering the period of a quarterly contract, say. This obviously enforces a connection between prices to avoid arbitrage possibilities. In a mathematical sense, this decomposition can be done arbitrarily, leading to an infinitesimal condition which makes it basically impossible to state reasonable models. We are going to investigate these questions, and propose an approach taken from LIBOR models in the interest rate market.

As already mentioned, a problem in the energy markets in particular is the high volumes of swaps trading, but comparably low liquidity in the option markets (at least the organized ones). For instance, in the fixed income markets, one may use option price data to estimate the (implied) parameters of the risk-neutral dynamics of the underlying asset. In the energy markets, this may be problematic, since there is a lack of trustable data in the option market. Thus, it seems more appropriate to fit the model to the observed forward prices in the market. This is of course very natural to do, however, it is difficult to find the appropriate measure change in order to characterise the jump process under the market measure P. We must find some natural measure for the market price of risk. This could in principle be found by minimising the distance between the theoretical and observed option prices, where the theoretical ones are deduced from the forward price dynamics after an Esscher transform. However, the data quality may seriously damage the estimates. On the other hand, this is the best estimate we may get from the market.

Note that the above discussion is particularly relevant for the HJM modelling approach we apply here. When using the spot market modelling approach, we have the advantage of estimating a market price of risk given by the "mismatch" between the spot and forward/swap market. Since both markets are rather liquid, we get a good estimate for the market price of risk. This counts in favour of the spot modelling approach. However, it is notoriously difficult to find good models for the spot dynamics in many markets, and the market price of risk may have a complicated stochastic behaviour. Further, the stylised facts of the market price of risk are still not well understood. So far, there are investigations hinting towards a negative market price of risk in the short end. For instance, as we already mentioned in Subsect. 1.5.2, [Cartea and Figueroa (2005)] estimate a negative market price of risk for the England and Wales electricity market, whereas [Ollmar (2003)] and [Weron (2005)] find evidence of a changing sign from negative in the short end to positive in the long end for Nord Pool contracts. Further, we recall that [Cartea and Williams (2006)] find that the UK gas market may have a price of risk changing sign in the short term and negative for long dated instruments.

6.2 HJM modelling of forwards

Assume that the forward dynamics under a risk-neutral probability Q is

$$f(t,\tau)=f(0,\tau)\exp\Big(\int_0^t a(u,\tau)\,du+\sum_{k=1}^p\int_0^t \sigma_k(u,\tau)\,dW_k(u)$$
$$+\sum_{j=1}^n\int_0^t \eta_j(u,\tau)\,dJ_j(t)\Big)\,. \tag{6.1}$$

Here, a, σ_k and η_j, $k=1,\ldots,p, j=1,\ldots,n$, are real-valued continuous functions on $[0,\tau]\times[0,\mathcal{T}]$, where $\mathcal{T}$ is an upper bound for the delivery times in the market. It is convenient to assume that the functions σ_k are positive, since they are describing the volatility of the forward price evolution. Furthermore, we assume that the initial forward curve $f(0,\tau)$ is a continuous function in τ. We also let W_k, $k=1,\ldots,p$ be independent Brownian motions, and J_j, $j=1,\ldots,n$ be independent II processes, which are assumed to be independent of the Brownian motions. The Poisson random measure of J_j is denoted $M_j(dt,dz)$, with the compensator measure $\nu_j(dz,dt)$.

We find the following risk-neutral dynamics of the forward price, together with a drift condition for $a(u,\tau)$ ensuring the martingale property.

Proposition 6.1 *Suppose for each $j=1,\ldots,n$, that the exponential integrability condition*

$$\int_0^{\mathcal{T}}\int_{|z|\geq 1}\exp(\eta_j(u,\tau)z)\,\nu_j(dz,du)<\infty\,,$$

holds for every $\tau\leq\mathcal{T}$. Under the drift condition

$$\int_0^t a(u,\tau)+\frac{1}{2}\sum_{k=1}^p\sigma_k^2(u,\tau)\,du+\sum_{j=1}^n\int_0^t\eta_j(u,\tau)\,d\gamma_j(u)$$
$$+\sum_{j=1}^n\int_0^t\int_{\mathbb{R}} e^{\eta_j(u,\tau)z}-1-\eta_j(u,\tau)z\mathit{1}_{|z|<1}\,\nu_j(dz,du)=0 \tag{6.2}$$

for every $t \leq \tau$ and $\tau \leq \mathcal{T}$, the forward price $f(t,\tau)$ in (6.1) *has the following dynamics for $t \leq \tau$*

$$\frac{df(t,\tau)}{f(t-,\tau)} = \sum_{k=1}^{p} \sigma_k(t,\tau)\, dW_k(t) + \sum_{j=1}^{n} \int_{\mathbb{R}} e^{\eta_j(t,\tau)z} - 1\, \widetilde{M}_j(dt,dz)\,.$$

Proof. Itô's Formula leads to

$$\begin{aligned}\frac{df(t,\tau)}{f(t-,\tau)} &= \left(a(t,\tau) + \frac{1}{2}\sum_{k=1}^{p}\sigma_k^2(t,\tau)\right) dt + \sum_{j=1}^{m} \eta_j(t,\tau)\, d\gamma_j(t)\\ &\quad + \sum_{j=1}^{n}\int_{\mathbb{R}} e^{\eta_j(t,\tau)z} - 1 - \eta_j(t,\tau) z 1_{|z|<1}\, \nu_j(dt,dz)\\ &\quad + \sum_{k=1}^{p} \sigma_i(t,\tau)\, dW_k(t) + \sum_{j=1}^{n}\int_{\mathbb{R}} e^{\eta_j(t,\tau)z} - 1\, \widetilde{M}_j(dt,dz)\,.\end{aligned}$$

We used the exponential integrability condition to merge the jump components into one integral. Hence, we read off the drift condition as the condition on $a(t,\tau)$ which ensures the martingale property of $f(t,\tau)$ under the risk-neutral probability. □

If the drift condition is not satisfied, the forward price dynamics will not be a martingale under the risk-neutral probability. This means that the forward price dynamics is not arbitrage-free.

Observe that the forward dynamics resembles very well the risk-neutral dynamics resulting from the geometric spot price model derived in Prop. 4.8 in the previous chapter. There we had

$$\sigma_k(t,\tau) = \sum_{i=1}^{m} \sigma_{ik}(t) \exp\left(-\int_t^{\tau} \alpha_i(u)\, du\right)$$

and

$$\eta_j(t,\tau) = \eta_j(t) \exp\left(-\int_t^{\tau} \beta_j(u)\, du\right)$$

for $k = 1,\ldots,p$ and $j = 1,\ldots,n$.

We now study the market dynamics of the forward price defined by (6.1). We change the probability measure using the Esscher transform as we did in the spot market modelling approach in Chapter 3. The only difference here is that we go the opposite way, namely from the risk-neutral to the market probability. We assume that the measure change can be expressed in terms of a market price of risk $(\widehat{\theta}, \widetilde{\theta})$. We can state the dynamics of $f(t,\tau)$ explicitly under P, as follows. Let

$$dW_k(t) = \widehat{\theta}_k(t)\,dt + dB_k(t)$$

and I_j is an II process with drift

$$\gamma_j(t) + \int_0^t \int_{|z|<1} z\left(e^{\widetilde{\theta}_j(u)z} - 1\right)\nu_j(dz, du)\,,$$

and jump measure $e^{\widetilde{\theta}_j(t)}\,\nu_j(dt, dz)$. Then

$$\begin{aligned} f(t,\tau) = f(0,\tau)\exp\Big(&\int_0^t a(s,\tau)\,ds + \sum_{k=1}^p \int_0^t \sigma_k(s,\tau)\widehat{\theta}_k(s)\,ds \\ &+\sum_{k=1}^p \int_0^t \sigma_k(s,\tau)\,dB_k(s) + \sum_{j=1}^n \int_0^t \eta_j(s,\tau)\,dI_j(s)\Big)\,. \end{aligned} \tag{6.3}$$

This is easily obtained from the Esscher transform and the characteristics of the jumps after the change of measure. Note that the drift $a(t,\tau)$ satisfies the drift condition in Prop. 6.1. Moreover, the market dynamics of f is stated in the following Proposition.

Proposition 6.2 *The dynamics of $f(t,\tau)$ under P is*

$$\begin{aligned} \frac{df(t,\tau)}{f(t-,\tau)} &= \sum_{k=1}^p \sigma_k(t,\tau)\widehat{\theta}_k(t)\,dt + \sum_{k=1}^p \sigma_k(t,\tau)\,dB_k(t) \\ &\quad+\sum_{j=1}^n \int_{|z|<1}\left(e^{\eta_j(t,\tau)z} - 1\right)\left(e^{\widetilde{\theta}_j z} - 1\right)\nu_j(dt,dz) \\ &\quad-\sum_{j=1}^n \int_{|z|\geq 1}\left(e^{\eta_j(t,\tau)z} - 1\right)\nu_j(dt,dz) \\ &\quad+\sum_{j=1}^n \int_{|z|<1}\left(e^{\eta_j(t,\tau)z} - 1\right)\widetilde{N}_j(dt,dz) \\ &\quad+\sum_{j=1}^n \int_{|z|\geq 1}\left(e^{\eta_j(t,\tau)z} - 1\right)N_j(dt,dz)\,. \end{aligned}$$

Here, N_j is the random measure associated to I_j, and $\widetilde{N}_j$ is its compensator.

Proof. Suppose $p = n = 1$. An application of Itô's Formula on (6.3) yields

$$\begin{aligned}\frac{df(t,\tau)}{f(t,\tau)} &= \left(a(t,\tau) + \sigma(t,\tau)\widehat{\theta}(t)\right)\,dt \\ &\quad + \sigma(t,\tau)\,dB(t) + \frac{1}{2}\sigma^2(t,\tau)\,dt + \eta(t,\tau)\,d\gamma(t) \\ &\quad + \int_{|z|<1} \eta(t,\tau)z\left(\mathrm{e}^{\widetilde{\theta}(t)z} - 1\right)\,\nu(dt,dz) \\ &\quad + \int_{|z|<1} \left(\mathrm{e}^{\eta(t,\tau)z} - 1\right)\,\widetilde{N}(dt,dz) \\ &\quad + \int_{|z|\geq 1} \left(\mathrm{e}^{\eta(t,\tau)z} - 1\right)\,N(dt,dz) \\ &\quad + \int_{|z|<1} \left(\mathrm{e}^{\eta(t,\tau)z} - 1 - \eta(t,\tau)z\right)\,\mathrm{e}^{\widetilde{\theta}(t)z}\,\nu(dt,dz)\,.\end{aligned}$$

The proof is complete after invoking the drift condition for $a(t,\tau)$ and reorganizing the integral terms. □

If the jump measures ν_j satisfy the exponential integrability condition

$$\int_0^T \int_{|z|\geq 1} \mathrm{e}^{(\eta_j(u,\tau)+\widetilde{\theta}_j(u))z}\,\nu_j(du,dz) < \infty \tag{6.4}$$

for every $\tau \leq T$ and $j = 1,\ldots,n$, then we can write the market dynamics of $f(t,\tau)$ as

$$\begin{aligned}\frac{df(t,\tau)}{f(t,\tau)} &= \sum_{k=1}^{p} \sigma_k(t,\tau)\widehat{\theta}_k(t)\,dt + \sum_{k=1}^{p} \sigma_k(t,\tau)\,dB_k(t) \\ &\quad + \sum_{j=1}^{n} \int_{\mathbb{R}} \left(\mathrm{e}^{\eta_j(t,\tau)z} - 1\right)\left(\mathrm{e}^{\widetilde{\theta}_j(t)z} - 1\right)\,\nu_j(dt,dz) \\ &\quad + \int_{\mathbb{R}} \left(\mathrm{e}^{\eta_j(t,\tau)z} - 1\right)\,\widetilde{N}_j(dt,dz)\,.\end{aligned} \tag{6.5}$$

From the forward price dynamics in Prop. 6.2, or the simpler form in (6.5), we find that the market dynamics of $f(t,\tau)$ has a drift given explicitly by the market price of risk $\theta = (\widehat{\theta},\widetilde{\theta})$. If we want to fit the forward price model to data, this drift will be the expected logreturn, and can thus be estimated by the empirical expected logreturns.

Let us consider two simple examples, where we suppose that $m = n = 1$. First, let us disregard jumps by assuming $J = 0$. Then the drift condition in Prop. 6.1 becomes

$$\int_0^t a(u,\tau) + \frac{1}{2}\sigma^2(u,\tau)\,du = 0\,,$$

for all $0 \le t \le \tau$. Thus, we find that

$$a(t,\tau) = -\frac{1}{2}\sigma^2(t,\tau)\,,$$

which we already knew from standard properties of geometric Brownian motion. Hence, the market dynamics of the forward price is

$$\frac{df(t,\tau)}{f(t,\tau)} = \sigma(t,\tau)\widehat{\theta}(t)\,dt + \sigma(t,\tau)\,dB(t)\,.$$

Next, consider a model where we also have jump risk. Suppose that the drift $d\gamma$ and the jump measure $\nu(du,dz)$ are both absolutely continuous with respect to the Lebesgue measure, that is, $d\gamma(u) = \gamma(u)\,du$ and $\nu(du,dz) = \nu(u,dz)\,du$ for two integrable (in u) functions $\gamma(u)$ and $\nu(u,dz)$, respectively. Then, the drift condition of Prop. 6.1 becomes

$$\begin{aligned} a(t,\tau) &+ \frac{1}{2}\sigma^2(t,\tau) + \eta(t,\tau)\gamma(t) \\ &+ \int_{\mathbb{R}} \left(e^{\eta(t,\tau)z} - 1 - \eta(t,\tau)z 1_{|z|<1}\right)\nu(t,dz) = 0\,, \end{aligned}$$

for all $0 \le t \le \tau \le \mathcal{T}$. Hence, the drift condition tells us that $a(t,\tau)$ becomes a function of the volatility $\sigma(t,\tau)$ and the jump volatility $\eta(t,\tau)$, together with the drift γ of the jump process.

6.3 HJM modelling of swaps

The electricity and gas markets trade in forward contracts having a delivery period, for which we here will use the common notion swaps. The owner of a swap contract with delivery over the time interval $[\tau_1,\tau_2]$ would receive a constant flow of the commodity over this period, against a fixed payment per unit. Our aim in this Section is to derive a price dynamics for such swap contracts based on the HJM approach.

Denote by $F(t,\tau_1,\tau_2)$ the price at time t for a swap contract where the underlying is delivered over the period $[\tau_1,\tau_2]$. The swap contract is usually traded over the time period $t \in [0,\tau_1)$. Typical delivery periods are days, weeks, months, quarters or years. In the electricity market, one encounters contracts having overlapping delivery periods. For instance, in the marketplace one may buy swaps with delivery in the three first months of a year, or a contract with delivery over the first quarter. There are also quarterly contracts over the whole year, together with a contract delivering over the year. Indeed, in order to avoid arbitrage, one needs to have certain relations between the prices of these contracts.

Recall from (4.3) the definition of the weight function $\widehat{w}(u,\tau_1,\tau_2)$ connecting forwards with swaps.

$$\widehat{w}(u,s,t)=\frac{w(u)}{\int_s^t w(v)\,dv},$$

where $w(u)=1$ when the settlement of the swap takes place at the end of the delivery period, while it is $w(u)=\exp(-ru)$ if the contract is settled continuously during the delivery period. Recall that the risk-free interest rate is r. The weight function becomes $\widehat{w}(u,s,t)=1/(t-s)$ in the former case, while it is

$$\widehat{w}(u,s,t)=\frac{re^{-ru}}{e^{-rs}-e^{-rt}},$$

in the latter. Consider the swap price $F(t,\tau_1,\tau_N)$ of a contract with delivery over $[\tau_1,\tau_N]$, and N contracts $F(t,\tau_k,\tau_{k+1})$ with delivery over $[\tau_k,\tau_{k+1}]$ for $k=1,\dots,n-1$. Assume that $\tau_1<\tau_2<\cdots<\tau_N$. Then, by appealing to arbitrage arguments, we find the following *no-arbitrage relation* between the swap prices,

$$F(t,\tau_1,\tau_N)=\sum_{k=1}^{N-1} w_k F(t,\tau_k,\tau_{k+1})\,. \tag{6.6}$$

Here, we use the notation

$$w_k=\frac{\int_{\tau_k}^{\tau_{k+1}} w(u)\,du}{\int_{\tau_1}^{\tau_N} w(u)\,du}, \tag{6.7}$$

for $k=1,\dots,N-1$. Any arbitrage-free model of the swap price needs to satisfy the condition (6.6), at least for those products traded in the market.

In a theoretical model one may ask for arbitrage-freeness for all possible delivery periods. This is in the spirit of the HJM approach, because it is often of interest to have a model which holds for arbitrary delivery periods. This leads to a stronger version of the no-arbitrage relation. Suppose $\tau_k=\tau_1+(k-1)\times\Delta$, with $\Delta=(\tau_2-\tau_1)/N$. Letting $N\to\infty$ and using (6.6), we reach the continuous version of the no-arbitrage condition

$$F(t,\tau_1,\tau_2) = \int_{\tau_1}^{\tau_2} \widehat{w}(u,\tau_1,\tau_2)F(t,u,u)\,du\,. \tag{6.8}$$

Observe that a swap contract starting and ending delivery at time u is the same as a forward contract. Thus, $F(t,u,u)$ is the forward price at time t of a contract with delivery at time $u \geq t$, or $F(t,u,u) = f(t,u)$ following the notation in the book. The implication of the continuous no-arbitrage condition (6.8) is that any model for the swaps valid for arbitrary delivery periods $[\tau_1,\tau_2]$, must come from a forward dynamics. We show that this is a very stringent condition which does not hold for the natural class of geometric Brownian motion models.

We now introduce the natural extension of the forward dynamics in (6.1) to the case of swap contracts. Suppose that the risk-neutral price dynamics of the swap is

$$\begin{aligned} F(t,\tau_1,\tau_2)) = F(0,\tau_1,\tau_2)\exp\Big(\int_0^t A(u,\tau_1,\tau_2)\,du \\ +\sum_{k=1}^{p}\int_0^t \Sigma_k(u,\tau_1,\tau_2)\,dW_k(u) \\ +\sum_{j=1}^{n}\int_0^t \Upsilon_j(u,\tau_1,\tau_2)\,dJ_j(t)\Big)\,. \end{aligned} \tag{6.9}$$

Here, $A(t,\tau_1,\tau_2)$, $\Sigma_k(t,\tau_1,\tau_2)$ and $\Upsilon_j(t,\tau_1,\tau_2)$, $i = k,\ldots,p, j = 1,\ldots,n$, are real-valued continuous functions where $0 \leq t \leq \tau_1 \leq \tau_2 \leq \mathcal{T}$, and $\mathcal{T}$ is an upper bound for the delivery times in the market. Further, we suppose that the functions Σ_k are positive and the initial forward curve $F(0,\tau_1,\tau_2)$ is a real-valued and continuous function for $0 \leq \tau_1 \leq \tau_2 \leq \mathcal{T}$. Since we need the explicit dynamics of the forward to be set under the risk-neutral probability Q, the price has to be a martingale. This entails a condition on A, which we state in the next Proposition together with the dynamics.

Proposition 6.3 *Suppose for each $j = 1,\ldots,n$, that the exponential integrability condition*

$$\int_0^{\tau_1}\int_{|z|\geq 1} \exp(\Upsilon_j(u,\tau_1,\tau_2)z)\,\nu_j(dz,du) < \infty$$

the marketplace one may buy swaps with delivery in the three first months of a year, or a contract with delivery over the first quarter. There are also quarterly contracts over the whole year, together with a contract delivering over the year. Indeed, in order to avoid arbitrage, one needs to have certain relations between the prices of these contracts.

Recall from (4.3) the definition of the weight function $\widehat{w}(u, \tau_1, \tau_2)$ connecting forwards with swaps.

$$\widehat{w}(u, s, t) = \frac{w(u)}{\int_s^t w(v)\, dv},$$

where $w(u) = 1$ when the settlement of the swap takes place at the end of the delivery period, while it is $w(u) = \exp(-ru)$ if the contract is settled continuously during the delivery period. Recall that the risk-free interest rate is r. The weight function becomes $\widehat{w}(u, s, t) = 1/(t - s)$ in the former case, while it is

$$\widehat{w}(u, s, t) = \frac{r\mathrm{e}^{-ru}}{\mathrm{e}^{-rs} - \mathrm{e}^{-rt}},$$

in the latter. Consider the swap price $F(t, \tau_1, \tau_N)$ of a contract with delivery over $[\tau_1, \tau_N]$, and N contracts $F(t, \tau_k, \tau_{k+1})$ with delivery over $[\tau_k, \tau_{k+1}]$ for $k = 1, \ldots, n - 1$. Assume that $\tau_1 < \tau_2 < \cdots < \tau_N$. Then, by appealing to arbitrage arguments, we find the following *no-arbitrage relation* between the swap prices,

$$F(t, \tau_1, \tau_N) = \sum_{k=1}^{N-1} w_k F(t, \tau_k, \tau_{k+1}) \,. \tag{6.6}$$

Here, we use the notation

$$w_k = \frac{\int_{\tau_k}^{\tau_{k+1}} w(u)\, du}{\int_{\tau_1}^{\tau_N} w(u)\, du}, \tag{6.7}$$

for $k = 1, \ldots, N - 1$. Any arbitrage-free model of the swap price needs to satisfy the condition (6.6), at least for those products traded in the market.

In a theoretical model one may ask for arbitrage-freeness for all possible delivery periods. This is in the spirit of the HJM approach, because it is often of interest to have a model which holds for arbitrary delivery periods. This leads to a stronger version of the no-arbitrage relation. Suppose $\tau_k = \tau_1 + (k - 1) \times \Delta$, with $\Delta = (\tau_2 - \tau_1)/N$. Letting $N \to \infty$ and using (6.6), we reach the continuous version of the no-arbitrage condition

$$F(t,\tau_1,\tau_2) = \int_{\tau_1}^{\tau_2} \widehat{w}(u,\tau_1,\tau_2)F(t,u,u)\,du\,. \tag{6.8}$$

Observe that a swap contract starting and ending delivery at time u is the same as a forward contract. Thus, $F(t,u,u)$ is the forward price at time t of a contract with delivery at time $u \geq t$, or $F(t,u,u) = f(t,u)$ following the notation in the book. The implication of the continuous no-arbitrage condition (6.8) is that any model for the swaps valid for arbitrary delivery periods $[\tau_1,\tau_2]$, must come from a forward dynamics. We show that this is a very stringent condition which does not hold for the natural class of geometric Brownian motion models.

We now introduce the natural extension of the forward dynamics in (6.1) to the case of swap contracts. Suppose that the risk-neutral price dynamics of the swap is

$$\begin{aligned} F(t,\tau_1,\tau_2)) = F(0,\tau_1,\tau_2)\exp\Big(\int_0^t A(u,\tau_1,\tau_2)\,du \\ +\sum_{k=1}^{p}\int_0^t \Sigma_k(u,\tau_1,\tau_2)\,dW_k(u) \\ +\sum_{j=1}^{n}\int_0^t \Upsilon_j(u,\tau_1,\tau_2)\,dJ_j(t)\Big)\,. \end{aligned} \tag{6.9}$$

Here, $A(t,\tau_1,\tau_2)$, $\Sigma_k(t,\tau_1,\tau_2)$ and $\Upsilon_j(t,\tau_1,\tau_2)$, $i = k,\ldots,p, j = 1,\ldots,n$, are real-valued continuous functions where $0 \leq t \leq \tau_1 \leq \tau_2 \leq \mathcal{T}$, and $\mathcal{T}$ is an upper bound for the delivery times in the market. Further, we suppose that the functions Σ_k are positive and the initial forward curve $F(0,\tau_1,\tau_2)$ is a real-valued and continuous function for $0 \leq \tau_1 \leq \tau_2 \leq \mathcal{T}$. Since we need the explicit dynamics of the forward to be set under the risk-neutral probability Q, the price has to be a martingale. This entails a condition on A, which we state in the next Proposition together with the dynamics.

Proposition 6.3 *Suppose for each $j = 1,\ldots,n$, that the exponential integrability condition*

$$\int_0^{\tau_1}\int_{|z|\geq 1} \exp(\Upsilon_j(u,\tau_1,\tau_2)z)\,\nu_j(dz,du) < \infty$$

holds for all $0 \le \tau_1 \le \tau_2 \le \mathcal{T}$. *Under the drift condition*

$$\int_0^t A(u,\tau_1,\tau_2) + \frac{1}{2}\sum_{k=1}^{p} \Sigma_k^2(u,\tau_1,\tau_2)\,du + \sum_{j=1}^{n}\int_0^t \Upsilon_j(u,\tau_1,\tau_2)\,d\gamma_j(u)$$
$$+ \int_0^t \int_{\mathbb{R}} \left(e^{\Upsilon_j(u,\tau_1,\tau_2)z} - 1 - \Upsilon_j(u,\tau_1,\tau_2) z \mathit{1}_{|z|<1}\right) \nu_j(dz,du) = 0 \tag{6.10}$$

for every $t \le \tau_1$, *the forward price* $F(t,\tau_1,\tau_2)$ *in* (6.9) *has the following dynamics for* $t \le \tau_1$,

$$\frac{dF(t,\tau_1,\tau_2)}{F(t-,\tau_1,\tau_2)} = \sum_{k=1}^{p} \Sigma_k(t,\tau_1,\tau_2)\,dW_k(t) + \sum_{j=1}^{n}\int_{\mathbb{R}} \left(e^{\Upsilon_j(t,\tau_1,\tau_2)z} - 1\right) \widetilde{M}_j(dt,dz)\,.$$

Proof. The proof is similar to the case of forwards in Prop. 6.1. Itô's Formula leads to

$$\begin{aligned}\frac{dF(t,\tau_1,\tau_2)}{F(t-,\tau_1,\tau_2)} &= \left(A(t,\tau_1,\tau_2) + \frac{1}{2}\sum_{k=1}^{p} \Sigma_k^2(t,\tau_1,\tau_2)\right) dt \\ &+ \sum_{j=1}^{m} \Upsilon_j(t,\tau_1,\tau_2)\,d\gamma_j(t) \\ &+ \sum_{j=1}^{n}\int_{\mathbb{R}} \left(e^{\Upsilon_j(t,\tau_1,\tau_2)z} - 1 - \Upsilon_j(t,\tau_1,\tau_2) z 1_{|z|<1}\right) \nu_j(dt,dz) \\ &+ \sum_{k=1}^{p} \Sigma_k(t,\tau_1,\tau_2)\,dW_k(t) \\ &+ \sum_{j=1}^{n}\int_{\mathbb{R}} \left(e^{\Upsilon_j(t,\tau_1,\tau_2)z} - 1\right) \widetilde{M}_j(dt,dz)\,,\end{aligned}$$

where the exponential integrability condition has been used to merge the jump components into one jump integral. The drift condition can now be read off the resulting dynamics to ensure the martingale property. □

As for forwards, the drift condition must hold in order to avoid arbitrage dynamics for the individual swap contracts.

The main question we want to answer is whether model (6.9) satisfies the continuous no-arbitrage relation (6.8). Unfortunately, the answer is negative for most of the interesting cases.

Lemma 6.1 *Let $n = 0$, and suppose that $A(u, \tau_1, \tau_2)$ and $\Sigma_k(u, \tau_1, \tau_2)$ are continuously differentiable in τ_2 for all $k = 1, \ldots, p$. Then, if $\partial_{\tau_2}\Sigma_k(u, \tau_1, \tau_2)$ is nonzero for u in a subset of positive measure of $[0, t]$ for at least one k, $k = 1, \ldots, p$, then the forward price dynamics* (6.9) *does not satisfy the continuous-time no-arbitrage relation* (6.8).

Proof. We restrict our attention to the case with a weight function $\widehat{w}(u, \tau_1, \tau_2) = 1/(\tau_2 - \tau_1)$ and $p = 1$. By differentiating both sides of (6.8) with respect to τ_2, and by appealing to the stochastic Fubini Theorem, we find that

$$F(t, \tau_1, \tau_2)\left(\frac{1}{\tau_2 - \tau_1} - \int_0^t \partial_{\tau_2} A(u, \tau_1, \tau_2)\, du - \int_0^t \partial_{\tau_2}\Sigma(u, \tau_1, \tau_2)\, dW(u)\right)$$
$$= \frac{1}{\tau_2 - \tau_1} F(t, \tau_2, \tau_2)\,.$$

The right-hand side of the equality above is positive, while the left-hand side may become negative since the Brownian motion integral attains all values on the real line as long as $\partial_{\tau_2}\Sigma$ is nonzero on a subset of $[0, t]$ of positive measure. The proof is complete. □

We conclude that the natural class of an exponential swap price dynamics will fail to satisfy the continuous version of the no-arbitrage condition when the volatility term structure depends on the end of delivery period. In most circumstances it is desirable and natural that the volatility depends on the delivery period of the swap. Thus, one needs to consider other models than the exponential class if arbitrage-freeness is to hold in general. A simple way to obtain such models is to generate them from forward contracts, which is the approach we consider in the next Subsection. An alternative path to the construction of swap models is inspired by the LIBOR models in fixed income theory (see, for example, [Brigo and Mercurio (2001)]), which will be the topic of Sect. 6.4.

6.3.1 *Swap models based on forwards*

In this Subsection we want to construct models for swaps based on the forward models studied in Sect. 6.2. In the electricity and gas markets, say, there is no trade with forwards for obvious reasons. However, one may still use the dynamics of such forwards as a building block, with the hope that they will induce reasonable models of the electricity and gas futures price dynamics which are feasible for further analysis.

Suppose that we model a risk-neutral forward price dynamics $f(t,\tau)$ as in (6.1), where the drift condition for $a(t,u)$ in Prop. 6.1 holds. The swap may be viewed as a continuous flow of forwards, as we recall from relation (4.1), that is

$$F(t,\tau_1,\tau_2)=\int_{\tau_1}^{\tau_2}\widehat{w}(u,\tau_1,\tau_2)f(t,u)\,du\,.$$

We know from Prop. 4.3 that $\lim_{\tau_2\downarrow\tau_1}F(t,\tau_1,\tau_2)=f(t,\tau_1)$, and thus trivially the continuous-time no-arbitrage condition (6.8) holds. Obviously, the no-arbitrage relation (6.6) also holds. In practice, the swap should be a weighted sum of the forwards, reflecting that the smallest delivery period is an hour (or that the spot is really an hourly delivery forward). We keep the integral formulation here for notational simplicity.

Let us study the implied price dynamics for the swap contract $F(t,\tau_1,\tau_2)$. To simplify the discussion, we suppose that $m=n=1$. Further, assume that the coefficient functions σ and η of $f(t,\tau)$ are continuously differentiable with respect to the second variable τ. From the dynamics of $f(t,\tau)$ in Prop. 6.1 (assuming the exponential integrability condition on the jump measure ν), we have

$$\begin{aligned}F(t,\tau_1,\tau_2)&=\int_{\tau_1}^{\tau_2}\widehat{w}(u,\tau_1,\tau_2)f(0,u)\,du\\&\quad+\int_{\tau_1}^{\tau_2}\int_0^t\widehat{w}(u,\tau_1,\tau_2)f(s,u)\sigma(s,u)\,dW(s)\,du\\&\quad+\int_{\tau_1}^{\tau_2}\int_0^t\widehat{w}(u,\tau_1,\tau_2)f(s-,u)\int_{\mathbb{R}}\left(\mathrm{e}^{\eta(s,u)z}-1\right)\widetilde{M}(dz,ds)\,du\,.\end{aligned}$$

Appealing to the stochastic Fubini Theorem and the classical Fubini-Tonelli Theorem, we find

$$\begin{aligned}F(t,\tau_1,\tau_2)&=\int_{\tau_1}^{\tau_2}\widehat{w}(u,\tau_1,\tau_2)f(0,u)\,du\\&\quad+\int_0^t\int_{\tau_1}^{\tau_2}\widehat{w}(u,\tau_1,\tau_2)\sigma(s,u)f(s,u)\,du\,dW(s)\\&\quad+\int_0^t\int_{\mathbb{R}}\int_{\tau_1}^{\tau_2}\widehat{w}(u,\tau_1,\tau_2)\left(\mathrm{e}^{\eta(s,u)z}-1\right)du\,\widetilde{M}(dz,ds)\,.\end{aligned}$$

Using integration by parts, it holds

$$\int_{\tau_1}^{\tau_2} \widehat{w}(u,\tau_1,\tau_2)\sigma(s,u)f(s,u)\,du = \sigma(s,\tau_2)F(s,\tau_1,\tau_2) - \int_{\tau_1}^{\tau_2} \partial_u\sigma(s,u)\int_{\tau_1}^{u} \widehat{w}(\tau,\tau_1,\tau_2)f(s,\tau)\,d\tau\,du\,,$$

where the notation $\partial_u\sigma(s,u) = \partial\sigma(s,u)/\partial u$ is used. Observe that $\widehat{w}(\tau,\tau_1,\tau_2)/\widehat{w}(\tau,\tau_1,u)$ is independent of τ. Hence,

$$\int_{\tau_1}^{\tau_2} \widehat{w}(u,\tau_1,\tau_2)\sigma(s,u)f(s,u)\,du = \sigma(s,\tau_2)F(s,\tau_1,\tau_2) - \int_{\tau_1}^{\tau_2} \partial_u\sigma(s,u)\frac{\widehat{w}(\tau,\tau_1,\tau_2)}{\widehat{w}(\tau,\tau_1,u)}F(s,\tau_1,u)\,du$$

and, similarly,

$$\int_{\tau_1}^{\tau_2} \widehat{w}(u,\tau_1,\tau_2)\left(e^{\eta(s,u)z}-1\right)f(s-,u)\,du = \left(e^{\eta(s,\tau_2)z}-1\right)F(s-,\tau_1,\tau_2) - \int_{\tau_1}^{\tau_2} z\eta(s,u)\partial_u\eta(s,u)\frac{\widehat{w}(\tau,\tau_1,\tau_2)}{\widehat{w}(\tau,\tau_1,u)}F(s-,\tau_1,u)\,du\,.$$

The dynamics of the forward price is therefore (in differential form)

$$\begin{aligned} dF(t,\tau_1,\tau_2) &= \sigma(t,u)F(t,\tau_1,\tau_2)\,dW(t) \\ &\quad + F(t-,\tau_1,\tau_2)\int_{\mathbb{R}}\left(e^{\eta(t,u)z}-1\right)\widetilde{M}(dz,dt) \\ &\quad - \int_{\tau_1}^{\tau_2} \partial_u\sigma(t,u)\frac{\widehat{w}(\tau,\tau_1,\tau_2)}{\widehat{w}(\tau,\tau_1,u)}F(t,\tau_1,u)\,du\,dW(t) \\ &\quad - \int_{\mathbb{R}}\int_{\tau_1}^{\tau_2} z\eta(t,u)\partial_u\eta(t,u)\frac{\widehat{w}(\tau,\tau_1,\tau_2)}{\widehat{w}(\tau,\tau_1,u)}F(t-,\tau_1,u)\,du\,\widetilde{M}(dz,dt)\,. \end{aligned}$$

In conclusion, we see that the dynamics $dF(t,\tau_1,\tau_2)$ depends on $F(t,\tau_1,\tau_2)$ and $F(t,\tau_1,u)$ for $u \in [\tau_1,\tau_2]$. In fact, we integrate $F(t,\tau_1,u)$ over all $u \in [\tau_1,\tau_2]$, weighted with the derivative of the coefficient functions σ and η. Thus, the dynamics of the swap is not a geometric Brownian motion as long as the derivatives of σ and η with respect to the second variable is non-vanishing. Even though we start out with a geometric Brownian motion for the forward, we end up with a non-Markovian stochastic dynamics for the swap contracts, involving all delivery periods contained in $[\tau_1,\tau_2]$. Such a dynamics is time consuming to simulate due to the complicated dependency on all swaps with shorter delivery periods, and moreover, it is hard to use it for data estimation or pricing of derivatives.

As a digression, we may consider the time dynamics of $F(t, \tau_1, \tau_2)$ as an infinite-dimensional stochastic process. With this interpretation we have in fact a multplicative structure on the dynamics. However, this requires that we consider the stochastic dynamics of the functional-valued stochastic process $F(t, \tau_1, \cdot)$. We refer to [DaPrato and Zabczyk (1992)] for more details on the theory for infinite-dimensional stochastic processes.

Consider the case when $\partial_u \sigma(t, u) = \partial_u \eta(t, u) = 0$, that is, with $\sigma(t, u) = \sigma(t)$ and $\eta(t, u) = \eta(t)$. Hence, we analyse the case when both diffusional and jump volatility of the forward are not dependent on the maturity of the contract. From the analysis above we find the stochastic dynamics to be

$$\frac{dF(t, \tau_1, \tau_2)}{F(t-, \tau_1, \tau_2)} = \sigma(t)\, dW(t) + \int_{\mathbb{R}} \left(e^{\eta(t)z} - 1 \right) \widetilde{M}(dz, dt) \,.$$

The volatility structure of the swap dynamics does not depend on the delivery period, which seems unrealistic. In mean-reverting markets, the volatility of forwards will depend explicitly on the maturity of the contract, which has the implication that the swaps cannot have a volatility which does not depend on the delivery period.

If we start out with a model for forwards, the question of estimating the parameters arises, given that the observed data is for the swap. As we have seen earlier, the dynamics of the swap contract becomes intractable. Also, by integrating the explicit representation of the swap in terms of forwards seems to be rather difficult if we want to estimate parameters. But the alternative is to turn the question around, and ask if we can *derive* forward price data from swaps. This can be done by a smoothing technique, where we find a smooth representation of the swap curve. From this curve, one may read off “forward data”. Moreover, by smoothing historical swap curves, we can construct historical price series, and thereby find data to use for estimation of volatility, jumps and the market price of risk. This is the topic of Chapter 7.

We now discuss an approximation approach suggested by [Bjerksund, Rasmussen and Stensland (2000)]. In the paper, the electricity market is considered, and a one-factor model is proposed for the forward dynamics without jumps with a specific volatility function given by

$$\sigma(t, \tau) = \frac{\sigma}{\tau - t + b} + c \,,$$

where σ, b and c are constants. This volatility function can create a sharp increase in volatility as time to maturity of the contract decreases. It is claimed in the paper that the exponentially damping volatility function

$$\sigma(t,\tau) = \sigma \exp(-\alpha(\tau - t))$$

implied by a mean reversion dynamics for the spot does not produce an increase in volatility which is sufficiently sharp in the short end of the curve. To have a tractable dynamics for the traded contracts in the electricity market, [Bjerksund, Rasmussen and Stensland (2000)] suggest to approximate the dynamics for $F(t,\tau_1,\tau_2)$ by

$$\frac{dF(t,\tau_1,\tau_2)}{F(t,\tau_1,\tau_2)} = \Sigma(t,\tau_1,\tau_2)\, dW(t)\,, \tag{6.11}$$

where Σ is the (weighted) average volatility of the forward over the delivery period, defined as

$$\Sigma(t,\tau_1,\tau_2) = \int_{\tau_1}^{\tau_2} \widehat{w}(u,\tau_1,\tau_2)\sigma(t,u)\, du\,. \tag{6.12}$$

Consider the case where $w(u) = 1$ and $\sigma(t,\tau) = \sigma \exp(-\alpha(\tau - t))$, that is, a volatility term structure coming from an exponential mean reversion model with constant volatility and speed of mean reversion α. The approximated volatility then becomes

$$\Sigma(t,\tau_1,\tau_2) = \frac{\sigma}{\alpha(\tau_2 - \tau_1)} \left\{ \mathrm{e}^{-\alpha(\tau_1 - t)} - \mathrm{e}^{-\alpha(\tau_2 - t)} \right\}.$$

If we consider the volatility specification of [Bjerksund, Rasmussen and Stensland (2000)], we are led to the expression

$$\Sigma(t,\tau_1,\tau_2) = \frac{\sigma}{\tau_2 - \tau_1} \ln\left(\frac{\tau_2 - t + b}{\tau_1 - t + b}\right) + c\,.$$

We will return to this model in Chapter 9 when considering option pricing and the Black-76 formula.

6.4 The market models

The LIBOR models in interest rate theory form a flexible class of dynamical models for LIBOR rates matching the implied volatility of captions or swaptions traded in the market. We propose a similar modelling approach for the swap price dynamics in the energy market, where the idea is to construct a dynamics for the *traded* contracts matching with the observed volatility term structure. Since the option markets on electricity and gas are rather thin, we want to estimate the model on the swap prices themselves. The difference from the HJM approach discussed above is that we consider models only for the products traded in the market, and thereby make the possible range of models much wider since we avoid the continuous-time no-arbitrage condition. We refer to this as *the market model.*

In the Nord Pool market (and other power markets) there is organized trade in electricity futures contracts with overlapping delivery periods. For example, we may buy futures contracts with delivery over each quarter next year, or one futures with delivery over the whole year. When introducing market models, we first single out the contracts which cannot be decomposed into contracts with smaller delivery periods. For example, we single out the quarterly contracts, and do not model the yearly contract directly, but rather as a sum of the quarterly ones (to be discussed below). These contracts will be called the *basic contracts.*

Let

$$\{[\tau_1^b, \tau_1^e], \dots, [\tau_C^b, \tau_C^e]\}$$

be a sequence of delivery periods for the different basic contracts, for $c = 1, \dots, C$. The only condition set for this sequence of delivery periods is that it should not be possible to decompose one period into a union of other delivery periods. In normal markets, we additionally have that we can order this sequence so that $\tau_c^e \leq \tau_{c+1}^b$, a typical example being monthly contracts, that is, contracts with delivery each month over the year. In this case, assuming that we are at the beginning of the year, $\tau_c^b = (c-1)/12$ and $\tau_c^e = c/12$, with $c = 1, \dots, C$ and with time measured in years.

We state the forward dynamics for each contract under the risk-neutral probability, in line with the HJM modelling approach discussed in the sections above. Denote by

$$F_c(t) \triangleq F(t, \tau_c^b, \tau_c^e) \tag{6.13}$$

for $c = 1, \dots, C$, and assume that the risk-neutral explicit dynamics is

$$F_c(t) = F_c(0) \exp\Big(\int_0^t A_c(u)\, du + \sum_{k=1}^{p} \int_0^t \Sigma_{c,k}(u)\, dW_k(u) + \sum_{j=1}^{n} \int_0^t \Upsilon_{c,j}(u)\, dJ_j(u) \Big)\,. \tag{6.14}$$

The functions $A_c, \Sigma_{c,k}$ and $\Upsilon_{c,j}$ are continuous real-valued functions on $[0, \tau_c^b]$, since we assume that trading of the contracts ends at the beginning of the delivery period. Now, we state the drift condition that ensures the swap price dynamics to be a martingale.

Proposition 6.4 *Assume that the exponential integrability condition for each $j = 1, \ldots, n$*

$$\int_0^{\tau_c^b} \int_{|z|\geq 1} \exp(\Upsilon_{c,j}(t)z)\, \nu_c(dz, dt) < \infty$$

holds. Under the drift condition

$$\int_0^t A_c(u) + \frac{1}{2}\sum_{k=1}^p \Sigma_{c,k}^2(u)\, du + \sum_{j=1}^n \int_0^t \Upsilon_{c,j}(u)\, d\gamma_j(u)$$
$$+ \sum_{j=1}^n \int_0^t \int_{\mathbb{R}} \left(e^{\Upsilon_{c,j}(u)z} - 1 - \Upsilon_{c,j}(u) z 1_{|z|<1}\right) \nu_j(dz, du) = 0 \quad (6.15)$$

for every $t \leq \tau_c^b$, the swap price $F_c(t)$ in (6.14) has for $t \leq \tau_c^b$ the following dynamics

$$\frac{dF_c(t)}{F_c(t)} = \sum_{k=1}^p \Sigma_{c,k}(t)\, dW_k(t) + \sum_{j=1}^n \int_{\mathbb{R}} \left(e^{\Upsilon_{c,j}(t)z} - 1\right) \widetilde{M}_j(dt, dz).$$

Proof. The proof is similar to the case of forwards in Prop. 6.1, and we will not repeat it here. □

We consider some examples of market models. First, in [Benth and Koekebakker (2005)], a simple one-factor model without jumps is considered,

$$\frac{dF_c(t)}{F_c(t)} = \sigma_c(t)\, dW(t), \quad (6.16)$$

where c labels the different contracts, and the volatility $\sigma_c(t)$ is a function explicitly dependent on the delivery period of the contract in question, $[\tau_c^b, \tau_c^e]$. Different specifications of the volatility is empirically investigated on a huge data set of prices collected from Nord Pool. The volatility specifications incorporated seasonality and maturity effect, where the preferred structure had the specification

$$\sigma_c(t) = \frac{1}{\tau_c^e - \tau_c^b} \int_{\tau_c^b}^{\tau_c^e} \sigma(t, u)\, du + s(t),$$

where $s(t)$ is a seasonality function defined as a truncated Fourier series and $\sigma(t,u) = \sigma \exp(-\alpha(u-t))$. To fit this volatility structure to observed data, we need to have the P dynamics available. In [Benth and Koekebakker (2005)] it was assumed a constant market price of risk, that is, a constant θ, leading to the P dynamics

$$\frac{dF_c(t)}{F_c(t)} = \theta\sigma_c(t)\,dt + \sigma_c(t)\,dB(t)\,.$$

Such a one-factor model is rather simplistic, and unlikely to capture all stylised facts of the electricity futures price dynamics. In view of the findings in [Frestad (2007a)], contracts with different lengths of delivery and delivery at different times of the year are not perfectly correlated, but show a rather complicated pattern of dependency. This calls for multi-factor models. [Kjaer and Ronn (2006)] use a forward model to study gas futures returns on NYMEX, where a non-stationary correlation structure is observed.

A two-factor model for the electricity futures prices at EEX is suggested by [Kiesel, Schindlmayer and Börger (2006)]. They model the basic contracts as

$$\frac{dF_c(t)}{F_c(t)} = \sigma_{c,1}(t)\,dW_1(t) + \sigma_{c,2}(t)\,dW_2(t)\,, \tag{6.17}$$

where c labels the different contracts, and

$$\sigma_{c,1}(t) = \sigma_1 \exp(-\alpha(\tau_c^b - t))\,, \tag{6.18}$$

$$\sigma_{c,2}(t) = \sigma_2\,. \tag{6.19}$$

The motivation for this model is that $\sigma_{c,1}$ mimics the volatility term structure arising from a mean reversion model (at least when considering forwards coming from a multiplicative spot model), while the second volatility models the non-stationary part. The volatility in the electricity futures contract will decay exponentially towards $\sigma_{c,2}$ with increasing time to maturity of the contract. They estimate the model on the implied volatility term structure of monthly, quarterly and yearly contracts. We refer to the paper [Kiesel, Schindlmayer and Börger (2006)] for more details on the calibration procedure and the results.

6.4.1 Modelling with jump processes

Empirical results in [Benth and Koekebakker (2005)], [Frestad, Benth and Koekebakker (2007)] and [Green (2006)] suggest that the logreturns of electricity futures prices are far from being normally distributed. This excludes the geometric Brownian motion models introduced above. In fact, the logreturns are heavy-tailed, rather symmetric, and peaked in the center of the distribution. The analysis in [Frestad, Benth and Koekebakker (2007)] and [Green (2006)] points towards the use of NIG models for the logreturns. We discuss such models in more detail here, where we aim to estimate the model on observed prices and not implicitly on option price data. We have models with prescribed distributional properties under the market probability P in mind, with the NIG distribution as the prime example.

In the rest of this subsection we discuss different modelling aspects for swap price dynamics going beyond the geometric Brownian motion. The focus is on models that can explain the stylised facts of logreturns of forward prices, to which NIG-based models belong. The discussion will not go into detail, since this is an area where there are still many unanswered questions. However, in Chapter 8, we present some empirical studies which argue in favour of using the NIG distribution for modelling the logreturns of electricity futures.

Consider first *one* swap contract specified by a given c. We defined our market models directly under the risk-neutral probability Q. If we want the logreturns of the swap price dynamics to have a specified distribution under the market probability P, we need to translate the dynamics by using the Esscher transform. We know that the transform essentially preserves the characteristics, and thus it is natural to assume that the Q dynamics of the swap price yields logreturns with the same distributional properties. Thus, suppose for the given contract that $p = 0$, $n = 1$ and $\Upsilon = 1$. Further, we assume that $A_c = 0$ and let J be a Lévy process such that $J(1)$ has the desired distribution. Knowing the distribution of $J(1)$ is equivalent to knowing the cumulant function ψ. The dynamics under Q of the forward price $F_c(t)$ is

$$F_c(t) = F_c(0) \exp\left(J(t)\right) ,$$

where $J(t)$ must satisfy a martingale condition. The cumulant function ψ associated with the Lévy process J has to be so that $\mathrm{e}^{t\psi(-\mathrm{i})} = 1$, or,

$$\widetilde{\psi}(-\mathrm{i}) = 0 . \tag{6.20}$$

This is an alternative way to state the martingale condition of the process $F_c(t)$ when modelling the distribution of J rather than a Lévy process with drift as in the general model. Introducing a market price of risk $\widetilde{\theta}(t)$, we get the complete characteristics for J under P from the characteristics under Q.

Returning to our prime example, the NIG distribution, we find that under a constant Esscher transform $\widetilde{\theta}$ of J, the assumption that $J(1) \sim \mathrm{NIG}(\alpha, \beta, \delta, \mu)$ under Q becomes $J(1) \sim \mathrm{NIG}(\alpha, \beta+\widetilde{\theta}, \delta, \mu)$ under P. Recall the explicit cumulant function for the NIG distribution in (2.31). The cumulant condition (6.20) thus becomes

$$\mu + \delta\left(\sqrt{\alpha^2 - \beta^2} - \sqrt{\alpha^2 + (\beta+1)^2}\right) = 0\,. \tag{6.21}$$

When performing a ML estimation, all the parameters $\alpha, \beta+\widetilde{\theta}, \delta$ and μ will be fitted to the data, leaving β and $\widetilde{\theta}$ unestimated. However, (6.21) yields a condition on β in terms of μ, δ and α. Thus, we can find the market price of risk by solving it for β.

We have several possible models which can be employed to model contracts F_c, $c = 1, \ldots, C$. We use a one-factor model in the sense that we state the dynamics

$$F_c(t) = F_c(0)\exp(\Upsilon_c J(t))$$

for constants Υ_c. In this case, the martingale condition becomes

$$\psi(-\mathrm{i}\Upsilon_c) = 0\,,$$

for all $c = 1\ldots, C$. This means, in particular for the NIG case, that

$$\mu\Upsilon_c + \delta\left(\sqrt{\alpha^2 - \beta^2} - \sqrt{\alpha^2 - (\beta+\Upsilon_c)^2}\right) = 0\,.$$

It is natural to let the "scaling" parameter Υ_c vary with the contract, since the distributional characteristics most likely will depend strongly on the delivery period, that is, the choice of c. However, varying Υ_c means that we get as many conditions for the martingale property as contracts, and as a result the parameters α, β, δ and μ will become dependent on c as well. Letting Υ_c be time-dependent only makes the situation worse.

To solve this, we introduce several independent Lévy processes J_j by stating a multi-factor model. Each contract depends on one or more of the factors in the following fashion. For $c = 1, \ldots, C$,

$$F_c(t) = F_c(0) \exp\left(\sum_{j=1}^{n} \Upsilon_{c,j} J_j(t)\right) \tag{6.22}$$

with constants $\Upsilon_{j,c}$. This is the analogue to a multi-factor model with Brownian motion where marginals and the correlation structure are fully specified. If we go back to the specific case of NIG, we lose the explicit form of the marginals. In general, a sum of independent NIG distributed random variables (as the terms in the sum in (6.22) will be), is *not* NIG distributed.[2] Hence, although (6.22) opens for modelling the correlation structure among the different contracts, we lose the attractive property of the explicit knowledge of the marginal distribution, and cannot be sure that the theoretical model has the desirable distribution. We can perform a numerical fitting, but this may become a rather complicated task.

An alternative is to assume that $\mathbf{J}(1) = (J_1(1), \ldots, J_C(1))$ is distributed according to a multivariate NIG distribution. The multivariate NIG distribution is defined via its cumulant function, which means that $\mathbf{J}(1)$ is $\mathrm{NIG}_C(\alpha, \beta, \delta, \mu, \Delta)$ if it has a cumulant function being equal to

$$\begin{aligned}\psi_{\mathrm{mNIG}}(\theta) &= \ln \mathbb{E}[\mathrm{e}^{\mathrm{i}\theta \mathbf{J}(1)}] \\ &= -\mathrm{i}\mu\theta' + \delta\left(\sqrt{\alpha^2 - \beta\Delta\beta'} - \sqrt{\alpha^2 - (\beta + \mathrm{i}\theta)\Delta(\beta + \mathrm{i}\theta)'}\right).\end{aligned} \tag{6.23}$$

Here, $\mu, \beta \in \mathbb{R}^C$, $\delta > 0$, $\alpha > 0$, and $\theta \in \mathbb{R}^C$. Finally, Δ is a positive definite matrix in $R^{C \times C}$ with determinant equal to one, and u' means the transpose of u. We refer the interested reader to [Rydberg (1997)] for a discussion of properties of the multivariate NIG distribution. The model becomes

$$F_c(t) = F_c(0) \exp(J_c(t)),$$

for $c = 1, \ldots, C$. We choose as many NIG variables as contracts. The parameters in the model need to satisfy a martingale condition in line with the one-dimensional case above. From the cumulant function (6.23), we get

$$\mathbf{e}_c\mu + \delta\left(\sqrt{\alpha^2 - \beta\Delta\beta'} - \sqrt{\alpha^2 - (\beta + \mathbf{e}_c)\Delta(\beta + \mathbf{e}_c)'}\right) = 0, \tag{6.24}$$

for $c = 1, \ldots, C$, and $\mathbf{e}_c$ being the cth unit vector in $\mathbb{R}^C$. The Esscher transform must in this case be a slightly generalized form of the univariate

[2]The convolution property holds for NIG distributions as long as the skewness and α are the same for each term.

transform discussed in Subsect 4.1.1. Let the density process of the Radon-Nikodym derivate of P with respect to Q^θ be (for constant $\widetilde{\theta} \in \mathbb{R}^C$)

$$\left.\frac{dP}{dQ}\right|_{\mathcal{F}_t} = \exp(\widetilde{\theta}\mathbf{J}'(t) - \psi_{\mathrm{mNIG}}(-\mathrm{i}\widetilde{\theta})t)\,. \tag{6.25}$$

It is easy to see that in this case the distribution of $\mathbf{J}(1)$ is a multivariate NIG, with the same parameters except for the skewness β, which under P becomes $\beta + \widetilde{\theta}$, analogous to the univariate case.

The multivariate NIG is feasible for ML estimation. However, it is questionable if it is flexible enough to capture the rather complicated dependency structure observed for futures prices in the electricity market. The multivariate GH distribution, and the particular case of NIG, was proposed and empirically analysed by [Börger *et al.* (2007)] as a joint model for electricity, gas, coal and CO2 returns. The authors apply their model to answer question related to risk management.

The drawback with the multivariate NIG is the high dimensionality which may cause numerical problems when fitting the likelihood function to a set of logreturn data. Following recent theory on copulas, we may keep to a much simpler path. First, we fit each contract marginally with the desired Lévy model. The next step is to introduce a dependency structure by using a so-called Lévy copula presented in [Kallsen and Tankov (2006)]. The Lévy copula creates a C-dimensional Lévy process $(J_1(t), \ldots, J_C(t))$ from C Lévy processes J_j with marginally defined distributions. This way of modelling a multivariate Lévy process is very flexible, since we can do it first marginally, and then model the dependency. Also, we get a more general dependency structure than in the multivariate NIG case above, since we are not restricted to use the dependency structure defined by the multivariate NIG. Instead we can choose from a wide variety of Lévy copula models. The drawback is that it is not straightforward statistically to estimate the empirical Lévy copula. The Esscher transform can be applied to obtain the market dynamics, however, it can also be that the dependency structure is transformed. Hence, even when having NIG marginals, not only the skewness of the marginals are changed, but also properties of the Lévy copula are.

Of course, when creating models for swap contracts traded in the market, we could state the model directly under P. This seems to be simpler because we are then in the situation where the data are observed. On the other hand, we are interested in pricing derivatives on these swap contracts, where the risk-neutral dynamics is required. As we saw earlier for the NIG case, it is more or less the same if the dynamics is stated under P or Q, as long as we refer to the Esscher transform to model the change in probability. The same holds for many other models, for which the Esscher transform is structure preserving (see Subsect. 4.1.2). The crucial question when stating the model under P is if the dynamics is arbitrage-free. For the complete market, we need to have the existence of an equivalent martingale measure so that the dynamics of each swap becomes a martingale process. This puts some restrictions on the number of jump processes compared with the contracts. In general, we need to have at least as many independent sources of risk as we have contracts to ensure arbitrage-freeness.

In Chapter 8, we come back to more detailed discussions on the statistical properties of electricity futures prices.

Chapter 7

Constructing Smooth Forward Curves in Electricity Markets

Representing forward prices by one continuous term structure curve is a necessary input when modelling the forward price dynamics or when marking an OTC product to the market. The participants in the electricity market are trading large volumes of financial contracts OTC, and many of these do not have a settlement period coinciding with any of those traded on the exchange. In marking-to-market, the investors need to combine market prices to reflect the "market" value of the OTC product. Hence, a smooth forward curve which can be utilised to price electricity futures having any settlement period will be an essential tool. In this chapter, we propose an algorithm for constructing a smooth curve of forward prices from swap contracts. This technology is also very useful in empirical work, and we apply it for analysing curve dynamics on Nord Pool. The theory and empirical findings in this chapter are an extension of the paper of [Benth, Koekebakker and Ollmar (2007)].

A term structure curve is also required if one wants to implement one of the many no-arbitrage term structure models for risk management or derivative pricing purposes. Following [Bjerksund, Rasmussen and Stensland (2000)] (see Sects. 6.2 and 6.3), we may model the price dynamics of forwards, and make an approximate dynamics for contracts exhibiting a delivery period. The initial condition for such a HJM approach when modelling forwards is a smooth curve describing today's forward prices. This curve must be extracted from the prices observed in the market, which are quoted for products having a settlement period. Thus, one needs methods to extract a smooth curve from quoted prices.

Fitting a yield curve to market data in a fixed income market has been studied for many years. The seminal paper in this field is [McCulloch (1971)]. A survey of different methods for constructing yield curves is provided in [Anderson and Deacon (1996)]. The two main approaches are either to fit a parametric function to the entire yield curve by regression, or to fit all observed yields with a spline. In this chapter we propose a mixture of the two approaches, assuming that the forward curve can be represented as the sum of a seasonal function and an adjustment function. In particular, we apply a maximum smoothness criterion in the specification of the adjustment term, first used in fixed income markets by [Adams and van Deventer (1994)] (see also [Forsgren (1998)] and [Lim and Xiao (2002)]). We cannot apply the methods developed by these authors directly for two reasons. First, our market data are not fixed point yields. Second, in the case of non-storability the cost-of-carry relationship between spot and contracts for future delivery no longer exists. This means that even financial prices may vary seasonally.

The information provided by the market may hide the seasonality if the settlement period is long. This means that we must specify a seasonal function based on more information than can be read off the market prices. There are many ways to do this, for example, based on spot price data which can be linked to forward prices. However, since there is no clear arbitrage-free connection between the spot prices and the forward curve in the electricity market, the choice of seasonal function will necessarily be ad hoc to some degree.

The maximum smoothness approach can easily be modified to allow for average based price inputs. When closing prices are provided as input data, the curve estimation is transformed into solving a linear system of equations. When confronted with bid-ask prices, we suggest an iterative procedure to compute a smooth forward curve that consistently price all products constrained by all bid-ask spreads. The algorithm is computationally efficient, and converges rapidly towards the optimal forward curve.

We have implemented our algorithm and tested the methodology on data collected from Nord Pool. We perform several studies with different specifications of the seasonality function to show the flexibility of our approach. As an application we perform an empirical study of historical electricity futures curves. These curves are used to estimate a forward curve model and a swap price market model. These are simple arithmetic models, and from a theoretical point of view they are identical, that is, the arithmetic dynamics of the forward curve implies a specific arithmetic electricity futures price dynamics, and vice versa. The specification of the seasonality function is affecting the volatility estimates. In the short end of the curve the model specifications produce similar results. But the volatility in the long end of the curve seems to be upward biased using a forward curve approach compared to the swap market model approach. Although we focus on electricity markets in this chapter, the methodology is clearly applicable for constructing smooth forward curves from gas futures prices as well. We use the terms electricity futures and swaps interchangeably in this chapter.

7.1 Swap and forward prices

The methodology we will describe in this chapter is related to [Fleten and Lemming (2003)], where they smoothen an electricity futures curve based on a bottom-up model called the MPS model, see [Botnen *et al.* (1992)]. The MPS model calculates weekly equilibrium prices and production quantities based on fundamental factors for demand and production, like, for example, temperature, fuel costs, snow levels, capacities, etc. The approach of [Fleten and Lemming (2003)] is non-parametric, in the sense that they derive a sequence of daily (or any other appropriate time resolution) forward prices minimising the distance in the least squares sense to the output from the MPS model. The optimisation is constrained on the bid-ask spreads of market prices, and the curve is appropriately smoothened by a penalty term. The MPS model may serve as a seasonal function in our context, however, we are not limited to such a specification. In fact, we may use any seasonal function, which gives more flexibility. In this way speculators may conduct quantitative analyses without profound resource based models using financial market data only. Further, we introduce an adjustment term in the smoothing procedure which we interpret as a market price of risk coming from the link between spot prices and forwards. Our methodology provides a flexible framework for combining a seasonal specification with smoothing techniques in line with existing work from fixed income markets, see [Adams and van Deventer (1994)].

7.1.1 *Basic relationships*

In Chapter 4 we derived the relationships between the spot price, forwards and swaps. Recall from Prop. 4.1 that the price of a swap contract at time

$t \leq \tau_1$, having delivery period $[\tau_1, \tau_2]$, is

$$F(t, \tau_1, \tau_2) = \int_{\tau_1}^{\tau_2} w(u, \tau_1, \tau_2) f(t, u)\, du\,, \tag{7.1}$$

where $w(u, \tau_1, \tau_2)$ is defined in (4.3). In practice, the contracts are not settled continuously over the delivery period, but at discrete times. Assuming settlement at N points in time $u_1 < u_2 < \ldots < u_N$, with $\tau_1 = u_1$, $\tau_2 = u_N$ and $\Delta_i \triangleq u_{i+1} - u_i$, the relationship becomes

$$F(t, \tau_1, \tau_2) = \sum_{i=1}^{N} w(u_i, \tau_1, \tau_2) f(t, u_i) \Delta_i\,. \tag{7.2}$$

In this chapter we are concerned with the extraction of $f(0, u)$ based on the observations of swap contracts, $F(0, \tau_1, \tau_2)$, for different delivery periods. We write $f(u)$ for $f(0, u)$ and $F(\tau_1, \tau_2)$ for $F(0, \tau_1, \tau_2)$ in the next section.

7.1.2 *A continuous seasonal forward curve*

Assume that we observe m contracts a given day. Furthermore let τ_s be the start of the settlement period for the contract with the shortest time to delivery, and denote by τ_e the end of the settlement period for the contract going longest into the future. In the following subsections we will specify a certain functional form for f that makes the relationship (7.1) or (7.2) hold for m contracts at a given point in time. We decompose the forward price into

$$f(u) = \Lambda(u) + \varepsilon(u)\,, \quad u \in [\tau_s, \tau_e] \tag{7.3}$$

for two continuous functions $\Lambda(u)$ and $\varepsilon(u)$. We interpret $\Lambda(u)$ to be the seasonality of the forward curve, and $\varepsilon(u)$ to be an adjustment function that captures the forward curve's deviation from the seasonality.

Since there is no unique arbitrage-free relationship between the spot price dynamics and the forward curve for many commodity markets like, for example, electricity, one may use the objective probability as the risk-neutral pricing measure. Thus, the forward price should simply be the predicted spot price, given by the seasonality function. However, it is natural to believe that the traders in the market are including a risk premium for the lack of a hedge of the forward, which is reflected as a deviation from the seasonality in the forward curve. In this sense, the adjustment function can be understood as the market price of risk, and it is quite natural to believe that this is a function of time to delivery (see, for example, [Benth *et al.* (2003)]). In the short end of the curve, the traders have much information about future price formations, based on, for instance, weather conditions, hydro reservoir fillings etc. The long end of the curve may be several years ahead, and obviously the market's view on risk is less sensitive to time. This defends not only a time-varying ε, but also that the adjustment function should be flat in the long end. Therefore, from now on we assume that

$$\varepsilon'(\tau_e) = 0 .$$

Note that the smoothness is calculated on the adjustment function $\varepsilon(u)$, and not on the forward function $f(u)$. The reason for this is to better retain seasonal patterns.

We may explain the decomposition (7.3) in further detail by relating the forward dynamics to a spot model. In [Koekebakker (2003)] and [Benth, Kallsen and Meyer-Brandis (2007)] different types of arithmetic models have been proposed for the electricity spot price dynamics, as discussed in Subsect. 3.2.2 of Chapter 3. In the simple case, we may represent the price dynamic in these models as

$$S(u) = \Lambda(u) + X(u) ,$$

for a deterministic seasonality function $\Lambda(u)$ and some OU process $X(u)$ as discussed in Chapter 3. Assuming such a model, we recall from Sect. 4.1 that

$$f(u) = \Lambda(u) + \mathbb{E}_Q\left[X(u)\right] .$$

The expectation is a risk-neutral one, introducing a correction to the seasonality function $\Lambda(u)$ in the representation of the forward prices. Hence, under an arithmetic model for the spot price dynamics, we obtain a decomposition like in (7.3). This also gives a mathematical argument for the dependency on delivery time in the adjustment function $\varepsilon(u)$. Further, observe that given a specification of $X(u)$ under the objective probability P (which will be the case after fitting $S(u)$ to the observed spot prices), we have to calculate the risk-neutral expectation. Using the Esscher transform as the risk-neutral probability with a constant θ, and assuming constant coefficients in the dynamics of X for simplicity, we find from Prop. 4.10

that

$$\mathbb{E}_\theta\left[X(u)\right] = \frac{\mu + \sigma\theta}{\alpha}(1 - \mathrm{e}^{-\alpha u}) + X(0)\mathrm{e}^{-\alpha u}\,,$$

which converges to $(\mu + \sigma\theta)/\alpha$ when u tends to infinity. This implies that ε should be flat in the long end, providing a rationale for the assumption $\varepsilon'(\tau_e) = 0$ stated above. Note that the link between an arithmetic spot dynamics and the forward curve may be used to suggest a specific functional form of ε (see Prop. 4.10, in particular the function $\Theta(t, u; \theta)$).

In passing, we note that using a geometric model for the spot price dynamics would naturally lead to a multiplicative decomposition of the forward curve like

$$f(u) = \Lambda(u)\varepsilon(u)\,.$$

It is possible to adjust our approach to cope with such a specification as well, however, it is not possible to do a logarithmic transform since we cannot represent the swap prices in terms of the logarithm of forward prices.

The function $\Lambda(u)$ may be interpreted as a subjective forward curve specified based on the market trader's beliefs. Then it becomes an exogenous prior function, and the adjustment function describes the degree of misspecification. For instance, one may use the bottom-up model of [Fleten and Lemming (2003)], which yields a non-parametric prior estimate of $\Lambda(u)$ based on fundamental modelling of demand and supply curves defining equilibrium prices. By further expanding their approach, we introduce the correction term ε accounting for a risk premium which may be parametrized and fitted to observed swap prices. On the other hand, letting $\Lambda(u)$ be parametrized, one can estimate it together with the correction term ε on market data directly, and in this way the beliefs of the trader are specified in a structural way, and not fixed by some a priori specified function.

We use a parametric form of $\Lambda(u)$, while the adjustment function is specified as a spline in order to have a perfectly matching smoothed forward curve to observed average-based financial data. Loosely speaking, we fit the parameters of the seasonal function using the least squares approach, and then use the spline specification of the adjustment function to ensure a perfect fit to the observed swap prices of the market. However, we may choose ε to be parametric as well. It could, for instance, be the function obtained from $\mathbb{E}_Q[X(u)]$ in the arithmetic modelling of the spot prices discussed above. In this case, the definition of ε is coming from the risk-neutral specification of $X(u)$. Next, we will explain in detail an algorithm for computing a forward curve based on the maximum smoothness criterion.

7.2 Maximum smooth forward curve

In [Adams and van Deventer (1994)] the maximum smoothness criterion is used to construct forward yield curves. We briefly describe their method in our context. Let $C_0^2([\tau_s, \tau_e])$ be the set of real-valued functions on the interval $[\tau_s, \tau_e]$ which are twice continuously differentiable with zero derivative in τ_e. Consider some subclass $\mathcal{C}$ of functions. Using smoothness of a function expressed as the mean square value of its second derivative, we define the smoothest possible forward curve on an interval $[\tau_s, \tau_e]$ as the one that minimises

$$\int_{\tau_s}^{\tau_e} \left[\varepsilon''(u)\right]^2 \, du \,,$$

over the set $\mathcal{C}$, if such exists. We define $\varepsilon \in \mathcal{C}$ to be this minimising function, and interpret the smoothest forward curve to be the one for which ε solves the minimisation problem above. The subclass $\mathcal{C}$ is chosen so that it is easy to take the swap price data into account, either by exact matching, or by a constraint on the bid-ask spread prices. In addition to be smooth, we want the adjustment function to be twice continuously differentiable and horizontal at time τ_e. It turns out that the class of polynomial spline functions of order four is appropriate for our purposes.

7.2.1 *A smooth forward curve constrained by closing prices*

Let

$$\mathcal{S}_p = \{(\tau_1^b, \tau_1^e), (\tau_2^b, \tau_2^e), \ldots, (\tau_m^b, \tau_m^e)\}$$

be a list of start and end dates for the settlement periods of m different swap contracts. To be able to handle overlapping settlement periods we construct a new list of dates, $\{\tau_0, \tau_1, \ldots, \tau_n\}$, where overlapping contracts are split into sub-periods. This is illustrated in Fig. 7.1. Note that $\tau_s = \tau_1^b = \tau_0$ and $\tau_e = \tau_m^e = \tau_n$.

As we can see from Fig. 7.1, the new list is basically the elements in $\mathcal{S}_p$ sorted in ascending order with any duplicated dates removed. The bid and ask prices for the forward contract $i \in \{1, \ldots, m\}$ are denoted F_i^B and F_i^A, respectively. For the time being, we concentrate on the closing prices.

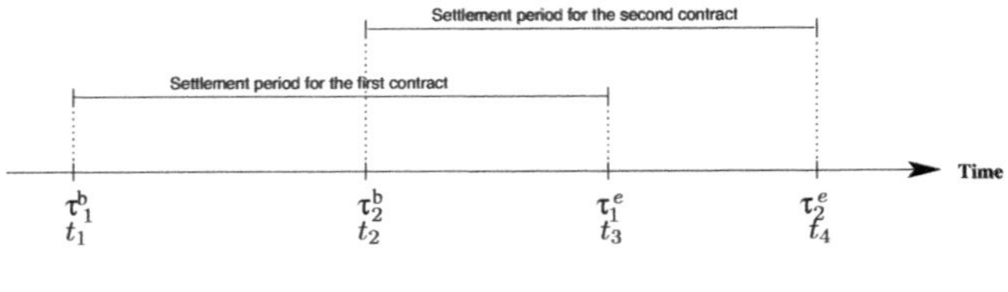

Fig. 7.1 Splitting overlapping swaps.

The adjustment functions ε will be chosen from the class $\mathcal{C}$, which, we assume, consists of polynomial splines of order four defined as

$$\varepsilon(u) = \begin{cases} a_1u^4 + b_1u^3 + c_1u^2 + d_1u + e_1 \,, & u \in [\tau_0, \tau_1] \,, \\ a_2u^4 + b_2u^3 + c_2u^2 + d_2u + e_2 \,, & u \in [\tau_1, \tau_2] \,, \\ \quad \vdots \\ a_nu^4 + b_nu^3 + c_nu^2 + d_nu + e_n \,, & u \in [\tau_{n-1}, \tau_n] \,. \end{cases}$$

Furthermore, the splines are assumed to be twice continuously differentiable and having a zero derivative in τ_n.

To find the parameters (here $\mathbf{x}'$ means the transpose of the vector $\mathbf{x}$)

$$\mathbf{x}' = [a_1\ b_1\ c_1\ d_1\ e_1 \quad a_2\ b_2\ c_2\ d_2\ e_2\ \ldots\ a_n\ b_n\ c_n\ d_n\ e_n] \,,$$

to the adjustment function, we solve the following equality constrained convex quadratic programming problem

$$\min_{\mathbf{x}} \int_{\tau_0}^{\tau_n} \left[\varepsilon''(u;\mathbf{x})\right]^2 du \,, \tag{7.4}$$

subject to the natural constraints in the connectivity and smoothness of derivatives at the knots, $j = 1, \ldots, n-1$,

$$(a_{j+1} - a_j)u_j^4 + (b_{j+1} - b_j)u_j^3 + (c_{j+1} - c_j)u_j^2 \\ + (d_{j+1} - d_j)u_j + e_{j+1} - e_j = 0 \,, \tag{7.5}$$

$$4(a_{j+1} - a_j)u_j^3 + 3(b_{j+1} - b_j)u_j^2 \\ + 2(c_{j+1} - c_j)u_j + d_{j+1} - d_j = 0 \,, \tag{7.6}$$

$$12(a_{j+1} - a_j)u_j^2 + 6(b_{j+1} - b_j)u_j + 2(c_{j+1} - c_j) = 0 \,, \tag{7.7}$$

and

$$\varepsilon'(u_n;\mathbf{x}) = 0 \,, \tag{7.8}$$

$$F_i^C = \int_{\tau_i^b}^{\tau_i^e} w(u, \tau_i^b, \tau_i^e)\, (\varepsilon(u) + s(u))\, du\,, \tag{7.9}$$

for $i = 1, \ldots, m$. In the empirical estimation below we use $w(u, \tau_1, \tau_2) = 1/(\tau_2 - \tau_1)$ even though the contracts traded on Nord Pool are marked-to-market in the delivery period. This approximation is very good for reasonable levels of the interest rate (see [Lucia and Schwartz (2002)]). In effect, we are assuming zero interest rate, arguing that the interest rate effect is less than the effect from both the seasonality and adjustment functions.

The minimisation problem has a total of $3n + m - 2$ constraints. By inserting the expression for $\varepsilon''(u)$ and integrating we can rewrite the minimisation problem (7.4) as

$$\min_{\mathbf{x}} \mathbf{x}'\mathbf{H}\mathbf{x}\,,$$

where

$$\mathbf{H} = \begin{bmatrix} \mathbf{h}_1 & & 0 \\ & \ddots & \\ 0 & & \mathbf{h}_n \end{bmatrix}, \; \mathbf{h}_j = \begin{bmatrix} \frac{144}{5}\Delta_j^5 & 18\Delta_j^4 & 8\Delta_j^3 & 0 & 0 \\ 18\Delta_j^4 & 12\Delta_j^3 & 6\Delta_j^2 & 0 & 0 \\ 8\Delta_j^3 & 6\Delta_j^2 & 4\Delta_j^1 & 0 & 0 \\ 0 & 0 & 0 & 0 & 0 \\ 0 & 0 & 0 & 0 & 0 \end{bmatrix}$$

and

$$\Delta_j^l = \tau_j^l - \tau_{j-1}^l\,.$$

Here, the vector $\mathbf{x}$ is $5n$-dimensional, while the symmetric matrix $\mathbf{H}$ has dimensions of $5n \times 5n$. All the constraints in (7.5)–(7.9) are linear in $\mathbf{x}$, and can be formulated in matrix form as $\mathbf{A}\mathbf{x} = \mathbf{b}$, where $\mathbf{A}$ is a $(3n+m-2) \times 5n$-dimensional matrix, and $\mathbf{b}$ is a $(3n+m-2)$-dimensional vector. We obtain an explicit solution by the Lagrange multiplier method. Let $\boldsymbol{\lambda}' = [\lambda_1, \lambda_2, \ldots, \lambda_{3n+m-2}]$ be the corresponding Lagrange multiplier vector to the constraints. We can now express (7.4) as the following unconstrained minimization problem

$$\min_{\mathbf{x},\lambda} \quad \mathbf{x}'\mathbf{H}\mathbf{x} + \lambda'(\mathbf{A}\mathbf{x} - \mathbf{b})\,. \tag{7.10}$$

The solution $[\mathbf{x}^*, \lambda^*]$ is thus obtained by solving the linear equation

$$\begin{bmatrix} \mathbf{2H} & \mathbf{A}' \\ \mathbf{A} & \mathbf{0} \end{bmatrix} \begin{bmatrix} \mathbf{x} \\ \lambda \end{bmatrix} = \begin{bmatrix} \mathbf{0} \\ \mathbf{b} \end{bmatrix}. \tag{7.11}$$

The dimension of the left matrix is $(8n+m-2) \times (8n+m-2)$. The solution vector and the rightmost vector both have the dimension $(8n+m-2)$.

Solving (7.11) numerically is standard, and can be done using various techniques. We have chosen to use QR factorisation. If n or m is large, one could improve the calculation speed by utilising the sparseness of the matrix.

7.2.2 *A smooth forward curve constrained by bid and ask spreads*

When the market is open for trading we do not observe exact prices but rather bid-ask spreads. We will now extend the model to handle bid-ask prices. We need to replace the fixed closing price constraint (7.9) with

$$F_i^B \leq \int_{\tau_i^b}^{\tau_i^e} w(u, \tau_i^b, \tau_i^e) \left(\varepsilon(u) + \Lambda(u)\right) du \leq F_i^A \qquad i = 1, \ldots, m .$$

The smooth forward function is constrained between the market bid-ask prices. Unfortunately, all of the constraints are no longer binding by equality and it is therefore not possible to use the fast and easy Lagrange multiplier method.

There exist several methods to solve this problem numerically. See [Judd (1998)] for a short description of some of the most commonly used algorithms. We suggest a method inspired by the active set approach. The strategy to solve the minimisation problem (7.10) is to use a pseudo-closing price which is iteratively changing within the boundaries of the bid-ask spread in the direction implied by the sign of the Lagrangian. The algorithm is outlined below.

(1) **Initialization.** Set the initial pseudo-closing prices, F_i^C, closest to the seasonal function and solve (7.10). Alternatively, $F_i^C = (F_i^A + F_i^B)/2$ can be used as initial values.
(2) **Start optimisation.** Let $\lambda_{3n-2}, \ldots, \lambda_{3n+m-2}$ be the closing price Lagrangians, and call the one with the largest absolute value λ^C. Adjust the pseudo-closing price according to

$$F_{new}^C = \begin{cases} F^A & \text{if } \lambda^C > 0 \text{ and } \lambda^A \geq 0\,, \\ F^C - \frac{\lambda^C (F^A - F^C)}{\lambda^A - \lambda^C} & \text{if } \lambda^C > 0 \text{ and } \lambda^A < 0\,, \\ F^C - \frac{\lambda^C (F^C - F^B)}{\lambda^C - \lambda^B} & \text{if } \lambda^C < 0 \text{ and } \lambda^B > 0\,, \\ F^B & \text{if } \lambda^C < 0 \text{ and } \lambda^B \leq 0\,, \end{cases}$$

where λ^A and λ^B denote the contract's Lagrangians with an average price equal to the ask price F^A and the bid price F^B, respectively.

(3) **Stopping criteria.** We apply two different stopping criteria. The first one is to stop if for each F_i the following is true: *(a)* either the average price F_i^C is equal to F^B and λ_i^C is still negative, or *(b)* the average price F_i^C is equal to F^A and λ_i^C is still positive. This means that it is not possible to improve the smoothness by changing F_i^C.
The second criterion is to stop when the improvement of the smoothness is below some percentage γ, that is,

$$1 - \frac{(\mathbf{x}'\mathbf{H}\mathbf{x})_k}{(\mathbf{x}'\mathbf{H}\mathbf{x})_{k-1}} < \gamma ,$$

where k is the iteration number. The minimisation ends when one of the two criteria is fulfilled. If neither of the stopping criteria is satisfied the algorithm returns to step 2.

The main advantage of this algorithm is the calculation speed. Convergence is usually obtained in m to $2m$ iterations. The reason for this rapid convergence is mainly due to the relative small bid-ask spread compared to the value of the adjustment function, that is, a small bid-ask spread usually implies that the bid or the ask constraint is binding and thereby reducing the number of constraints with inequalities.

7.3 Putting the algorithm to work

In this section we apply our algorithm to real world data. In examples we use electricity futures price data collected from Nord Pool (see Chapter 1 for a detailed description of this market). In the first example we compute a smooth curve on a particular day. In the second example we compute the term structure of volatility from historical data. Using a simple arithmetic model we are able to illustrate spot, forward and swap price modelling in this market.

7.3.1 *Nord Pool example I: A smooth curve*

In this example we study the effect of the choice of seasonal function on the smoothed forward curve. The input data is from 4 May 2005, and

Table 7.1 Market data from Nord Pool, 4 May 2005

Ticker	Start date	End date	Bid	Ask
ENOW19-05	2005-05-09	2005-05-15	252.5	253.00
ENOW20-05	2005-05-16	2005-05-22	248.75	250.00
ENOW21-05	2005-05-23	2005-05-29	253.25	257.00
ENOW22-05	2005-05-30	2005-06-05	252.00	255.00
ENOW23-05	2005-06-06	2005-06-12	252.00	255.00
ENOW24-05	2005-06-13	2005-06-19	252.00	255.00
ENOMJUN-05	2005-06-01	2005-06-30	252.00	254.50
ENOMJUL-05	2005-07-01	2005-07-31	236.00	236.50
ENOMAUG-05	2005-08-01	2005-08-31	256.00	258.00
ENOMSEP-05	2005-09-01	2005-09-30	260.00	263.00
ENOMOCT-05	2005-10-01	2005-10-31	263.00	269.00
ENOMNOV-05	2005-11-01	2005-11-30	275.00	277.00
FWV2-05	2005-10-01	2005-12-31	276.00	276.50
ENOQ1-06	2006-01-01	2006-03-31	280.50	283.50
ENOQ2-06	2006-04-01	2006-06-30	235.00	240.00
ENOQ3-06	2006-07-01	2006-09-30	230.00	233.00
ENOQ4-06	2006-10-01	2006-12-31	251.00	259.00
ENOQ1-07	2007-01-01	2007-03-31	266.50	272.00
ENOQ2-07	2007-04-01	2007-06-30	220.50	228.50
ENOQ3-07	2007-07-01	2007-09-30	218.50	226.50
ENOQ4-07	2007-10-01	2007-12-31	250.00	250.50
ENOYR-06	2006-01-01	2006-12-31	249.00	249.50
ENOYR-07	2007-01-01	2007-12-31	240.50	241.00
ENOYR-08	2008-01-01	2008-12-31	240.00	244.00

is reported in Table 7.1.[1] The six weekly contracts ("ENOWxxxx") are futures-style contracts and the rest of the contracts are forward-style, that is, no marking-to-market. Recall that in this book we have denoted all these contracts electricity futures no matter if they are forward- or futures-style.

The seasonality is most easily seen from the quarterly contracts in both 2006 and 2007 where the first quarter (Q1) has the highest price and the third quarter (Q3) the lowest. To visualise the effect of the seasonal function on a smooth forward curve we try three different specifications:

(1) zero seasonality, $\Lambda(u) = 0$,
(2) a simple trigonometric function,

$$\Lambda(u) = 145.732 + 29.735 \times \cos\left((u + 6.691)\frac{2\pi}{365}\right),$$

[1] Quarterly and yearly contracts are traded in EUR. The NOK conversion is done via the spot NOK/EUR currency rate on the actual day. See Sect. 4.2 for a discussion on currency conversion.

(3) a spot prognosis from a bottom-up model.

The first alternative has no seasonality. In markets trading swap contracts and showing no seasonality, this is a natural candidate (one may think of the oil market as having no seasonality). Seasonality is a prominent feature for electricity prices at Nord Pool, however, we include the specification for comparison. The second specification is a spot-based estimate of the seasonality in the Nord Pool market taken from [Lucia and Schwartz (2002)], where the parameter values are estimates from their one-factor arithmetic model. The third example is a prognosis supplied by Agder Energi AS on 4 May 2005 (see lower panel of Fig. 7.2), based on their internal bottom-up model.

The smooth forward curves are extracted from $m = 21$ contracts and are represented by a spline consisting of $n = 32$ polynomials. The algorithm converged after about 28 iterations, depending on the seasonal function we used. The seasonal functions are plotted in the lower panel of Fig. 7.2, and the corresponding smoothed forward curves are plotted in the upper panel. Note that the three forward curves are all plausible, in the sense that they can all reproduce electricity futures contracts within the bid-ask prices reported in Table 7.1.

We have deliberately chosen very different seasonal functions. The trigonometric function is much smoother and with a regular wave pattern compared to the bottom-up prognosis. They also differ in level, with the zero alternative as an extreme. The effect of the three different specifications on the shape of the forward curve varies with maturity. All three seasonal functions produce more or less identical forward curves for maturities less than a year. The reason is simple. Since the delivery periods for short term contracts are short, the marking-to-market constraint on the forward curve completely dominates the deterministic seasonality part of the forward curve function. For maturities between one and two years (roughly between 400 and 800 days to maturity), the bottom-up prognosis differs somewhat from the two others. The four quarterly contracts force the curve with flat seasonality to fluctuate over the yearly cycle, making it indistinguishable from the curve with trigonometric seasonality. The market price covering the rightmost end of the curve is from a yearly contract alone. The curve without seasonality flattens. Overall, this example shows that the choice of seasonal function has little effect on the smooth forward curve in the short end. With longer maturities the forward curve can acquire various shapes depending on the seasonality function supplied by the

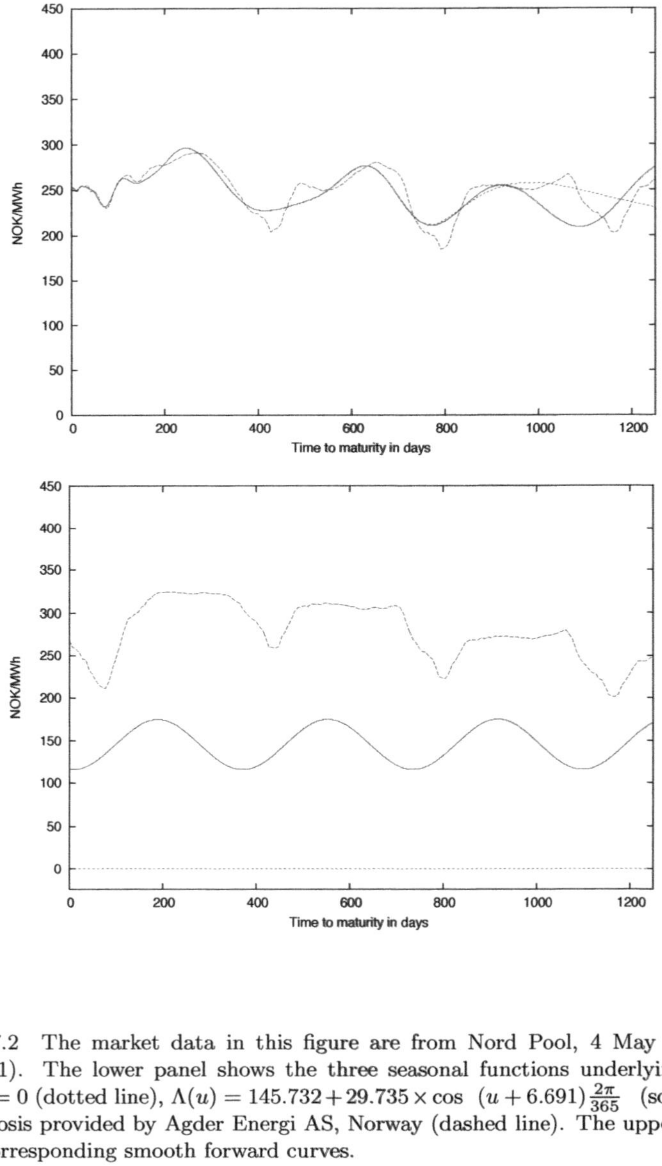

Fig. 7.2 The market data in this figure are from Nord Pool, 4 May 2005 (see Table 7.1). The lower panel shows the three seasonal functions underlying the curves: $\Lambda(u) = 0$ (dotted line), $\Lambda(u) = 145.732 + 29.735 \times \cos\ (u + 6.691)\frac{2\pi}{365}$ (solid line) and a prognosis provided by Agder Energi AS, Norway (dashed line). The upper panel shows the corresponding smooth forward curves.

user. In the long end, with price information from only one yearly contract, the seasonal function becomes imperative in shaping the forward curve.

7.3.2 *Nord Pool example II: Preparing a data set and analysing volatility*

Even in highly liquid futures markets one will often need to estimate prices for other maturity dates than observed in the market. In this subsection an investigation of volatility based on historical data is conducted. It is shown that the way the panel data set is prepared affects the estimated volatility dynamics. Note that the analysis is for illustrative purposes only, a full scale empirical analysis is beyond the scope of this book.

We use the first two seasonal functions defined above, zero seasonality ($\Lambda(u) = 0$) and yearly variation estimated in [Lucia and Schwartz (2002)] ($\Lambda(u) = 145.732 + 29.735 \times \cos\left((u + 6.691)\frac{2\pi}{365}\right)$). On a given date t_i, $i = 1, ..., N$, we apply our algorithm to compute the forward curve with maximum smoothness using all reported swap contracts on Nord Pool constrained by their closing bid-ask spreads. The sample period starts 2 January 2001, and ends 29 April 2005, a total of $N = 1076$ trading days. Hence, we extract a total of 1076 smooth forward curves.

We next extract panel data sets in two ways for each choice of seasonal function; a panel of forward prices and a panel of electricity futures prices. From each day's computed forward curve, we pick a cross section of forward and electricity futures prices. The cross sections of forward prices are computed for a set of delivery times u_j, $j = 1, \ldots, M$, while for the electricity futures we use the relationship in (7.1) to produce cross sections of M contracts, having delivery periods $\{\tau_j^b, \tau_j^e\}$, $j = 1, \ldots, M$. In the empirical study, we use $M = 22$, a number of contracts which roughly mimics the situation in the real market.

The maturities of the forwards and the delivery periods are given in Table 7.2. The columns τ_j^b and τ_j^e are the beginning and end of the delivery periods for non-overlapping electricity futures contracts, respectively. The first contract starts the delivery period in seven days. It is a weekly contract, that is, having a seven days delivery period. The first seven contracts are all weekly contracts, the next 10 contracts are monthly contracts (30 days of delivery), then four quarterly contracts (90 days of delivery), and finally a yearly contract (360 days of delivery). This structure of delivery periods mimics roughly the actual traded contracts in this market, with delivery periods increasing with time to delivery. The column u_j provides the maturities of the forwards. The maturities are chosen as the midpoints in the delivery periods of the electricity futures. Hence, the term structure spans a little less than three years (1076 days) in the data set of electricity

Table 7.2 Maturities of forwards and delivery periods for electricity futures (in days)

m	u_j	τ_j^b	τ_j^e
1	10.5	7	14
2	17.5	14	21
3	24.5	21	28
4	31.5	28	35
5	38.5	35	42
6	45.5	42	49
7	52.5	49	56
8	72	56	86
9	101	86	116
10	131	116	146
11	161	146	176
12	191	176	206
13	221	206	236
14	251	236	266
15	281	266	296
16	311	296	326
17	341	326	356
18	401	356	446
19	491	446	536
20	581	536	626
21	671	626	716
22	896	716	1076

futures contracts, and a little less than 2.5 years (896 days) of forwards. Since we consider two different contract types and two choices of seasonal functions, we have four different data sets to investigate.

We now turn our attention to applying the smooth data set for modelling of the forward dynamics. The authors [Lucia and Schwartz (2002)] investigated one- and two-factor models for the spot price of electricity, and fitted them to Nord Pool data. They studied both arithmetic and geometric models, and concluded that the arithmetic class of models had the best fit to the price observations. We will therefore follow a forward based version of the arithmetic model. The main reason why we consider this class, is the simple relationship between the dynamics of the forward and swap, which becomes useful in empirical analysis of the volatility term structure.

Following the HJM modelling paradigm discussed in Chapter 6, we assume that there exists a continuum of forwards $f(t, u)$, with $t \leq u \leq T$, and dynamics under the risk-neutral probability given by

$$df(t,u) = \sigma(t,u)\,dW(t)\,, \tag{7.12}$$

where $W(t)$ is a standard Brownian motion, and $\sigma(t,u)$ is a continuous and bounded function of time t and maturity u. This is an arithmetic model for the forward curve evolution yielding normally distributed forward prices.

Alternatively, we can model the swap contracts directly. Let the average-based forward contract for $0 \leq t \leq \tau_1 < \tau_2$ be given by

$$dF(t,\tau_1,\tau_2) = \Sigma(t,\tau_1,\tau_2)\,dW(t)\,. \tag{7.13}$$

Using the relation (7.1), a straightforward application of the stochastic Fubini Theorem yields that

$$\Sigma(t,\tau_1,\tau_2) = \int_{\tau_1}^{\tau_2} w(u,\tau_1,\tau_2)\sigma(t,u)\,du\,, \tag{7.14}$$

which links the volatility term structures of the two models.

The forward and swap price models, (7.12) and (7.13), respectively, describe the stochastic evolution under an equivalent martingale measure, and not under the real world measure where data are observed. Although there may be a risk premium in the market that cause forward prices to exhibit non-zero drift terms, the diffusion terms are equal under both measures. Hence, the volatility functions Σ and σ can be estimated from real world data. As noted by [Cortazar and Schwartz (1994)], this is only strictly correct when observations are sampled continuously. In our analysis we use daily observations as a proxy to a continuously sampled data set. Let $f(t_i,u_j)$ denote the forward price at date t_i with maturity u_j, where $t_i \leq u_j$. Our discrete approximations of models (7.12) and (7.13) are

$$df(t_i,u_j) \approx f(t_i,u_j) - f(t_{i-1},u_j) = x^f_{i,j}\,,$$

and

$$dF(t_i,\tau^s_j,\tau^e_j) \approx F(t_i,\tau^b_j,\tau^e_j) - F(t_{i-1},\tau^b_j,\tau^e_j) = x^F_{i,j}\,,$$

respectively, where $i = 1,...,N$ and $j = 1,...,M$. We prepare two different data sets

$$\mathbf{X}^k_{(N\times M)} = \begin{bmatrix} x^k_{1,1} & x^k_{1,2} & \cdots & x^k_{1,M} \\ x^k_{2,1} & x^k_{2,2} & \cdots & x^k_{2,M} \\ \vdots & \vdots & \ddots & \vdots \\ x^k_{N,1} & x^k_{N,2} & \cdots & x^k_{N,M} \end{bmatrix}$$

where $k = f, F$.

Let us investigate the term structure of volatility. We compute for each column in our data set

$$\widehat{\Sigma}_j = \sqrt{\frac{1}{N-1}\sum_{i=1}^{N}\left(x_{i,j}^F - \overline{x}_j^F\right)^2},$$

for the electricity futures price data, and

$$\widehat{\sigma}_j = \sqrt{\frac{1}{N-1}\sum_{i=1}^{N}\left(x_{i,j}^f - \overline{x}_j^f\right)^2},$$

for the forward price data, where $\overline{x}_j$ denotes the average daily price differences.

In Fig. 7.3 we plot the estimated volatility curves based on the four different data sets. In the upper panel the daily volatility based on fixed delivery seasonal and non-seasonal prices is presented. Perhaps unsurprisingly, the volatility estimates are very similar in the short end of the term structure. Only in the last year the estimates diverge. The data set based on the zero seasonal function produces higher volatility compared to the data set based on the trigonometric seasonal function. Also, note that both volatility curves rise during the last year (albeit this is more pronounced for the data set extracted with zero seasonality). This is hard to justify theoretically. If volatility increases in time to maturity, the underlying stochastic process exhibits some sort of explosive behaviour in the sense of becoming more and more volatile. The increase in volatility in the long end for the zero seasonality data set is more likely the result of misspecification. Ignoring seasonality will result in an unrealistic forward curve in the long end (cf. Fig. 7.2). Over time such an unrealistic curve will necessarily fluctuate quite a lot to fit the seasonal market prices. Therefore, the volatility in the long end is prone to an upward bias for misspecified volatility.

In the mid panel the volatility estimates for the seasonal and non-seasonal data sets of electricity futures contracts are plotted. We illustrate the estimates by using bar plots, where the width of each bar corresponds to the delivery period of that particular contract. Only for the last delivery period the two estimates differ. The data set prepared without seasonality shows an increasing tendency, but this is not the case for the data with

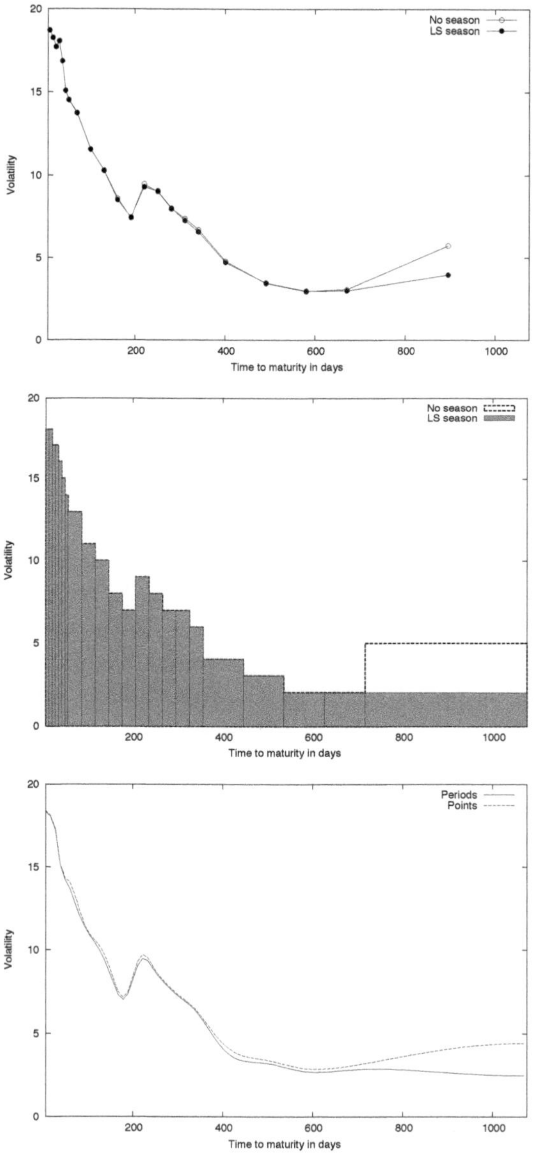

Fig. 7.3 The upper panel shows the volatility estimates for the forward price differences. The mid panel shows the volatility estimates for the electricity futures contracts. The bottom panel presents a smoothed forward volatility curve implied from electricity futures contracts compared to a smoothed forward curve computed from forward price differences, both using the trigonometric seasonality function.

a trigonometric seasonality function. It seems that working with electricity futures contracts directly reduces the problem with seasonality induced volatility bias. This is quite natural, since our data set mimics the actual delivery periods for market traded contracts. In the long end we only consider the volatility of a yearly electricity futures contract, not the volatility at a given point on the forward curve in the distant future. The former is naturally less sensitive to the seasonal specification.

We can make a more direct comparison of the $\widehat{\Sigma}$ or $\widehat{\sigma}$ using the relationship in (7.14). From our estimate of $\widehat{\Sigma}$ we can compute the implied volatility for the forward curve during the delivery period. We use $w(t, \tau_1, \tau_2) = 1/(\tau_2 - \tau_1)$, so that (7.14) becomes $\Sigma(t, \tau_1, \tau_2) = \int_{\tau_1}^{\tau_2} \sigma(t, u)\, du/(\tau_2 - \tau_1)$. In this model the volatility of the electricity futures is simply the average volatility of the forwards during the delivery period. Assuming that the implied forward volatility is smooth, we can use our proposed algorithm (with zero seasonality) to compute a smooth volatility curve. We simply replace the electricity futures prices with estimated electricity futures volatilities in the derivations of the previous section. The result is a maximum smooth implied volatility function. In the bottom plot of Fig. 7.3 a volatility curve implied by $\widehat{\Sigma}$ and a smoothed version of $\widehat{\sigma}$ is plotted. Both curves are based on data sets with trigonometric seasonality. Between one and two years to maturity, the forward approach leads to $5 - 10\%$ higher volatility for $\widehat{\sigma}$ compared to the estimate implied by $\widehat{\Sigma}$. Between two and three years to maturity this number increases steadily to more than 50% of the volatility estimate implied by the electricity futures prices.

Summing up, the specification of the seasonal function, $\Lambda(u)$, affects the appearance of the forward curve when the market information is restricted to swap contracts with long delivery periods. In the Nordic electricity market, this is the case when time to delivery is more than two years. If seasonality is ignored, volatility estimates seem to be biased upwards in the very long end. The volatility estimate is affected by how the historical data set is extracted. It seems that modelling swap prices is more promising than the forward curve approach, since volatility estimates seem to be less dependent on the specification of seasonality.

We focused on a simple arithmetic model, since this model allows comparison between the forward curve and swap price modelling. In a geometric model with maturity dependent volatility (for instance, in a geometric mean-reverting process), this relationship becomes complex and intractable. Direct modelling of the traded swaps is recommended. In the next chapter we apply our algorithm to a more extensive study of the term structure of volatility, based on a multi-factor market model of geometric type.

Chapter 8

Modelling of the Electricity Futures Market

In this chapter we conduct an empirical study of financial electricity contracts traded on Nord Pool. Empirical investigations of forward curve models in commodity markets have been done by, among others, [Cortazar and Schwartz (1994)] and [Clewlow and Strickland (2000)]. [Cortazar and Schwartz (1994)] studied the term structure of copper futures prices using principal component analysis (PCA) and found that three factors were able to explain 99% of the term structure movements. [Clewlow and Strickland (2000)] investigated the term structure of NYMEX oil futures and found that three factors explained 98.4% of the total price variation in the 1998-2000 period. The first factor (explaining 91% of total variation) shifted the whole curve in one direction. They termed this a *shifting* factor. The second factor, called the *tilting* factor, moved the short and long term contracts in opposite directions. The third factor, coined the *bending* factor, influenced the short and long end in opposite direction of the midrange of the term structure.[1]

In the paper by [Koekebakker and Ollmar (2005)] Nord Pool data was analysed using PCA techniques. The authors computed smooth forward curves using the technique described in Chapter 7. Their data set consisted of fixed delivery forward contracts (points on the forward curve) that mimicked the term structure of actual traded electricity futures. This means

[1]The multi-factor forward approach by [Heath, Jarrow and Morton (1992)] was originally developed for interest rate markets. Empirical work on factor dynamics in fixed income securities markets have been conducted by [Steeley (1990)], [Litterman and Sceinkman (1991)] and [Dybvig (1997)]. The results in these studies are quite similar to the work reported from the commodity markets. Typically, three factors explain 95%-98% of the total variation in the forward curve. In this chapter we use the names shift, tilt and bend for these factors, following [Clewlow and Strickland (2000)]. This is not entirely consistent with other literature, where the names of the same factors are shift, slope and curvature (see, for instance, [Lord and Pelsser (2007)]).

picking more points in the short end than in the long end of the curve in a manner similar to the procedure in Chapter 7, Subsect. 7.3.2. The results from the PCA analysis using data for the period 1995-2001 showed that the first three factors accounted for 80% of the price variation. Only the first two factors (shifting and bending) seem to be common across all maturities. It was also less straightforward to interpret the factor loadings in terms of a bending and tilting factor. In order to explain more than 98% of the variation in the empirical covariance matrix, more than 10 factors were needed. Also, the authors reported evidence that factors explaining a large proportion of the return variations in the long end of the curve, seemed to have very low explanatory power in the short end of the curve. The authors concluded that there might be a more complex factor structure in this market than others, and that some parts of the curve are exposed to some unique risk factors that other parts of the curve are not exposed to.

In line with the findings in [Koekebakker and Ollmar (2005)], [Audet *et al.* (2004)] suggested a simple model where each contract is driven by a Brownian motion, and the return on this contract is correlated with other contracts along the term structure by a negative exponential function. Their specification implies a complete market, and the suggested correlation function implies that contracts with delivery periods far apart are less correlated than those close together. They estimated their model in the short end only (four closest weekly contracts). [Frestad (2007b)] investigated empirically the model of [Audet *et al.* (2004)] using contracts across the complete term structure. The author found that the negative exponential function is too simple to explain the correlation structure across a broader set of contracts in this market. [Frestad (2007a)] further investigated the idea of common and unique risk factors at Nord Pool. In the proposed incomplete market model, electricity futures price returns are driven by some Brownian motions common to all traded contracts, together with a unique Brownian motion to each maturity. The author also tested for the number of common factors producing the return correlation matrix, and found that two or three factors are common to all electricity futures contracts. By dividing the data set into sub-periods, three factors accounted for 62.3% (1997–1999), 75.8% (2000–2002) or 69.1% (2003–2005) of the risk. The first three factors can be identified as shift, bend and tilt, although the tilting factor is less identifiable than the first two.

In this chapter, we perform PCA on daily Nord Pool electricity futures price data for the period (2001–2006). We prepare data sets of electricity futures prices following the lines described in Chapter 7. Based on these data, we re-establish the results of [Koekebakker and Ollmar (2005)] and [Frestad (2007a)] for the total market. We then take a less ambitious approach and analyse individual market segments. That is, weekly, monthly, quarterly and yearly electricity futures prices are given individual treatment. The results then become more transparent. The first three factors are quite clearly identified as shifting, tilting and bending factors within each segment. We also investigate the possibility of a parametric market model for each individual market segment. In a final section, we analyse empirically the distributional properties of the return data in view of heavy-tailed distributions.

8.1 The Nord Pool market and financial contracts

In this section we describe relevant issues for the market structure and the data available at Nord Pool. The content partly overlaps with the thorough discussion of the Nord Pool market in Chapter 1.

We obtained daily closing prices for all electricity futures contracts traded at Nord Pool from 2 January 2001 until 1 December 2006.[2] Information regarding settlement periods was also obtained from Nord Pool's database. The power contracts refer to 1 MW load during every hour (base load) for a given delivery period. The contracts are settled against the realised system prices in the delivery period. The trading period stops when the contracts enter the delivery period. The size and trading period vary considerably for the contracts available. We will give a brief description of weekly, monthly, seasonal and yearly contracts below.

The weekly contracts are specified with a delivery period of seven days (168 hours). The delivery period starts Sunday at midnight and ends midnight the following Sunday. The contracts with delivery the following week are lasting until the preceding Friday. Earlier, four new weekly contracts were introduced every fourth Monday, meaning that a maximum of seven and a minimum of four weekly contracts were traded at any given time. This has changed. Now a new weekly contract is introduced in the long end, as the contract in the short end enters the delivery period.

Block contracts had four-week delivery periods, but they do not exist anymore. These contracts were not traded in the month prior to delivery. They were broken up into separate weekly contracts. Since these contracts had delivery periods of 28 days (four weeks), each year was divided into 13

[2]Due to low liquidity, daily contracts are excluded from the data set.

block contracts, 10 of which traded simultaneously. But 13 blocks do not exactly add up to one year ($4 \times 28 \times 13 = 364$), and therefore the December block contract had one day longer delivery period than the others (two days extra in a leap year). Since 2003, no new block contracts have been introduced in the market. They have been replaced by monthly contracts with delivery periods consistent with the days in the particular months. The new contract structure on Nord Pool makes the financial contracts more similar to contracts traded at other power exchanges.

The seasonal contracts have also changed. Earlier, each year was divided into three seasons: V1 - late winter (1 January – 30 April), S0 - summer (1 May – 30 September) and V2 - early winter (1 October – 31 December). The setup of three seasons has been replaced with the more common four season system, with three months for each season. The first quarterly contracts were listed 2 January 2004 for each quarter of the year 2006. Now quarterly contracts have replaced all old seasonal contracts. There are between 8 and 11 quarterly contracts traded at any time. A new contract is introduced in the long end as the closest one enters the delivery period. Currently, quarterly contracts span more than two years.

Finally, the market trades in yearly contracts. As of 2007,[3] yearly contracts for the following five years are available for trading (2008, ..., 2012). These contracts have delivery periods of $24 \times 365 = 8760$ hours (8784 hours in a leap year). Each new contract is traded for five years, until it expires in late December prior to the start of the delivery period 1 January. In the beginning of January each year a new yearly contract is introduced with delivery period starting in five years. Yearly contracts were introduced in 1998. Then only three yearly contracts were traded. In our data set we therefore only use three yearly contracts, as contract starting delivery in four and five years have a very limited history.

Prior to 2003 all contracts traded at Nord Pool were denominated in NOK. It was decided to change denomination to EUR. In 2003 all new long term contracts were listed in EUR. This transition is now complete, and all contracts are currently denominated in EUR.

8.2 Preparing data sets

In [Benth and Koekebakker (2005)], closing prices of actual electricity futures prices were used when estimating their one-factor model. This ap-

[3] The time of writing this book.

proach becomes impractical when more advanced multi-factor models are considered. In this chapter, we follow the approach that is standard in the literature on estimating dynamic term structure models, where yield curves are estimated from real world fixed income assets (typically treasury bonds). The major advantage in working with estimated data, is that on each day the term structure can be specified to have a fixed delivery structure.

We wish to preserve the market's delivery structure of each day throughout the sample period for our data set. This is accomplished in the following way.

(1) All electricity futures prices at Nord Pool in the period 2 January 2001 until 1 December 2006 (1479 trading days) are collected (except contracts with 24 hour delivery period). Zero coupon bond prices in NOK and EUR are collected from Reuters.
(2) During the sample period Nord Pool has made a complete transition from contracts denominated in NOK to EUR. All electricity futures prices are converted to EUR using the formula in Prop. 4.5.
(3) We estimate a smooth curve for each day in the sample period using the maximum smoothness approach described in Chapter 7.
(4) Finally, we use the smooth forward curve to compute electricity futures prices with delivery structure specified in (8.1).

We prepare five different data sets. The first four are individual data sets for weekly, monthly, quarterly and yearly contracts, respectively. We introduce a sequence of delivery periods for the different contracts

$$\left\{\left[\tau_1^b, \tau_1^e\right], ..., \left[\tau_C^b, \tau_C^e\right]\right\} , \tag{8.1}$$

where the following conditions apply:

(1) $\tau_c^b - \tau_c^e = \Delta$, for $c = 1, \ldots, C$, with Δ being constant.
(2) $\tau_{c+1}^b = \tau_c^e$, for $c = 1, \ldots, C$.

The first condition implies that the data consist of contracts with delivery period of equal length. This modelling approach is well known from so-called market models in interest rate markets. The ingredient in such models is the term structure of swap or LIBOR rates with a particular tenor. Rates for different tenors are typically not modelled simultaneously. Applying this idea to the Nord Pool market, we model weekly, monthly, quarterly and yearly contracts separately. The second condition implies

Table 8.1 Delivery period structure for market models with time measured in years

	τ_c^b	τ_c^e	C
W_c - Week	$(c-1)/52$	$c/52$	6
M_c - Month	$(c-1)/12$	$c/12$	6
Q_c - Quarter	$(c-1)/4$	$c/4$	8
Y_c - Year	$c-1$	c	3

that the market is such that the end of the delivery period for one contract coincides with the start of the delivery period for the contract which is next in line along the term structure. This condition ensures that the contracts are non-overlapping (recall the discussion in Sect. 6.4) and mimicking the real world contracts traded at Nord Pool.

For each data set we have $\tau_c^b = (c-1)/\Delta$ and $\tau_c^e = c/\Delta$ for $c = 1, ..., C$. The first contract starts delivery immediately. The next contract along the term structure trades for a period Δ until it enters the delivery period. The third contract along the term structure trades for 2Δ, etc. The delivery structure for our four data sets is summarised in Table 8.1. The number of contracts included in each data set is roughly the same as the actual contracts traded on Nord Pool during the sample period.

The fifth data set represents the total market at Nord Pool. This is derived by extracting data from the four previous sets in the following way. We combine different delivery periods by including the first four weekly contracts $(W_1, ..., W_4)$, the five last monthly contracts $(M_2, ..., M_6)$, the last six quarterly contracts $(Q_3, ..., Q_8)$ and the last yearly contract (Y_3). This setup of the term structure has non-overlapping contracts, but delivery periods are of varying sizes.[4] This is an appropriate model if, for instance, it is used for calculating Value at Risk for a portfolio of contracts with different delivery periods.

8.3 Descriptive statistics

The upper panel of Fig. 8.1 shows prices of weekly, monthly, quarterly and yearly contracts with immediate delivery. The period from October 2002 to February 2003 catches the eye. This period is sometimes referred to as the

[4]Note that the condition of non-overlapping contracts is the only crucial one from a theoretical point of view (see discussion in Sect. 6.3).

Scandinavian power crisis. The hydrological balance was poor as of late summer of 2002, and it got worse during the autumn. This caused very high prices. Then in early 2003, the reservoir levels improved, and prices dropped again. It seems that the market did not expect such a quick drop in prices from the peak, as both the price for month and quarter followed the week price. We see that the weekly electricity futures demonstrate bigger variation than the monthly, quarterly and yearly contracts.

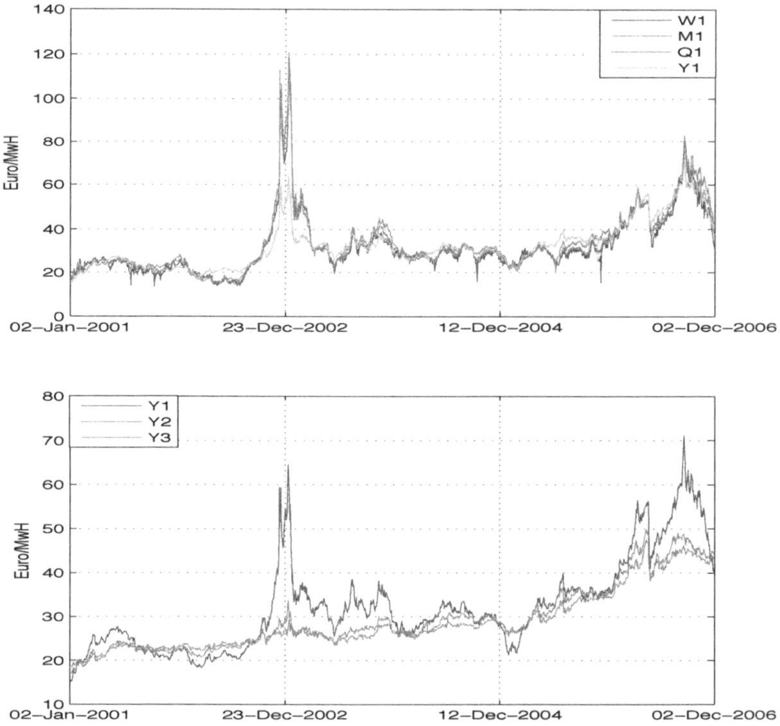

Fig. 8.1 Time series of electricity futures price data. Contracts with immediate delivery and varying delivery period - week (W_1), month (M_1), quarter (Q_1) and year (Y_1) on the upper panel. Contracts with varying time to delivery and yearly delivery period - year (Y_1), year (Y_2) and (Y_3) on the bottom panel.

In the bottom panel of Fig. 8.1 we plot time series of yearly contracts only (Y_1 with immediate delivery, Y_2 with delivery in one year and Y_3 with delivery in two years). It is clear from the plot that Y_1 exhibits much more variability than the other two contracts.

Table 8.2 Descriptive statistics of price levels from 2 January 2001 untill 1 December 2006, a total of 1,479 trading days

	Mean	Variance	Skewness	Kurtosis
Week				
W_1	31.71	179.62	2.02	8.94
W_2	32.83	205.07	2.25	10.61
W_3	33.24	212.72	2.11	9.45
W_4	33.50	214.47	2.00	8.72
W_5	33.62	213.94	1.93	8.29
W_6	33.66	209.98	1.82	7.60
Month				
M_1	32.95	202.27	2.07	9.17
M_2	33.71	205.14	1.70	6.89
M_3	33.92	195.36	1.45	5.50
M_4	33.82	188.37	1.38	5.09
M_5	33.49	172.88	1.32	5.02
M_6	32.89	150.64	1.32	5.25
Quarter				
Q_1	33.52	197.18	1.72	6.98
Q_2	33.40	165.26	1.30	4.92
Q_3	31.69	106.68	1.10	3.96
Q_4	30.36	69.95	0.96	3.52
Q_5	29.79	69.65	0.83	2.86
Q_6	29.52	79.26	0.90	3.31
Q_7	29.24	68.01	1.07	3.54
Q_8	29.11	51.66	1.04	3.35
Year				
Y_1	32.23	110.04	1.10	3.83
Y_2	29.41	58.88	1.01	3.09
Y_3	28.75	46.22	1.14	3.29

In Table 8.2 we report descriptive statistics for electricity futures price levels. We find that distributions of price levels are right skewed (positive skewness) and have excess kurtosis. Both kurtosis and skewness decrease as time to delivery and length of delivery period increase. The mean of the prices indicates a positive risk premium in the short end. A positive risk premium means that an electricity futures contract will be an upward biased estimate of the realised spot price, as the contract enters the delivery period, the electricity futures and the realised spot price converge. Hence, a positive risk premium implies an electricity futures price downward trending towards delivery. From Table 8.2 we see that the mean price of weekly electricity futures are lower for contracts close to delivery, indicating a positive risk premium in the short end of the term structure. The opposite is true for quarterly and yearly contracts, indicating a negative risk premium in the long end of the term structure. The explanation to this changing risk premium might be that there are different hedging needs along the term structure. Electricity producers are naturally short in the market, since they want to sell (some of the) expected future production, and in this way reducing the variability of future cash flow. The natural hedging horizon is the next 2–3 years, and monthly contracts are seldom used for this type of hedging. The seller side is thus bigger than the buyer side from a hedging perspective. Industry players with high power consumption are naturally on the purchase side in the financial market. But many of the biggest players have entered long term OTC deals (10 year contracts) to get predictable power costs over a longer period than Nord Pool offers. More demand than supply for electricity futures in the long end implies depressed prices, and a negative premium.

In the short end of the term structure the story might be somewhat different. Households are almost exclusively buying power in the spot market. Utilities sell spot power to end users. A typical spot contract can be changed by the utility every fortnight. In a sense, the household contracts are non-overlapping 14-day electricity futures contracts. Also, prices are independent of quantity. Hence, utilities perform price and demand prediction on a short term basis. They buy contracts to hedge short-term risk in their end user portfolio. If this explanation holds any merit, the buy side is bigger than the sell side for short-term contracts, with a negative risk premium as a natural implication.

In Table 8.3 we report descriptive statistics for electricity futures price returns. Skewness is positive only in the short end and negative for all other contracts. It is relatively low overall, suggesting a fairly symmetric electricity futures price return distribution. Kurtosis is in general high, but clearly declining as delivery period and time to maturity increase. Volatility is also decreasing in time to delivery and the length of the delivery period. For instance, the contract Y_3 with two years until delivery, has volatility close to half of Y_1. The delivery period for Y_1 starts immediately. The volatility of Y_3 is very close to the volatility of Y_2, an indication that volatility converges to a long run level of about 20%. In the very short end, the weekly contracts show volatility from about 55% to about 60%. The exception is W_1, the one week electricity futures with immediate delivery. It has an estimated volatility of 90%. There are reasons to believe that this estimate is not completely reliable. Recall that the electricity futures prices are extracted using an estimated forward curve. The forward curve is linked to

Table 8.3 Descriptive statistics of electricity futures price returns. Volatility is annualized using 250 trading days a year.

	Mean	Volatility	Skewness	Kurtosis
Week				
W_1	0.00	90.2 %	0.21	25.01
W_2	0.00	59.4 %	0.60	19.76
W_3	0.00	56.2 %	0.01	25.39
W_4	0.00	55.8 %	−0.28	26.95
W_5	0.00	56.5 %	−0.12	29.48
W_6	0.00	54.7 %	−0.35	26.02
Month				
M_1	0.00	52.6 %	0.40	22.02
M_2	0.00	52.1 %	−0.92	22.54
M_3	0.00	50.3 %	−1.48	19.47
M_4	0.00	48.3 %	−1.34	18.42
M_5	0.00	48.1 %	−1.00	14.54
M_6	0.00	47.5 %	−0.86	13.18
Quarter				
Q_1	0.00	48.9 %	−0.76	22.32
Q_2	0.00	45.1 %	−1.06	15.12
Q_3	0.00	49.7 %	−0.41	39.43
Q_4	0.00	36.4 %	−0.22	20.22
Q_5	0.00	30.6 %	−0.58	12.93
Q_6	0.00	29.3 %	−0.84	14.21
Q_7	0.00	29.4 %	−1.19	17.96
Q_8	0.00	25.0 %	−0.32	9.24
Year				
Y_1	0.00	38.2 %	−0.52	14.32
Y_2	0.00	22.3 %	−0.94	12.26
Y_3	0.00	21.3 %	−0.38	8.28

the spot price in the very short end (the daily average system price). The daily average system price can be considered as an electricity futures contract with immediate delivery and 24 hours delivery period. The traded weekly contract with delivery the following week has (on average) four days until delivery. All daily contracts are excluded from the sample due to low liquidity. Because of the way we have prepared our data, this means that the weekly contract with immediate delivery is roughly estimated as the average of the daily system price and the closest weekly contract. Therefore, the volatility of W_1 is closer to spot price volatility than to electricity futures price volatility. If we had used daily contracts in our curve generation (and ignored the fact that these contracts have low liquidity), this

Table 8.4 Volatility decomposed by season. Volatility is annualized assuming 250 trading days a year. "Const. vol" is the average volatility when all observations in the sample period are used. The rest of the table shows volatility calculated for individual months.

	Const. vol	Jan	Feb	Mar	Apr	May	Jun	Jul	Aug	Sep	Oct	Nov	Dec
Week													
W_1	90 %	102 %	75 %	62 %	60 %	84 %	79 %	103 %	74 %	98 %	86 %	135 %	97 %
W_2	59 %	105 %	70 %	40 %	38 %	44 %	45 %	40 %	50 %	35 %	52 %	42 %	106 %
W_3	56 %	100 %	46 %	44 %	39 %	40 %	39 %	42 %	53 %	36 %	46 %	42 %	105 %
W_4	56 %	93 %	46 %	44 %	44 %	38 %	40 %	42 %	52 %	36 %	44 %	46 %	106 %
W_5	56 %	91 %	46 %	45 %	46 %	42 %	41 %	39 %	51 %	36 %	44 %	49 %	111 %
W_6	55 %	90 %	44 %	44 %	47 %	41 %	40 %	36 %	51 %	36 %	44 %	49 %	101 %
Month													
M_1	53 %	90 %	47 %	39 %	38 %	37 %	42 %	40 %	48 %	35 %	43 %	46 %	94 %
M_2	52 %	90 %	44 %	44 %	46 %	38 %	37 %	35 %	51 %	36 %	42 %	45 %	88 %
M_3	50 %	94 %	47 %	42 %	42 %	36 %	33 %	36 %	50 %	37 %	40 %	46 %	69 %
M_4	48 %	84 %	43 %	41 %	45 %	35 %	29 %	36 %	47 %	35 %	39 %	49 %	70 %
M_5	48 %	82 %	39 %	44 %	45 %	34 %	32 %	35 %	43 %	35 %	39 %	48 %	75 %
M_6	47 %	80 %	38 %	37 %	42 %	34 %	28 %	32 %	42 %	34 %	44 %	47 %	81 %
Quarter													
Q_1	49 %	89 %	47 %	39 %	40 %	32 %	33 %	33 %	48 %	33 %	40 %	40 %	81 %
Q_2	45 %	78 %	37 %	37 %	41 %	32 %	28 %	32 %	42 %	32 %	41 %	47 %	69 %
Q_3	50 %	69 %	41 %	33 %	39 %	30 %	24 %	39 %	40 %	35 %	38 %	87 %	81 %
Q_4	36 %	43 %	28 %	35 %	41 %	27 %	25 %	36 %	26 %	26 %	29 %	35 %	72 %
Q_5	31 %	48 %	24 %	27 %	38 %	29 %	24 %	28 %	26 %	19 %	30 %	24 %	40 %
Q_6	29 %	48 %	29 %	30 %	32 %	31 %	22 %	23 %	19 %	30 %	21 %	18 %	33 %
Q_7	29 %	56 %	32 %	25 %	24 %	24 %	20 %	26 %	21 %	23 %	24 %	23 %	36 %
Q_8	25 %	35 %	20 %	20 %	27 %	22 %	20 %	26 %	25 %	31 %	22 %	21 %	24 %
Year													
Y_1	38 %	65 %	30 %	32 %	36 %	27 %	24 %	29 %	37 %	26 %	33 %	34 %	65 %
Y_2	22 %	37 %	22 %	22 %	26 %	22 %	16 %	20 %	18 %	17 %	19 %	15 %	26 %
Y_3	21 %	33 %	22 %	20 %	30 %	23 %	20 %	21 %	15 %	16 %	15 %	15 %	17 %

volatility estimate would probably be somewhat lower.[5] In the following sections we investigate empirically the term structure dynamics both with and without W_1 in the sample. In this way the effect of W_1 on the rest of the term structure can be assessed.

The rest of the descriptive statistics is given in Table 8.4. The first column repeats the volatility estimates from Table 8.3 for each contract. The 12 columns to the right present volatility estimates by month. Volatility seems to change over the course of the year. And for most of the contracts, December and January seem to be by far the most volatile months, indicating a non-symmetric seasonal effect. Also note that W_1 differs somewhat from this pattern, with peak volatility in November and July. We take this as further evidence that W_1 is potentially influenced by other factors than the rest of the weekly contracts, due to its close resemblance with the spot price.

8.4 A market model for electricity futures

We consider a simple market model as discussed in Sect. 6.4, and recall it together with some notations. Assume that market participants trade C different electricity futures contracts with non-overlapping delivery periods. The price at time t for an electricity futures with delivery period $\left[\tau_c^b, \tau_c^e\right]$ is denoted by $F_c(t) = F_c(t, \tau_c^b, \tau_c^e)$, $c = 1, \dots, C$. Assume that under the real world measure the price dynamics of $F_c(t)$ is lognormal, that is,

$$F_c(t) = F_c(0) \exp\left(\int_0^t A_c(u)du + \sum_{k=1}^{p} \int_0^t \Sigma_{c,k}(u)\, dB_k(u)\right), \qquad (8.2)$$

for $t < \tau_c^b$, with B_k, $k = 1, ..., p$ being independent Brownian motions and $\Sigma_{c,k}$ and A_c continuous functions on $[0, \tau_c^e]$. We assume that $p \le C$, that is, the number of traded contracts is at least as many as we have Brownian motions driving the swap price dynamics. This implies in particular that the market is complete. We further note that the proposed model does not deal with any idiosyncratic risk. If $p < C$, (8.2) may allow for arbitrage opportunities. Including more Brownian motions will model the idiosyncratic risk, and also remove the possibility of arbitrage in our model. Our focus in the coming empirical study of the model (8.2) is on common risk, maturity

[5] [Benth and Koekebakker (2005)] include low liquidity daily contracts to their sample. Their argument is that including the daily contracts yields more correct short-term volatility estimates.

and seasonality effects in the volatility, thus the assumption $p \leq C$.

The logreturn over the period $[t_{n-1}, t_n]$ of the contract F_c is defined as

$$x_{n,c} \triangleq \ln\left(\frac{F_c(t_n)}{F_c(t_{n-1})}\right).$$

Using (8.2), we have

$$x_{n,c} = \int_{t_{n-1}}^{t_n} A_c(u)\,du + \sum_{k=1}^{p} \int_{t_{n-1}}^{t_n} \Sigma_{c,k}(u)\,dB_k(u)\,. \tag{8.3}$$

Hence, the logreturns are normally distributed under the real world probability. By the Girsanov transform,[6] the drift will be altered under an equivalent martingale measure, while the volatility remains unchanged.

With $N+1$ trading days in our sample, the $N \times C$ data matrix $\mathbf{X}_{N\times C}$ is specified as

$$\mathbf{X}_{N\times C} = \begin{bmatrix} x_{1,1} & x_{1,2} & \cdots & x_{1,C} \\ x_{2,1} & x_{2,2} & \cdots & x_{2,C} \\ \vdots & \vdots & \ddots & \vdots \\ x_{N,1} & x_{N,2} & \cdots & x_{N,C} \end{bmatrix}.$$

In the next subsections we will analyse the factor dynamics, first by PCA, and next by estimating a multi-factor parametric model for electricity futures price returns.

8.5 Principal component analysis

In this section we will investigate the changes in the term structure of electricity futures price returns. PCA is utilised for the identification of structure within a set of interrelated variables. It establishes dimensions within the data, and serves as a data reduction technique. The aim is to determine factors (that is, principal components) in order to explain as much of the total variation in the data as possible.

We have a total of N observations of C return series, and collect time series of each contract in N-dimensional vectors $\mathbf{x}_1$, $\mathbf{x}_2, ..., \mathbf{x}_C$. The data matrix $\mathbf{X}_{N\times C}$ is then

$$\mathbf{X}_{N\times C} = \left[\mathbf{x}_1\ \mathbf{x}_2\ \cdots\ \mathbf{x}_C\right]. \tag{8.4}$$

[6]We assume here that the functions A_c and $\Sigma_{c,k}$, $k = 1, \ldots, p$ are sufficiently regular to admit a valid measure change.

The corresponding sample covariance matrix of dimension $C \times C$ is denoted $\boldsymbol{\Omega}$. The orthogonal decomposition of the covariance matrix is

$$\boldsymbol{\Omega} = \mathbf{P} \boldsymbol{\Lambda} \mathbf{P}', \tag{8.5}$$

where

$$\mathbf{P} = \left[\mathbf{p}_1 \ \mathbf{p}_2 \cdots \mathbf{p}_C \right] = \begin{bmatrix} p_{11} & p_{12} & \cdots & p_{1C} \\ p_{21} & p_{22} & \cdots & p_{2C} \\ \vdots & \vdots & \ddots & \vdots \\ p_{C1} & p_{C2} & \cdots & p_{CC} \end{bmatrix},$$

and $\boldsymbol{\Lambda}$ is a diagonal matrix with the eigenvalues $\lambda_1, \lambda_2, ..., \lambda_C$ on the diagonal. The matrix $\mathbf{P}$ is orthogonal, with ith column, $\mathbf{p}_i$, being the eigenvector corresponding to λ_i. $\mathbf{P}'$ denotes the transpose of $\mathbf{P}$. The matrix $\mathbf{Z} = \mathbf{XP}$ is called the matrix of principal components, while $\mathbf{P}$ the matrix of factor loadings. The eigenvectors on the diagonal of $\boldsymbol{\Lambda}$ are by convention ordered so that $\lambda_1 \geq \lambda_2 \geq ... \geq \lambda_C$. To explain all the variation in $\mathbf{X}$, we need C principal components. Since the objective of our analysis is to explain as much as possible of the covariance structure with just a few factors, we approximate the theoretical covariance matrix in (8.5) using only the first $M < C$ eigenvalues in $\boldsymbol{\Lambda}$ while putting the remaining equal to zero. The proportion of total variance accounted for by the first M factors is

$$\text{Cumulative contribution of first } M \text{ factors} = \frac{\sum_{i=1}^{M} \lambda_i}{\sum_{i=1}^{C} \lambda_i}.$$

The M factors should explain a "big" part of the total covariance of the underlying variables. In empirical studies, one is typically choosing M so that around 95% of the variation is explained.

Now we present the results from the PCA. First, we consider the complete data set. Recall that it consists of the 16 contracts representing the total market, $\{W_1, ..., W_4, M_2, ..., M_6, Q_3, ..., Q_8, Y_3\}$. Our results are comparable to the investigations in [Koekebakker and Ollmar (2005)] and [Frestad (2007a)]. Next, we analyse dynamics within each particular market segment (week, month, quarter and year). The descriptive statistics showed evidence of seasonality. Therefore all the return series have been normalised prior to the PCA analysis. Each return series is sorted according to the observation month, and then normalised by subtracting the mean of the series and dividing by the standard deviation.

8.5.1 *Principal component analysis of the total data set*

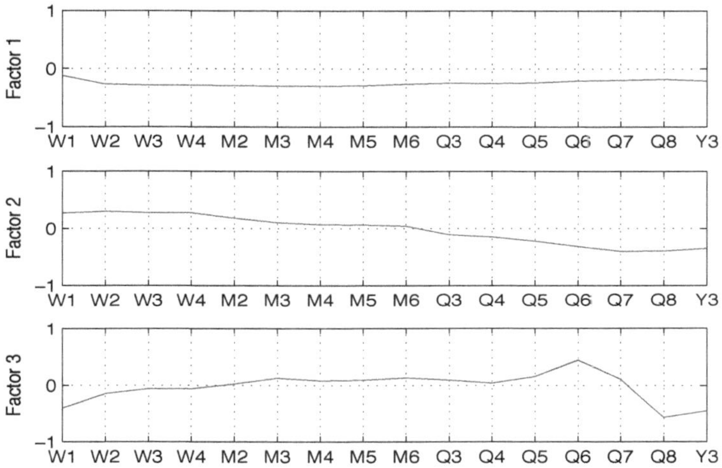

Fig. 8.2 First three factor loadings for the total data set.

In Fig. 8.2 we present factor loadings across contracts for the full data set. We see that the first factor can clearly be identified as a shifting factor. A shock to this factor shifts all contracts in the same direction. Factor two is the tilting factor. A shock to this factor moves weekly and monthly contracts in one direction, and the quarterly and yearly contracts in the opposite direction. The third factor is less clear, but it might perhaps be interpreted as a bending factor. The factor loadings change sign twice along the term structure, shifting the short and the long ends in one direction and the middle contracts in the opposite one. However, factor loadings are close to zero for most contracts, except W_1 in the short end, and Q_6, Q_8 and Y_3 in the long end. This can hardly be called a common factor. In Table 8.5 we report the variance explained by each factor individually, and the cumulative effect of adding extra factors. We note that the three first factors explain 70% of the total variance, which is in line with [Frestad (2007a)]. To reach 95% explained variance, 10 factors are needed. This is in line with findings in [Koekebakker and Ollmar (2005)]. We would like to emphasise that we do not find a 10-factor model for this market appropriate, rather the opposite. The factors starting already with the third one explain

Table 8.5 Individual and cumluative variance explained from PCA for the total data set

#	%-explained	%-cumulative
1	54 %	54 %
2	10 %	64 %
3	6 %	70 %
4	5 %	76 %
5	5 %	80 %
6	4 %	85 %
7	4 %	88 %
8	3 %	91 %
9	2 %	93 %
10	1 %	95 %
11	1 %	96 %
12	1 %	98 %
13	1 %	99 %
14	1 %	99 %
15	0 %	100 %
16	0 %	100 %

variations only in a small part of the term structure.

In Table 8.6 we report the correlation matrix for normalised electricity futures price returns. We see that contracts with delivery periods far apart have correlations lower than those close together. There are some indications for the short-term contracts that correlations seem to depend on the length of the delivery period. For instance, the correlation between W_2 and W_4 is 0.80, while the correlation between W_2 and M_2 (the first monthly contract) drops to 0.69. The Y_3 has low correlation with all contracts except the Q_8. When presenting descriptive statistics we noted that the first weekly contract with immediate delivery (W_1) is potentially influenced by other factors than the rest of the weekly contracts, due to its strong relation to the spot price. This indicates that the spot price, which is mostly influencing the dynamics of W_1, is driven by different factors than the other financial contracts. Our results may be interpreted as a warning against estimating spot price models (using, for instance, system price data), and then using the estimated model to derive the dynamics of electricity futures prices and the value of option prices.

In the next subsection we have a less ambitious agenda. Instead of modelling all electricity futures contracts in the market, we divide the market according to the length of the delivery period, and investigate the factor dynamics within each segment.

Table 8.6 Correlation matrix for total data set

	W_1	W_2	W_3	W_4	M_2	M_3	M_4	M_5	M_6	Q_3	Q_4	Q_5	Q_6	Q_7	Q_8	Y_3
W_1	1															
W_2	0.44	1														
W_3	0.31	0.87	1													
W_4	0.34	0.80	0.93	1												
M_2	0.35	0.69	0.76	0.78	1											
M_3	0.25	0.66	0.73	0.74	0.83	1										
M_4	0.32	0.61	0.68	0.68	0.76	0.86	1									
M_5	0.28	0.59	0.65	0.65	0.72	0.73	0.84	1								
M_6	0.19	0.53	0.59	0.60	0.61	0.66	0.65	0.83	1							
Q_3	0.17	0.48	0.53	0.53	0.55	0.56	0.57	0.56	0.55	1						
Q_4	0.22	0.48	0.52	0.53	0.59	0.58	0.57	0.54	0.50	0.65	1					
Q_5	0.20	0.42	0.45	0.46	0.54	0.57	0.55	0.53	0.49	0.46	0.67	1				
Q_6	0.09	0.34	0.38	0.37	0.42	0.49	0.48	0.45	0.40	0.45	0.41	0.60	1			
Q_7	0.12	0.32	0.35	0.35	0.36	0.42	0.44	0.41	0.36	0.41	0.44	0.39	0.67	1		
Q_8	0.14	0.32	0.33	0.33	0.33	0.36	0.38	0.37	0.32	0.35	0.39	0.37	0.23	0.53	1	
Y_3	0.15	0.33	0.37	0.37	0.41	0.43	0.45	0.45	0.42	0.44	0.46	0.48	0.35	0.35	0.70	1

8.5.2 *Principal component analysis for individual market segments*

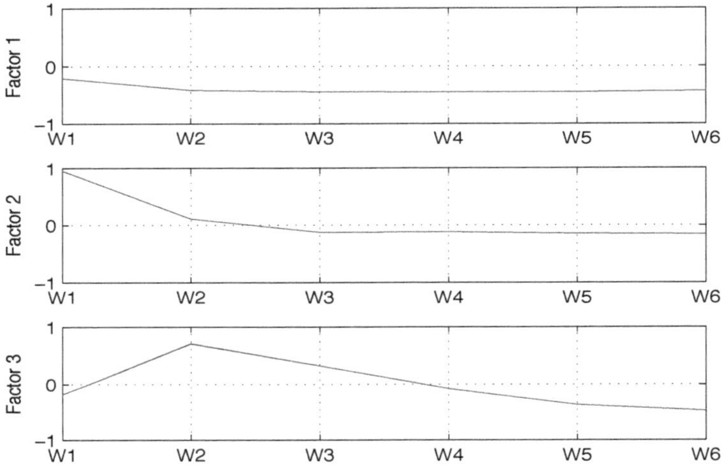

Fig. 8.3 First three factor loadings for weekly contracts.

In Fig. 8.3 we plotted factor loadings for the six weekly contracts. At first glance, it seems that the shapes of the first three factors correspond to shifting, bending and tilting. But knowing that there are low correlations between W_1 and the other weekly contracts, we go into more detail. Note that the first factor affects W_1 less than the other contracts. The second factor has an effect on W_1, and nearly zero effect on all the other contracts, while the third factor is a bending factor for all contracts except W_1, being basically not affected at all by the third factor. In the top panel of Table 8.7 we report the variance explained by the individual factors for the weekly electricity futures returns, and the cumulative variance for adding extra factors. The first factor explains 76% of the total variance, the second explains 14% and the third one adds another 6%, resulting in 96% of explained variance. Excluding W_1 from the sample, gives 88%, 7% and 3% of explained variance for the three first factors (98% in total). In the upper left panel of Fig. 8.4 we plot the first three factor loadings for weekly electricity futures returns, with W_1 excluded. The shifting, bending and tilting factors become very clear.

Table 8.7 Factor analysis and correlation matrices for weekly and monthly contracts

#	%-explained	%-cumulative	Correlation matrix – weekly						
				W_1	W_2	W_3	W_4	W_5	W_6
1	76 %	76 %	W_1	1					
2	14 %	91 %	W_2	0.44	1				
3	6 %	97 %	W_3	0.31	0.87	1			
4	2 %	99 %	W_4	0.34	0.80	0.93	1		
5	1 %	100 %	W_5	0.33	0.75	0.87	0.95	1	
6	0 %	100 %	W_6	0.31	0.72	0.83	0.88	0.95	1

#	%-explained	%-cumulative	Correlation matrix – monthly						
				M_1	M_2	M_3	M_4	M_5	M_6
1	79.1 %	79.1 %	M_1	1					
2	8.7 %	87.8 %	M_2	0.80	1				
3	4.7 %	92.5 %	M_3	0.74	0.84	1			
4	3.7 %	96.2 %	M_4	0.69	0.75	0.83	1		
5	2.3 %	98.5 %	M_5	0.67	0.70	0.76	0.86	1	
6	1.5 %	100.0 %	M_6	0.63	0.66	0.72	0.73	0.84	1

Table 8.8 Factor analysis and correlation matrices for quarterly and yearly contracts

#	%-explained	%-cumulative	Correlation matrix – quarterly								
				Q_1	Q_2	Q_3	Q_4	Q_5	Q_6	Q_7	Q_8
1	58.7 %	58.7 %	Q_1	1							
2	12.5 %	71.2 %	Q_2	0.80	1						
3	8.1 %	79.3 %	Q_3	0.59	0.67	1					
4	6.9 %	86.1 %	Q_4	0.56	0.60	0.64	1				
5	5.6 %	91.7 %	Q_5	0.58	0.60	0.48	0.65	1			
6	4.0 %	95.7 %	Q_6	0.52	0.57	0.46	0.46	0.67	1		
7	2.3 %	98.0 %	Q_7	0.40	0.50	0.38	0.45	0.41	0.60	1	
8	2.0 %	100.0 %	Q_8	0.36	0.44	0.36	0.41	0.44	0.39	0.67	1

#	%-explained	%-cumulative	Correlation matrix – yearly			
				Y_1	Y_2	Y_3
1	72 %	72 %	Y_1	1		
2	18 %	90 %	Y_2	0.70	1	
3	10 %	100 %	Y_3	0.50	0.52	1

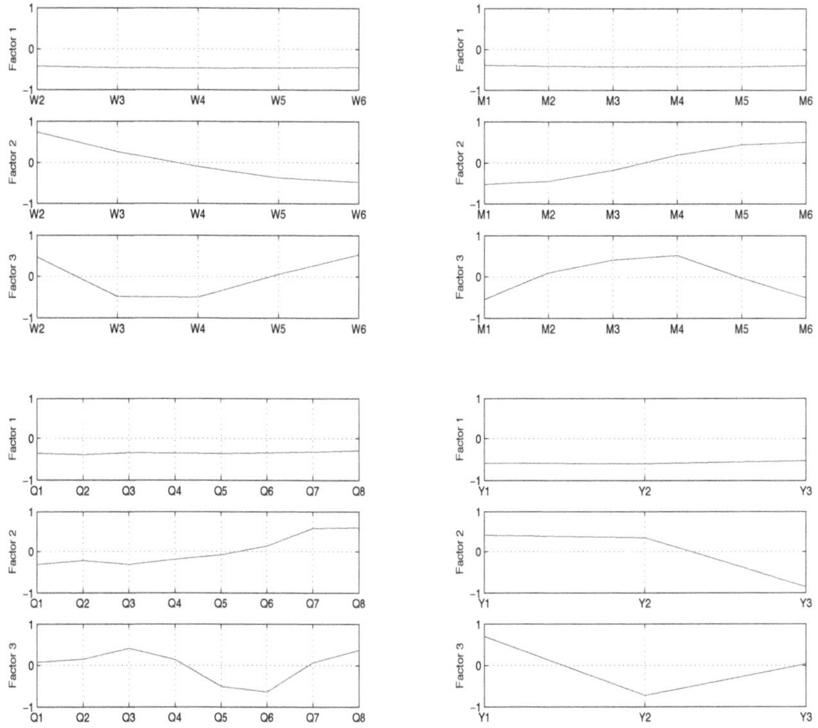

Fig. 8.4 First three factor loadings for the four different market segments: weekly contracts (upper left), monthly contracts (upper right), quarterly contracts (lower left), yearly contracts (lower right). The weekly contract with immediate delivery W_1 is excluded from the sample (see text for explanation).

In Fig. 8.4 we also plot the first three factor loadings for the other market segments; monthly contracts (upper left), quarterly contracts (lower left) and yearly contracts (lower right). For all market segments the first factor can be identified as a shifting factor, the second as a bending factor, and the third as a tilting factor. In the bottom panel of Table 8.7 and in Table 8.8 we report the variance explained by the individual factors and the cumulative effect for adding extra factors for the monthly, quarterly and yearly electricity futures returns. Since there are only three yearly contracts, three factors naturally add up to 100%. But we see from the correlation matrix that the contracts have fairly low correlations, and especially the contract in the far end of the curve is not strongly correlated with the two contracts closer to delivery. It is interesting to note that the correlation between Y_1 and Y_2 is 0.70, while the correlation between Y_2 and Y_3 is only 0.52. From the time series plot in Fig. 8.1 it looks like Y_2 and Y_3 are much stronger correlated than Y_1 and Y_2. This deceit is due only to the fact that the volatility of Y_1 is so much higher than that of Y_2 and Y_3. For the quarterly contracts, the first three factors explain in total 79.3% of the variance. This is the lowest number across all market segments. There are two reasons for this. First, there are eight contracts in total, and more contracts will typically require more factors to account for the variation. Secondly, the eight non-overlapping quarterly contracts span two full years of the term structure. In a term structure where contracts are not strongly correlated due to different risk factors along the term structure, a low dimensional factor model will deteriorate in performance as the total span of the term structure increases (see [Koekebakker and Ollmar (2005)] and [Frestad (2007a)]). For the six monthly contracts, with a span of only six months, the first three factors explain 92.5% of the total variance.

8.6 Estimating a parametric multi-factor market model

In this section we estimate a parametric market model for each market segment. For the weekly contracts we exclude W_1 from the analysis. Further, we assume that the dynamics of the electricity futures price for each market segment can be described by (8.2) with $p = 3$, that is, a three-factor model. In addition, we assume that $A_c(u)$ is constant for each c. Such a specification corresponds to deterministic market prices of risk which might be different across the contracts. This is of course a simplification, but our main interest lies in the volatility dynamics, and not in the nature of the market price of risk.

We also assume that the factor volatilities $\Sigma_{k,c}$ can be factorised into a common seasonal function $\sigma_S(t)$ (with t representing the time of year) and a maturity dependent function $\widetilde{\sigma}_k(\tau_c^b - t)$. The latter function depends on the time to the start of the delivery period, $\tau_c^b - t$. Thus, the factor volatilities can be represented as

$$\Sigma_{k,c}(t) = \sigma_S(t)\widetilde{\sigma}_k(\tau_c^b - t), \tag{8.6}$$

for $k = 1, 2, 3$. Note that $\Sigma_{k,c}$ implicitly depends on the delivery period as well, since we perform an empirical analysis for each market segment, where all contracts within a segment have the same length of delivery (week, month, quarter or year). The seasonal volatility is assumed to have the form

$$\sigma_S^2(t) = c_1 + \sum_{l=1}^{L}\left[c_{2l}\cos\left(\frac{2l\pi t}{365}\right) + c_{2l+1}\sin\left(\frac{2l\pi t}{365}\right)\right]. \tag{8.7}$$

If $L = 1$, the seasonal variation is symmetric, meaning that a peak in seasonal volatility produces an equally low variance exactly six months later. By increasing L, we allow for asymmetric seasonal variance. For reasons that will become clearer later, we chose $L = 4$ in the estimation procedure. The maturity function is specified as

$$\widetilde{\sigma}_k(\tau_c^b - t) = \sigma_0 + \left(\sigma_1 + \sigma_2(\tau_c^b - t)\right)\mathrm{e}^{-\kappa(\tau_c^b - t)}. \tag{8.8}$$

This form is the same for all factors, and it is chosen for its simplicity and flexibility. The functional form in (8.8) is chosen to allow for bends and humps in the term structure of volatility. Other functions could be used instead. One alternative specification is

$$\widetilde{\sigma}_k(\tau_c^b - t) = \sigma_0 + \sigma_1\mathrm{e}^{-\kappa_1(\tau_c^b - t)} + \sigma_2\mathrm{e}^{-\kappa_2(\tau_c^b - t)}.$$

This specification is used by [De Jong, Driessen and Pelsser (2004)] to model the volatility term structure in fixed income securities. A humped term structure of volatility can be accomplished by allowing both positive and negative values for the parameters σ_0, σ_1 and σ_2. Our choice (8.8) is adopted from the popular Nelson-Siegel model for the yield curve in interest rate theory (see, for example, [James and Webber (2000)]).

We can now compute daily normalised logreturns (with time measured in days) as

$$\overline{x}_{n,c} = \left(x_{n,c} - A_c\right)/\sigma_S(t_{n-1}). \tag{8.9}$$

The market price of risk A_c corresponds to the estimated average logreturn. From (8.2) and the assumed structure of the factor volatilities, the normalised logreturns become independent and centered normally distributed, with approximative factor volatilities given by $\widetilde{\sigma}_k(\tau_c^b - t_{n-1})$. An alternative version of this model can be expressed in terms of principal components in continuous time (see, for instance, Ch. 16 in [James and Webber (2000)]), yielding that the empirical factor volatilities can be written as $\sqrt{\lambda_k}p_k$, where λ_k are the eigenvalues and $p_k(\tau_c^b - t_n)$ are the eigenvectors of the covariance matrix of normalised logreturns.

We estimate the model in four steps.

(1) Estimate the deterministic seasonal volatility $\sigma_S(t)$.
(2) Normalise electricity futures logreturns using (8.9).
(3) Compute principal components from the normalised returns.
(4) For each factor, estimate the parameters of the maturity function $\widetilde{\sigma}_k$ from the empirical factor volatilities obtained via the values achieved in the previous step.

The results from the empirical analysis are discussed in the following subsections.

8.6.1 *Seasonal volatility*

To estimate the seasonal volatility σ_S, we first find the empirical mean and volatility for each contract c, denoted by $\widehat{m}_c$ and $\widehat{\sigma}_c$, respectively. Next, each price logreturn series are normalised (ignoring seasonality) using

$$\varepsilon_c(t) = \left(x_c(t) - \widehat{m}_c\right) / \widehat{\sigma}_c \,. \tag{8.10}$$

Obviously, $\varepsilon_c(t)$ will have an unconditional variance $\mathbb{E}\left[\varepsilon_c^2(t)\right] = 1$.

The deterministic variance is assumed to be constant across maturities. We therefore compute a series of average normalised squared returns

$$\bar{\varepsilon}_c^2(t) = \frac{1}{C}\sum_{c=1}^{C} \varepsilon_c^2(t) \,. \tag{8.11}$$

The parameters of σ_S are estimated by minimising, in a least squares sense, the difference between theoretical and empirical variance, that is, by finding the $\widehat{\mathbf{c}}_S$ which solves

$$\min_{\mathbf{c}_S} \frac{1}{T}\sum_{t=1}^{T} \left(\sigma_S^2(t) - \bar{\varepsilon}_c^2(t)\right)^2 ,$$

where T is the sample size and $\mathbf{c}_S = (c_1, c_2, ..., c_{2M+1})$ is the vector of parameters.

To account for asymmetric variance (high variance in December and January), we experimented with different values of L, and settled for $L = 4$ for all market segments as a reasonable choice. A lower value of L did not capture the asymmetry well, whereas a higher one did not give significantly better fit. This resulted in nine parameters to estimate for each data set. The parameter estimates are given in Table 8.9 and the variance functions are plotted in Fig. 8.5. We see that the asymmetric seasonality is clearly

Table 8.9 Fitted parameters for the seasonal volatility given in (8.7)

	Week	Month	Quarter	Year
c_1	1.0434	1.0420	1.0400	1.0380
c_2	0.8511	0.8007	0.6260	0.4686
c_3	−0.0786	−0.0430	−0.0098	0.2255
c_4	0.6640	0.5596	0.4474	0.3525
c_5	−0.1023	−0.0129	−0.0813	−0.0190
c_6	0.4270	0.4989	0.4541	0.4760
c_7	−0.1578	−0.0905	−0.1451	−0.0645
c_8	0.0863	0.1945	0.2396	0.2926
c_9	−0.0713	0.0563	0.0292	0.2109

present for all market segments. For weekly contracts, for example, the variance peaks in late December. From March until October, the variance is low and fairly constant, while it rises to levels close to six times higher in the peak period. We notice the same pattern in all segments. We observe a high peaking variance around the turn of the year, sharply rising and decreasing in the shoulder months. In the period from early spring to late autumn the variance is moderate, however, with a seasonal pattern. Interestingly, there is a volatility hump around April and a smaller one around August for all segments. The shape of the seasonal volatility shares many similar characteristics with the temperature variance, which we analyse in Chapter 10. Since temperature is one of the main factors driving electricity demand in Scandinavia, this is most likely not a coincidence. However, the coupling between temperature and prices is complicated, and we have not investigated this in any further detail (see, for example, [Vehviläinen and Pyykkönen (2005)] for a spot model with temperature dependence). [Benth and Koekebakker (2005)] consider a variance specification using only $L = 1$, imposing a symmetric structure. The results presented here suggest that such a specification is too simplistic.

8.6.2 *Maturity volatilities*

Next we estimate the parameters of the maturity volatilities $\widetilde{\sigma}_k(\tau_c^b - t)$.

First, we normalise logreturns using

$$\overline{x}_{n,c} = (x_{n,c} - \widehat{m}_c) / \widehat{\sigma}_S(t_{n-1}), \tag{8.12}$$

where $\widehat{\sigma}_S$ is the seasonal volatility obtained from the estimated parameter values $\widehat{\mathbf{c}}_S$ in Subsect. 8.6.1. Next, we estimate parameters by minimising,

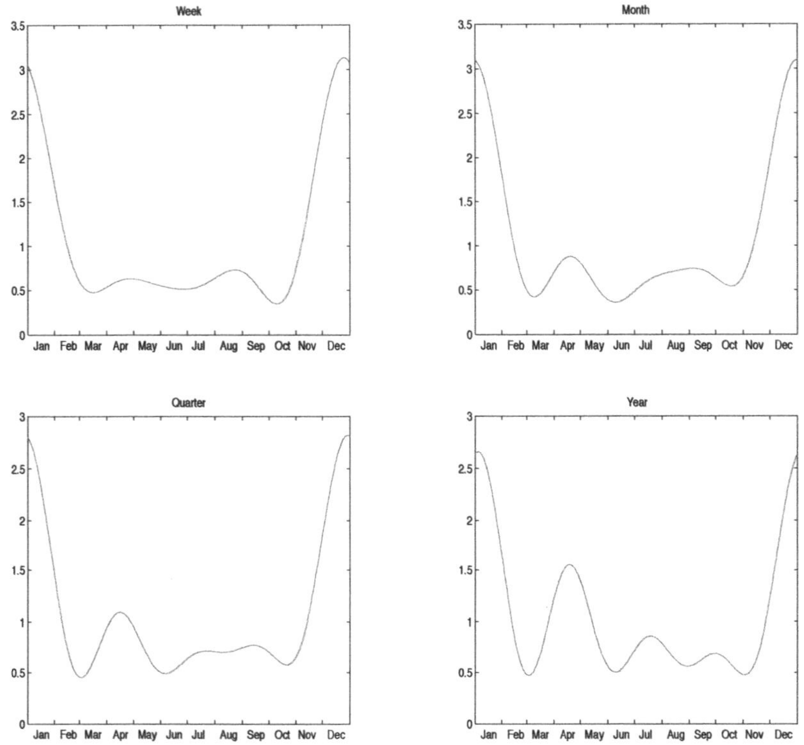

Fig. 8.5 Estimated seasonal functions for four market segments: weekly contracts (upper left), monthly contracts (upper right), quarterly contracts (lower left) and yearly contracts (lower right). W_1 is excluded from the weekly data set.

in a least squares sense, the difference between theoretical and empirical maturity volatilities obtained from a PCA of $\overline{x}_{n,c}$, that is,

$$\min_{\mathbf{m}_k} \frac{1}{C} \sum_{c=1}^{C} \left(\widetilde{\sigma}_k(\tau_c^b - t_n) - \sqrt{\widehat{\lambda}_k} \widehat{p}_k(\tau_c^b - t_n) \right)^2 .$$

Here, $\widehat{\lambda}_k$ are the empirical eigenvalues and $\widehat{p}_k(\tau_c^b - t_n)$ are the empirical eigenvectors. Furthermore, $\mathbf{m}_k = (\sigma_0, \sigma_1, \sigma_2, \kappa)$ is the vector of parameters for the maturity function.

The estimation routine is applied to three empirical volatility functions per data set, resulting in 12 parameters for each market segment. The parameter estimates are given in Table 8.10. Plots of the fitted volatility functions $\widehat{\sigma}_k^2(\tau_c^b - t)$ are presented in Fig. 8.6 for each factor and market

Table 8.10 Parameter estimates for a three-factor model. The function is the same for all three factors, and it is given in (8.8)

	σ_0	σ_1	σ_2	κ
Week				
Factor 1	−2.83	3.32	13.28	3.25
Factor 2	6.08	−5.56	−37.41	3.99
Factor 3	0.30	0.73	−57.85	33.23
Month				
Factor 1	−3.83	4.28	2.65	0.58
Factor 2	17.67	−17.88	4.63	−0.21
Factor 3	−26.85	26.67	21.27	0.68
Quarter				
Factor 1	0.02	0.35	0.00	0.50
Factor 2	13.47	−13.65	0.72	−0.04
Factor 3	−28.62	28.55	3.50	0.11
Year				
Factor 1	0.17	0.17	−0.07	1.00
Factor 2	0.21	−0.34	−1.34	1.87
Factor 3	−0.06	−0.09	1.13	1.94

segment.

In Fig. 8.6 we included the (discrete) empirical volatility for the first three factors, $\sqrt{\widehat{\lambda}_k \widehat{p}_k(\tau_c^b - t_n)}$, $k = 1, 2, 3$. The overall volatility is given both by the sum of the three fitted parametric volatility functions, $\sqrt{\sum_{k=1}^{3} \widehat{\sigma}_k^2(\tau_c^b - t)}$, and as the total volatility in terms of the standard deviation of the normalised returns $\overline{x}_{n,c}$. The empirical volatilities are presented as triangles (first factor), circles (second factor), squares (third factor) and diamonds (overall volatility). The fitted parametric volatility is presented as dashed line (factor 1), dashed-dotted line (factor 2), dotted line (factor 3) and solid line (overall volatility). For the weekly contracts, the parametric functions and the empirical volatility functions more or less coincide. The overall volatility shows that the three functions are sufficient to explain almost all of the total variance. Note that the overall volatility is almost flat, there is basically no maturity effect within the weekly segment. For the monthly electricity futures, overall volatility is slightly higher in the short end compared to the long end (five months to delivery). The parametric function is slightly off for the third factor, and the three-factor parametric

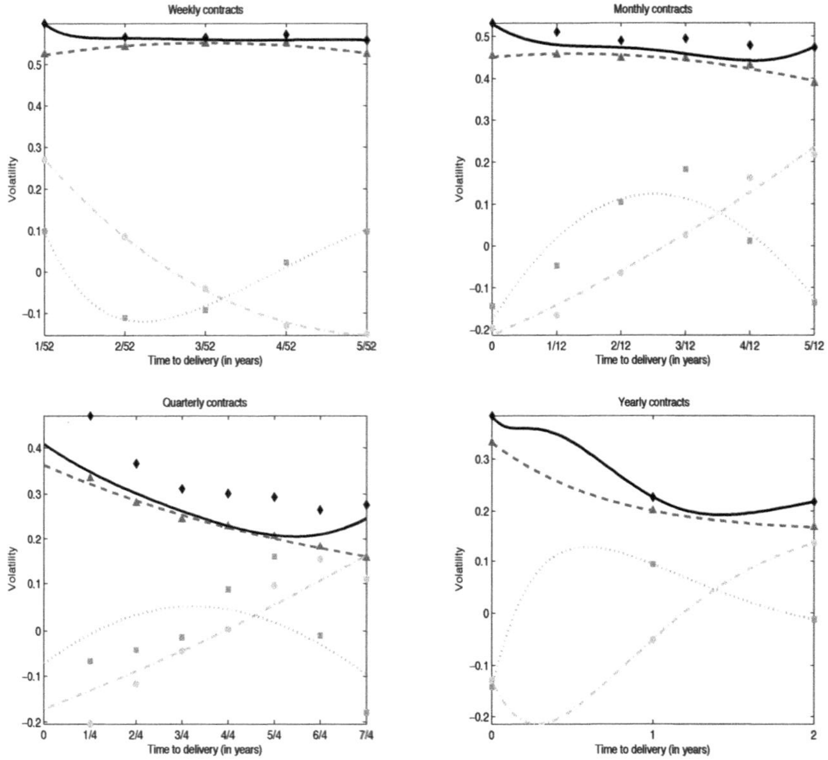

Fig. 8.6 Estimated maturity functions in a three-factor model: weekly contracts (upper left), monthly contracts (upper right), quarterly contracts (lower left) and yearly contracts (lower right). The first factor is the dashed line (triangles mark empirical volatility), the second factor is the dashed-dotted line (circles mark empirical volatility), the third factor is the dotted line (empirical volatility is marked by squares) and overall volatility is the solid line (empirical volatility is marked by diamonds). W_1 is excluded from the weekly data set.

overall volatility is a bit lower than the overall empirical volatility. For the quarterly contracts, the parametric function has problems fitting the second factor and serious trouble fitting the third factor. The maturity effect is evident in this market segment. The three empirical functions underestimate the total empirical volatility. For the yearly contracts, with only three maturities, the parametric function becomes too flexible, creating shapes that are not justified by the data.

The parametric multi-factor model for the electricity futures volatility indicates a clear seasonality effect. Moreover, it explains for some segments the maturity dependent volatility reasonably well by a simple three-factor structure. However, in particular for quarterly contracts, the data are far more complicated than the model is able to explain. Hence, more complex models are called for.

So far we have assumed conditional normally distributed logreturns. In the final section we challenge this assumption.

8.7 Normalised logreturns and heavy tails

In this Chapter we started out quite ambitiously with building a model that instantaneously incorporates all market segments (contracts with different delivery period length). Then we pursued a less ambitious task, by isolating each market segment and estimating multi-factor models for each segment (week, month, quarter or year). In this final section, we will not attempt to model joint dynamics at all, but instead focus on *single* contracts. That is, we pick a single contract with a specified time to delivery and a specified delivery period. We analyse the distributional properties of normalised logreturns, and show that they are far from being Gaussian. Clear signs of heavy tails are detected in all market segments, and we show that the NIG distribution models the stylised facts of the normalised logreturns in an excellent way. Our results are in line with the studies of [Frestad, Benth and Koekebakker (2007)].

Since we do not assume any particular parametric form for seasonality or maturity, we apply a different normalising routine than the one in the previous Section. For each contract in the market, we transform logreturns according to

$$x_{n,c} = (x_{n,c} - m_{S,c}) \,/ \sigma_{S,c} \,,$$

where $m_{S,c}$ and $\sigma_{S,c}$ are (seasonal) mean and standard deviation for $c = 1, ..., C$, respectively. We assume that $m_{S,c}$ and $\sigma_{S,c}$ are constant within each month. Therefore each data series is sorted according to month, normalised by subtracting the mean and dividing by standard deviation. Since we do this for each contract, both the seasonality and the maturity effects are removed from the data. In Table 8.11 the estimated parameters for the NIG distribution are presented for the six weekly contracts, starting with immediate delivery, and then delivery starting next week, two weeks later and so on. In addition, we include the estimates of the shape triangle parameters (defined in (2.32)), where in particular we observe that ξ is ranging from 0.76 to 0.85, implying that the electricity futures logreturns are far from being normal. It is worth noticing that the first weekly contract have the highest ξ, whereas the others vary at the significantly lower levels between 0.76 and 0.79. These levels are more in line with what we find for the contracts with longer delivery period. This is again a sign that the first weekly contract has some distinct features not corresponding with the features of other contracts (see Sect. 8.3 above). The χ parameters are close to zero, telling us that the distributions are close to symmetrical. We do not detect any dependence of the parameters on the time to delivery.

Table 8.11 NIG parameters estimated for weekly contracts

Week	μ	α	β	δ	ξ	χ
W1	0.013	0.62	−0.014	0.62	0.85	−0.019
W2	−0.005	0.85	0.005	0.84	0.76	0.004
W3	0.008	0.83	−0.008	0.82	0.77	−0.007
W4	0.016	0.84	−0.016	0.82	0.77	−0.015
W5	0.022	0.79	−0.022	0.78	0.79	−0.022
W6	0.025	0.80	−0.026	0.78	0.78	−0.025

The estimates for the monthly contracts are found in Table 8.12. We find estimates being similar to the weekly ones, with maybe lower ξ parameter in the shape triangle. The parameter χ is close to zero for all six months, a reflection of hardly any skewness in the data. As a consequence of the data normalisation, the estimates of μ are close to zero. It seems to be a tendency of increasing α with the month (that is, time to maturity), whereas ξ seems to be slightly decreasing with the month. It is not easy to tell the maturity effect on δ, where estimates vary a bit up and down in an unclear pattern.

Table 8.12 NIG parameters estimated for monthly contracts

Month	μ	α	β	δ	ξ	χ
M1	−0.075	0.82	0.076	0.81	0.78	0.072
M2	0.037	0.78	−0.037	0.78	0.79	−0.036
M3	0.019	0.89	−0.019	0.90	0.74	−0.016
M4	0.087	0.91	−0.089	0.89	0.74	−0.072
M5	−0.044	0.91	0.045	0.89	0.74	0.037
M6	0.066	0.95	−0.067	0.93	0.73	−0.051

The results for the quarterly contracts are presented in Table 8.13.

Again, we observe a slight decrease in ξ, reflecting a tendency towards normality, although, very small. Notice the similarity in estimates for the second and sixth quarters. Even though the data are normalised to remove the seasonality effects, there can still be some seasonality in ξ left. Also

Table 8.13 NIG parameters estimated for quarterly contracts

Quarter	μ	α	β	δ	ξ	χ
Q1	−0.018	0.84	0.019	0.84	0.77	0.017
Q2	0.031	0.90	−0.031	0.88	0.75	−0.026
Q3	0.051	0.81	−0.053	0.79	0.78	−0.051
Q4	0.094	0.94	−0.096	0.91	0.73	−0.075
Q5	0.054	0.96	−0.055	0.95	0.72	0.041
Q6	0.035	0.89	−0.036	0.86	0.75	−0.030
Q7	0.088	0.88	−0.089	0.87	0.75	−0.076
Q8	0.022	0.97	−0.022	0.96	0.72	−0.016

for the yearly contracts there is a tendency towards a smaller ξ, in comparison to the contracts with quartely delivery period. It is hard to draw

Table 8.14 NIG parameters estimated for yearly contracts

Year	μ	α	β	δ	ξ	χ
Y1	0.035	0.95	−0.035	0.94	0.73	−0.027
Y2	0.105	1.10	−0.108	1.06	0.68	−0.067
Y3	0.100	0.99	−0.101	0.97	0.71	−0.073

some conclusions regarding the dependency of the NIG parameters on the length of delivery period, except that there is a tendency towards decreasing values of ξ as delivery period increases. This is a sign of convergence towards normality, however at a very slow rate. A priori, one could expect that contracts with long delivery periods (like yearly contracts) are close to lognormally distributed since they can be understood as an aggregation of the spot over a long time interval, smoothing out the jump risk. This does not seem to be the case when investigating the data.

In the panel plot depicted in Fig. 8.7, we show the fitted NIG distribution together with the empirical and standard normal. The chosen contracts are W2 (top left), M2 (top right), Q2 (bottom left) and Y2 (bottom right). We see that the tails are heavy, and that the NIG distribution is superior to the normal in fitting the data along length of the delivery period and time to delivery. The center of the empirical distribution is more peaky than captured by the NIG. Trying the more flexible GH distribution did not improve significantly the lack of fit. The density plots for the other

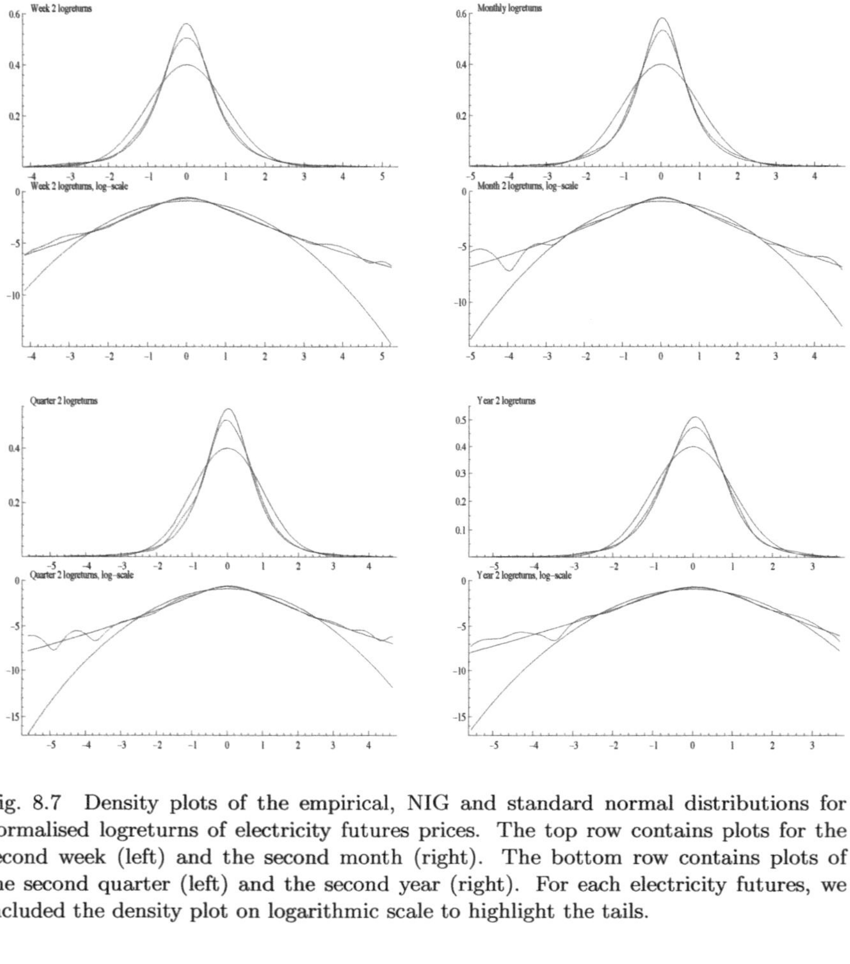

Fig. 8.7 Density plots of the empirical, NIG and standard normal distributions for normalised logreturns of electricity futures prices. The top row contains plots for the second week (left) and the second month (right). The bottom row contains plots of the second quarter (left) and the second year (right). For each electricity futures, we included the density plot on logarithmic scale to highlight the tails.

electricity futures are similar.

The empirical analysis suggests a market model including jump processes. Following the discussion in Subsect. 6.4.1, we may in a given segment choose the one-factor dynamics for contract c as

$$F_c(t) = F_c(0)\exp(\Upsilon_c J(t)) ,$$

where J is a NIG Lévy process. The scaling factor Υ_c can be interpreted as a volatility structure. It may be hard to determine *one* set of parameters for J

and scaling functions Υ_c which matches the estimated NIG distributions for the logreturns in question. Also, it is unlikely that we have a close to perfect dependency between the contracts, as discussed by [Frestad (2007b)]. The alternative is to model the electricity futures price dynamics by

$$F_c(t) = F_c(0)\exp(J_c(t)),$$

with one NIG Lévy process J_c per contract c. This makes it simple to estimate the characteristics of J_c directly from data. The next step then is of course to introduce a possible dependency structure on the contracts. If we choose a multivariate NIG, we would need to estimate the distribution parameters on all data, a difficult numerical task taking into account the dimension of the market and amount of data available. A copula structure is another possibility, as discussed in Subsect. 6.4.1.

8.8 Final remarks

In this chapter we conducted an empirical investigation of market models at Nord Pool. This research is, at the time of writing this book, still in its infancy, and we will probably see a lot of new developments.

There seems to be low correlation between the very short end of the term structure (spot price), and the financial contracts trading at the exchange. If spot price models are to be used, they should be estimated on the traded contracts, and volatility must be estimated on implied volatility or empirical term structure volatility. But then a market model approach seems to be far superior to spot price models. However, a market model approach does not solve all our problems. Modelling all contracts simultaneously is a daunting task. Financial electricity contracts seem to behave more idiosyncratic than what we usually see in other commodity markets. Modelling different parts of the term structure individually may be a better idea than modelling all contracts simultaneously in the market. Of course this does not help us if we need a model for the whole market (which is the case, for example, when analysing portfolio Value at Risk for a trading department). From an empirical perspective, the models investigated in this Chapter are still a long way from being satisfactory term structure models.

Chapter 9

Pricing and Hedging of Energy Options

Options constitute a very important asset class in energy markets. There exists an abundance of OTC products, designed to hedge or speculate on different events in the spot and forward markets. These options can be highly exotic, but at the exchanges more standardised plain vanilla options are traded. We will consider the pricing of European call and put options. In the energy markets, such options are usually written on forward contracts, and we analyse the option price dynamics for contracts written on both forwards and swaps. Further, we separate between geometric and arithmetic models for the underlying spot, and the case of jumps or no jumps.

The question of hedging is also discussed, more specifically in the so-called complete case where there exists a unique replication strategy to any option. We derive the delta hedge for several different models and options. When the underlying dynamics is driven by jump processes, there will in general not exist any perfect hedge. In this situation the market is usually referred to as being incomplete. We will briefly mention how one can approach the development of partial hedges in incomplete markets.

We include a section where we consider two types of exotic options frequently encountered in the energy markets. Spread options are contracts written on the difference of two commodities, like, for instance, the spark spread which is the difference of electricity and gas price. Further, we look at an average-type option (Asian option) written on a spot price. We price such options based on the arithmetic spot model. Asian-type options were traded at the Nord Pool power exchange in the 1990s, while spread options are hugely traded OTC.

The last Section contains a case study where we use a direct modelling approach to obtain semi-explicit prices for spark spread options. We base the model on a jump diffusion dynamics, for which we can use the Fourier transform to analyse the option price dynamics. Data from the UK are used in the case study. Our method provides fast and accurate prices of spark spread options.

9.1 Pricing and hedging options on forwards and swaps

We analyse pricing and hedging of call and put options written on forwards and swaps. We treat the case when the forward dynamics does not have any jump components in the dynamics separately. In this situation we can derive explicit pricing formulas for the plain vanilla options being slight extensions of the Black-76 Formula (see [Black (1976)]). When the forward dynamics has jumps, the question of pricing becomes more delicate, and we resort to the use of Fourier techniques. We distinguish our analysis between options on forwards and swaps, and arithmetic and geometric models.

9.1.1 *The case of no jumps – the Black-76 Formula*

Consider a call option written on a forward contract, where the exercise time is T and strike price is K. The forward contract has maturity $\tau \geq T$, and we suppose the risk-neutral dynamics to be

$$\frac{df(t,\tau)}{f(t,\tau)} = \sum_{k=1}^{p} \sigma_k(t,\tau)\, dW_k(t)\,, \tag{9.1}$$

where W_k are p independent Brownian motions under the risk-neutral probability Q. Recalling the forward price dynamics resulting from a geometric spot model derived in Prop. 4.8, we have

$$\sigma_k(t,\tau) = \sum_{i=1}^{m} \sigma_{ik}(t) \exp\left(-\int_t^{\tau} \alpha_i(u)\, du\right)\,.$$

The speeds of mean reversion are described by the functions α_i, and the spot volatilities by σ_{ik}. The forward dynamics in (9.1) can also stem from the direct modelling of the forward price curve as analysed in Chapter 6.

The following Proposition states the price of a call option, known as the Black-76 Formula (see [Black (1976)]).

Proposition 9.1 *The price of a call option at time $t \leq T$, written on a forward with delivery at time τ, where the option has exercise time $T \leq \tau$*

and strike price K, is

$$C(t;T,K,\tau) = e^{-r(T-t)}\left\{f(t,\tau)\Phi(d_1) - K\Phi(d_2)\right\}.$$

Here,

$$d_1 = d_2 + \sqrt{\sum_{k=1}^{p}\int_t^T \sigma_k^2(u,\tau)\,du}\,,$$

$$d_2 = \frac{\ln(f(t,\tau)/K) - 0.5\sum_{k=1}^{p}\int_t^T \sigma_k^2(u,\tau)\,du}{\sqrt{\sum_{k=1}^{p}\int_t^T \sigma_k^2(u,\tau)\,du}}\,,$$

and Φ is the cumulative standard normal probability distribution function.

Proof. Consider the case $p = 1$. We have that,

$$\ln f(T,\tau) \stackrel{d}{=} \ln f(t,\tau) - \int_t^T \sigma^2(u,\tau)\,du + X\sqrt{\int_t^T \sigma^2(u,\tau)\,du}\,,$$

where X is a standard normally distributed random variable. From general option theory, the price is defined as the present expected payoff, with expectation taken under the risk-neutral probability. Hence,

$$\begin{aligned}
&C(t;T,K,\tau)\\
&\quad = \mathrm{e}^{-r(T-t)}\mathbb{E}_Q\left[\max\left(f(T,\tau) - K, 0\right) \mid \mathcal{F}_t\right]\\
&\quad = \mathrm{e}^{-r(T-t)}\mathbb{E}\left[\max\left(\mathrm{e}^{\ln f(t,\tau) - 0.5\int_t^T \sigma^2(u,\tau)\,du + X\sqrt{\int_t^T \sigma^2(u,\tau)\,du}} - K, 0\right)\right].
\end{aligned}$$

Observe that we get a positive payoff from the option only when $X > -d_2$. Using this, a standard calculation exploiting the properties of normal random variables yields the option pricing formula. □

We now turn our attention to the question of hedging the call option on the forward. From option theory, the delta hedging strategy is defined as follows

$$\Delta(t;T,K,\tau) \triangleq \frac{\partial C(t;T,K,\tau)}{\partial f(t,\tau)}\,. \tag{9.2}$$

The delta hedge gives the number of forwards one should have at all times up to the exercise in a hedge of the call option. The strategy is derived in the following Proposition.

Proposition 9.2 *The delta hedge of the call option written on a forward with maturity at time τ, and where the option has exercise time $T \leq \tau$ and strike K, is given as*

$$\Delta(t;T,K,\tau) = e^{-r(T-t)}\Phi(d_1)\,,$$

where Φ and d_1 are defined in Prop. 9.1.

Proof. A differentiation leads to

$$\Delta(t;T,K,\tau) = \mathrm{e}^{-r(T-t)}\left\{\Phi(d_1) + f(t,\tau)\Phi'(d_1)\frac{\partial d_1}{\partial f} - K\Phi'(d_2)\frac{\partial d_2}{\partial f}\right\},$$

where d_2 is defined in Prop. 9.1. The partial derivatives of d_1 and d_2 with respect to f coincide, and a further calculation using that

$$d_1 = d_2 + \sqrt{\int_t^T \sigma^2(u,\tau)\,du}\,,$$

leads to the desired formula. □

We recognise the price and hedging strategy as the analogue of a call option in the Black & Scholes context. The only difference is that the forward dynamics is a martingale in the risk-neutral setting, while it is the discounted asset price which is a martingale in the Black & Scholes framework. This leads to some minor modifications of the price and hedge in the case of forward options. Remark further that we have many sources of risk (that is, many Brownian motions) in the forward dynamics. This can imply problems with the completeness of the market, but not in the case of the forward option analysed here, since we could merge all the Brownian motion terms into one single. We emphasise that for the purpose of option pricing, it is no need to use a forward price model with several factors, however, when modelling the forward curve dynamics it may be desirable in order to have the flexibility of introducing correlation among different forwards. If we collapse the model into one single diffusion term, we have in fact nearly perfectly correlated prices across the forward curve.

Let us look at an example where the forward price dynamics comes from a Schwartz model with constant volatility and speed of mean reversion. Assuming $p = m = 1$, we find

$$\sigma(u,\tau) = \sigma\mathrm{e}^{-\alpha(\tau-u)}\,.$$

Thus, the aggregated volatility to be inserted into the Black-76 Formula becomes

$$\int_t^T \sigma^2(u,\tau)\,du = \frac{\sigma^2}{2\alpha}\left(\mathrm{e}^{-2\alpha(\tau-T)} - \mathrm{e}^{-2\alpha(\tau-t)}\right).$$

Not surprisingly, this aggregated volatility increases with the exercise time, and decreases with the maturity of the forward. Hence, if the maturity of the forward is far into the future, the aggregated volatility will be relatively low if exercise of the option is close in time. The aggregated volatility is decreasing with an increasing speed of mean reversion α.

Consider a call option written on a swap contract. Suppose that the delivery period is $[\tau_1, \tau_2]$, and recall the forward dynamics given by the market models discussed in Subsect. 6.4

$$\frac{dF(t,\tau_1,\tau_2)}{F(t,\tau_1,\tau_2)} = \sum_{k=1}^{p} \Sigma_k(t,\tau_1,\tau_2)\,dW_k(t)\,. \tag{9.3}$$

Note that we use the general notation for the direct modelling of swaps, rather than the specific notation for the market models, since we here want to emphasise the delivery period. Following the case of options on forwards above, the following result is reached.

Proposition 9.3 *Suppose a call option written on a swap contract with delivery period $[\tau_1, \tau_2]$ has exercise time $T \leq \tau_1$ and strike K. The option price at time t is then given as*

$$C(t;T,K,\tau_1,\tau_2) = e^{-r(T-t)}\left\{F(t,\tau_1,\tau_2)\Phi(d_1) - K\Phi(d_2)\right\},$$

where

$$d_1 = d_2 + \sqrt{\sum_{k=1}^{p}\int_t^T \Sigma_k^2(s,\tau_1,\tau_2)\,ds}\,,$$

$$d_2 = \frac{\ln(F(t,\tau_1,\tau_2)/K) - 0.5\sum_{k=1}^{p}\int_t^T \Sigma_k^2(s,\tau_1,\tau_2)\,ds}{\sqrt{\sum_{k=1}^{p}\int_t^T \Sigma_k^2(s,\tau_1,\tau_2)\,ds}}\,.$$

The delta hedge of the option is

$$\Delta(t;T,K,\tau_1,\tau_2) = e^{-r(T-t)}\Phi(d_1)\,.$$

Here, Φ is the cumulative standard normal probability distribution function.

Proof. The proof is analogous to the case of options on forwards, see Props. 9.1 and 9.2. □

Let us turn our attention to the model of [Bjerksund, Rasmussen and Stensland (2000)] discussed in Subsect. 6.3.1. There we started out with a geometric Brownian motion model for a forward, and derived a volatility for a swap as the (weighted) average volatility. Specifically, we get that the swap has volatility

$$\Sigma(t,\tau_1,\tau_2)=\int_{\tau_1}^{\tau_2}\widehat{w}(u,\tau_1,\tau_2)\sigma(t,u)\,du\,,$$

with $\sigma(t,u)$ being the volatility of the forward. Thus, for the Black-76 Formula in the Proposition above we must calculate the aggregated volatility

$$\int_t^T \Sigma^2(s,\tau_1,\tau_2)\,ds=\int_t^T\{\int_{\tau_1}^{\tau_2}\widehat{w}(u,\tau_1,\tau_2)\sigma(s,u)\,du\}^2\,ds\,.$$

Suppose for simplicity $\widehat{w}(u,\tau_1,\tau_2)=1/(\tau_2-\tau_1)$, and consider first the specification

$$\sigma(t,u)=\sigma\mathrm{e}^{-\alpha(u-t)}\,.$$

This leads to the volatility

$$\Sigma(t,\tau_1,\tau_2)=\frac{\sigma}{\alpha(\tau_2-\tau_1)}\left\{\mathrm{e}^{-\alpha(\tau_1-t)}-\mathrm{e}^{-\alpha(\tau_2-t)}\right\}\,.$$

Hence, the aggregated volatility to be used in the Black-76 Formula is the following

$$\begin{aligned}\int_t^T \Sigma^2(s,\tau_1,\tau_2)\,ds&=\frac{\sigma^2}{\alpha^2(\tau_2-\tau_1)^2}\left(\mathrm{e}^{-\alpha\tau_1}-\mathrm{e}^{-\alpha\tau_2}\right)^2\int_t^T\mathrm{e}^{2\alpha s}\,ds\\&=\frac{\sigma^2}{2\alpha^3(\tau_2-\tau_1)^2}\left(\mathrm{e}^{-\alpha\tau_1}-\mathrm{e}^{-\alpha\tau_2}\right)^2\left(\mathrm{e}^{2\alpha T}-\mathrm{e}^{2\alpha t}\right)\,.\end{aligned}$$

Moving on to the volatility

$$\sigma(t,u)=\frac{\sigma}{u-t+b}+c\,,$$

used by [Bjerksund, Rasmussen and Stensland (2000)], we have

$$\Sigma(t,\tau_1,\tau_2)=\frac{\sigma}{\tau_2-\tau_1}\ln\left(\frac{\tau_2-t+b}{\tau_1-t+b}\right)+c\,.$$

Next, let us calculate the integral of $\Sigma^2(t, \tau_1, \tau_2)$ up to the exercise time T.

$$\begin{aligned}\int_t^T \Sigma^2(s, \tau_1, \tau_2)\, ds = {} & \frac{\sigma^2}{(\tau_2 - \tau_1)^2} \int_t^T \left(\ln\left(\frac{\tau_2 - s + b}{\tau_1 - s + b}\right)\right)^2 ds \\ & + \frac{2\sigma c}{\tau_2 - \tau_1} \int_t^T \ln\left(\frac{\tau_2 - s + b}{\tau_1 - s + b}\right) ds \\ & + c^2(T - t)\,. \end{aligned} \tag{9.4}$$

The first integral is

$$\int_t^T \left(\ln\left(\frac{\tau_2 - s + b}{\tau_1 - s + b}\right)\right)^2 ds = v(T) - v(t)\,, \tag{9.5}$$

where, for $s \leq \tau_1$,

$$\begin{aligned} v(s) = {} & (\tau_2 + b - s)(\ln(\tau_2 + b - s))^2 \\ & - 2(\tau_2 + b - s)\ln(\tau_2 + b - s)\ln(\tau_1 + b - s) \\ & + 4\sigma \ln(\tau_2 - \tau_1)\ln\left(\frac{\tau_1 + b - s}{\tau_2 - \tau_1}\right) \\ & - 2(\tau_2 - \tau_1)\mathrm{dilog}\left(\frac{\tau_2 + b - s}{\tau_2 - \tau_1}\right) \\ & + (\tau_1 + b - s)(\ln(\tau_1 + b - s))^2 - 2(\tau_2 - \tau_1)\,. \end{aligned} \tag{9.6}$$

The dilogarithm function is defined for $x \geq 0$ (see, for example, [Abramowitz and Stegun (1972), p. 1004]) as

$$\mathrm{dilog}(x) = -\int_1^x \frac{\ln(s)}{s - 1}\, ds\,, \tag{9.7}$$

which, for numerical purposes, may be approximated using

$$\mathrm{dilog}(x) \approx \begin{cases} \sum_{k=1}^n \frac{(x-1)^k}{k^2}\,, & 0 \leq x \leq 1\,, \\ -\frac{1}{2}(\ln(x))^2 - \sum_{k=1}^n \frac{((1/x)-1)^k}{k^2}\,, & x > 1\,, \end{cases}$$

for a sufficiently large positive integer n. The second integral in (9.4) is

$$\begin{aligned}\int_t^T \ln\left(\frac{\tau_2 - s + b}{\tau_1 - s + b}\right) ds = {} & (\tau_2 + b - T)\ln(\tau_2 + b - T) \\ & - (\tau_1 + b - T)\ln(\tau_1 + b - T) \\ & - (\tau_2 + b - t)\ln(\tau_2 + b - t) \\ & + (\tau_1 + b - t)\ln(\tau_1 + b - t)\,. \end{aligned} \tag{9.8}$$

As we can see, the aggregated volatility to be inserted into the Black-76 Formula is rather complicated in the approximation suggested by [Bjerksund, Rasmussen and Stensland (2000)], and numerical evaluation of the dilogarithm function is required.

We next discuss some problems related to pricing of options for the market models presented in Subsect. 6.4. Say that we want to price and hedge a call option written on a swap with delivery over the next quarter. We have seen in Subsect. 6.4 that the market models give the dynamics for the monthly contracts, and that the quarterly swap dynamics is the weighted sum of the corresponding three monthly swaps. The option on the quarterly swap therefore becomes an option on the weighted sum of three monthly swaps. Hence, we have in fact a basket option. We cannot write down any Black-76 Formula for the price of such an option directly, since it is written on the weighted average of three geometric Brownian motions, and the average of lognormal random variables does not have any simple distribution. One way out of this is to use Monte Carlo simulation techniques. However, one may ask for an approximation procedure. We discuss it next.

Suppose we want to price a call option written on a swap contract which can be represented as the sum of other, basic contracts, each modelled by a geometric Brownian motion. That is, assume we have the delivery period $[\tau_1, \tau_2]$ for the swap contract of interest, and that we can find basic swap contracts $F_c(t)$, $c = 1, \dots, N$, with disjoint delivery periods $[\tau_c^b, \tau_c^e]$ with their union equal to $[\tau_1, \tau_2]$. From the no-arbitrage principle (6.6), we can write

$$F(t, \tau_1, \tau_2) = \sum_{c=1}^{N} w_c F_c(t)$$

where the weights w_c are defined in (6.7). Further, we assume that the dynamics of $F_c(t)$ is defined by (6.14) without jumps (that is, $\Upsilon_{c,j} = 0$) and the drift condition in Prop. 6.4 holds. Thus,

$$F_c(t) = F_c(0) \exp\left(-\frac{1}{2} \sum_{k=1}^{p} \int_0^t \Sigma_{c,k}^2(u)\, du + \sum_{k=1}^{p} \int_0^t \Sigma_{c,k}(u)\, dW_k(u) \right),$$

for $c = 1, \dots, N$. If the exercise time of the call option is T with strike K, its fair value is given by

$$C(t; T, K, \tau_1, \tau_2) = \mathrm{e}^{-r(T-t)} \mathbb{E}\left[\max\left(F(T, \tau_1, \tau_2) - K, 0 \right) \mid \mathcal{F}_t \right]$$

$$= e^{-r(T-t)}\mathbb{E}\left[\max\left(\sum_{c=1}^{N} w_c F_c^{t,x_c}(T) - K, 0\right)\right]\Bigg|_{x_c=F_c(t)} .$$

The expectation is with respect to the risk-neutral probability, under which we have stated the market model for $F_c(t)$. Moreover, the notation $F_c^{t,x}(T)$ means that $F_c(t) = x$. Thus, we need to calculate an expectation of a function of a sum of lognormal variables. In [Levy (1992)] and [Brigo and Mercurio (2001)] it is suggested to approximate this sum with a lognormal variable. This entails in finding a lognormal variable X such that its mean and variance coincide with that of $\sum_{c=1}^{N} w_c F_c^{t,x_c}(T)$. Hence, supposing that

$$\ln X \sim \mathcal{N}(m, s^2)$$

where m and s^2 are defined in a following way

$$\mathbb{E}[X] \triangleq \mathbb{E}\left[\sum_{c=1}^{N} w_c F_c^{t,x_c}(T)\right] ,$$

$$\mathrm{Var}[X] \triangleq \mathrm{Var}\left(\sum_{c=1}^{N} w_c F_c^{t,x_c}(T)\right) ,$$

the expectation is given as

$$\mathbb{E}\left[\sum_{c=1}^{N} w_c F_c^{t,x_c}(T)\right] = \sum_{c=1}^{N} w_c x_c . \tag{9.9}$$

We calculate the variance using the Itô isometry, which leads to the expression

$$\begin{aligned}
&\mathrm{Var}\left[\sum_{c=1}^{N} w_c F_c^{t,x_c}(T)\right] \\
&\quad = \sum_{c=1}^{N} w_c^2 \mathrm{Var}[F_c^{t,x_c}(T)] + 2 \sum_{\substack{c,c'=1\\ c<c'}}^{N} w_c w_{c'} \mathrm{Cov}[F_c^{t,x_c}(T), F_{c'}^{t,x_{c'}}(T)] \\
&\quad = \sum_{c=1}^{N} w_c^2 x_c^2 \left(\exp\left(\sum_{k=1}^{p} \int_t^T \Sigma_{c,k}^2(u)\, du\right) - 1\right) \qquad (9.10) \\
&\qquad + \sum_{\substack{c,c'=1\\ c<c'}}^{N} w_c w_{c'} x_c x_{c'} \left(\exp\left(\sum_{k=1}^{p} \int_t^T \Sigma_{c,k}(u)\Sigma_{c',k}(u)\, du\right) - 1\right) .
\end{aligned}$$

Since, by definition of X,

$$\mathrm{E}[X] = \mathrm{e}^{m+0.5s^2}, \qquad \mathrm{Var}[X] = \mathrm{e}^{2m+s^2}\left(\mathrm{e}^{s^2} - 1\right),$$

we find, after matching with (9.9) and (9.10), that

$$s^2 = \ln\left(1 + \frac{\mathrm{Var}\left[\sum_{c=1}^N w_c F_c^{t,x_c}(T)\right]}{\left(\sum_{c=1}^N w_c x_c\right)^2}\right), \tag{9.11}$$

and

$$m = \ln\left(\sum_{c=1}^N w_c x_c\right) - \frac{1}{2}s^2. \tag{9.12}$$

An approximate value of the call option price is therefore

$$C(t;T,K,\tau_1,\tau_2) \approx \mathrm{e}^{-r(T-t)}\mathbb{E}\left[\max(X-K,0)\right]\Big|_{x_c=F_c(t),c=1,\ldots,N},$$

which we can explicitly calculate using the same arguments as for the Black-76 Formula. The result is the approximation

$$C(t;T,K,\tau_1,\tau_2) \approx \mathrm{e}^{-r(T-t)}\left\{F(t,\tau_1,\tau_2)\Phi(d_1) - K\Phi(d_2)\right\}, \tag{9.13}$$

where

$$d_1 = d_2 + s, \tag{9.14}$$

$$d_1 = \frac{\ln(F(t,\tau_1,\tau_2)/K) - 0.5s^2}{s}, \tag{9.15}$$

and Φ the cumulative standard normal probability distribution function. Here we used the equality

$$m + \frac{1}{2}s^2 = \sum_{k=1}^p w_c F_c(t) = F(t,\tau_1,\tau_2),$$

with s^2 given by (9.11). The accuracy of this approximation has been studied by [Levy (1992)] for currency exchange rates and [Brigo and Liinev (2005)] for interest rates. See also [Henriksen (2007)] for a general study of this approximation in view of spread options, and further [Benth and Henriksen (2008)] for a corresponding analysis of the NIG case.

9.1.2 *The case of jumps*

When considering the case of jumps in the forward price dynamics, we do not, in general, have available an explicit distribution function for the price of the underlying asset at the exercise time. However, we can express the distribution in terms of the cumulant functions of the jump processes driving the dynamics, and we shall exploit this fact to derive the call option price using Fourier analysis. Before proceeding, we recall the Fourier transform of a function $g \in L^1(\mathbb{R})$ as

$$\widehat{g}(y) = \int_{\mathbb{R}} g(x)\mathrm{e}^{-\mathrm{i}yx}\,dx\,. \tag{9.16}$$

The inverse Fourier transform is

$$g(x) = \frac{1}{2\pi}\int_{\mathbb{R}} \widehat{g}(y)\mathrm{e}^{\mathrm{i}yx}\,dy\,. \tag{9.17}$$

These expressions will be frequently used in the derivations below. The reader should note that we use a minus sign in the exponent in the definition (9.16), which may not correspond to the most common definition of the transform (see however [Folland (1984)]). Avoiding to have a minus sign in (9.17) turns out to be convenient when relating the Fourier transform with the characteristic function of a random variable.

Consider a forward price dynamics $f(t,\tau)$ like in (6.1) for a contract with delivery at time τ. We observe that we can write this dynamics as

$$f(t,\tau) = h(t,\tau)\exp\left(\sum_{k=1}^{p}\int_0^t \sigma_k(u,\tau)\,dW_k(u) + \sum_{j=1}^{n}\int_0^t \eta_j(u,\tau)\,dJ_j(t)\right)\,, \tag{9.18}$$

with h being a deterministic function defined as

$$h(t,\tau) = f(0,\tau)\exp\left(\int_0^t a(u,\tau)\,du\right)\,.$$

The function $a(u,\tau)$ satisfies the drift condition stated in Prop. 6.1, where we suppose the integrability conditions to hold.[1] We further note that the forward price dynamics has the form stated in (9.18) also when derived from a geometric spot model as analysed in Chapter 4. This can be shown

[1] We will assume a stronger condition shortly.

by using Prop. 4.8 with

$$\sigma_k(u,\tau) = \sum_{i=1}^{m} \sigma_{ik}(u) \exp\left(-\int_u^{\tau} \alpha_i(v)\, dv\right),$$

$$\eta_j(u,\tau) = \eta_j(u) \exp\left(-\int_u^{\tau} \beta_j(v)\, dv\right),$$

and a rather technical and long expression for the deterministic function $h(t,\tau)$ that we refrain from expressing explicitly here. Note that h is a positive and continuous function with respect to t. We introduce the following shorthand notation

$$Z(t,\tau) \triangleq \sum_{k=1}^{p} \int_0^t \sigma_k(u,\tau)\, dW_k(u) + \sum_{j=1}^{n} \eta_j(u,\tau)\, dJ_j(u).$$

Thus, we can write the forward price compactly as $f(t,\tau) = h(t,\tau)\exp(Z(t,\tau))$.

Before calculating the price of a call option with strike K at exercise time $T \leq \tau$, we state the following Lemma representing the payoff function in terms of Fourier transform.

Lemma 9.1 *Define for $a > 1$ the function*

$$g_{T,\tau}(x) \triangleq e^{-ax} \max\left(h(T,\tau)e^x - K, 0\right).$$

Then we have

$$\widehat{g}_{T,\tau}(y) = \frac{K}{(a-1+iy)(a+iy)} \left(\frac{K}{h(T,\tau)}\right)^{-(a+iy)},$$

where $\widehat{g}_{T,\tau}$ denotes the Fourier transform of $g_{T,\tau}$.

Proof. Using the definition of the Fourier transform (9.16), we have

$$\begin{aligned}\widehat{g}_{T,\tau}(y) &= \int_{\mathbb{R}} \mathrm{e}^{-ax} \max(h(T,\tau)\mathrm{e}^x - K, 0)\mathrm{e}^{-iyx}\, dx \\ &= \int_{\ln(K/h(T,\tau))}^{\infty} (h(T,\tau)\mathrm{e}^x - K)\mathrm{e}^{-(a+iy)x}\, dx.\end{aligned}$$

The result follows from a straightforward calculation. □

The reason for considering a damping factor $\exp(-ax)$ in front of the payoff function of the call is that the latter does not define an integrable function on the real line. We need this property when applying the Fourier approach to option pricing. The price of a call written on a forward contract is derived in the next Proposition.

Proposition 9.4 *Suppose that there exist constants c_j such that for some constant $a > 1$, Condition G holds for $J_j(t)$ with $c_j \geq a \sup_{0 \leq u \leq T} |\eta_j(u, \tau)|$, $j = 1, \ldots, n$. Then the price $C(t; K, T, \tau)$ at time $t \leq T$ of a call option with strike price K at exercise time T, written on a forward with delivery at time $\tau \geq T$, is given as*

$$C(t; T, K, \tau) = e^{-r(T-t)} \frac{1}{2\pi} \int_{\mathbb{R}} \widehat{g}_{T,\tau}(y) \Psi(t, T, \tau) e^{(a+iy)Z(t,\tau)} \, dy \,,$$

where

$$\ln \Psi(t, T, \tau) = \frac{1}{2}(a + iy)^2 \sum_{k=1}^{p} \int_t^T \sigma_k^2(u, \tau) \, du + \sum_{j=1}^{n} \psi_j(t, T; (y - ia)\eta_j(\cdot, \tau)) \,.$$

Proof. Appealing to the inverse Fourier transform (9.17), we get from Lemma 9.1 that

$$\mathbb{E}\left[\max\left(h(T, \tau)e^{Z(T,\tau)} - K, 0\right) \mid \mathcal{F}_t\right] = \frac{1}{2\pi} \int_{\mathbb{R}} \widehat{g}_{T,\tau}(y) \mathbb{E}\left[e^{(a+iy)Z(T,\tau)} \mid \mathcal{F}_t\right] \, dy \,.$$

This is valid because of the integrability condition on the jump measures yielding exponential integrability of the jump terms in the dynamics of $f(t, \tau)$. We calculate the conditional expectation of the right-hand side. By independence of the processes W_k and J_j, and the definition of Z, we have

$$\mathbb{E}\left[e^{(a+iy)Z(T,\tau)} \mid \mathcal{F}_t\right] = e^{(a+iy)Z(t,\tau)} \prod_{k=1}^{p} \mathbb{E}\left[e^{(a+iy) \int_t^T \sigma_k(u,\tau) \, dW_k(u)}\right] \times \prod_{j=1}^{n} \mathbb{E}\left[e^{(a+iy) \int_t^T \eta_j(u,\tau) \, dJ_j(u)}\right] \,.$$

By appealing to the definition of the cumulants of W_k and J_j, the result follows. $\square$

The expression for $C(t; T, K, \tau)$ is suitable for numerical valuation using the fast Fourier transform (FFT). The FFT approach was suggested by [Carr and Madan (1998)] to price options on assets which are not governed by a geometric Brownian motion, but where the characterstic functions are available. Note that we need to perform an integration in the valuation of ψ_j, which is not always analytically possible. Hence, a FFT approach also involves a numerical integration in order to have the cumulants readily available for valuation. Prices of put options can be derived through the put-call parity.

Next, let us consider the case of options on swaps. We concentrate on the class of market models introduced in (6.14), defined as

$$F_c(t) = F_c(0)\exp\Big(\int_0^t A_c(u)\,du + \sum_{k=1}^p \int_0^t \Sigma_{c,k}(u)\,dW_k(u) + \sum_{j=1}^n \int_0^t \Upsilon_{c,j}(u)\,dJ_j(u)\Big),$$

with A_c satisfying the drift condition in Prop. 6.4. This dynamics is of a geometric type and may be fitted into the above calculations of prices for forward options. We recall the expression for $f(t,\tau)$ stated in (9.18), and substitute 'τ' with 'c' to obtain

$$h(t,c) = F_c(0)\exp\left(\int_0^t A_c(u)\,du\right),$$
$$\sigma_k(t,c) = \Sigma_{c,k}(t),$$

and

$$\eta_j(t,c) = \Upsilon_{c,j}(t).$$

By using these notational definitions in Prop. 9.4 we find the price of a call option with strike K at exercise time T written on the contract with dynamics $F_c(t)$ to be equal

$$C(t;T,K,c) = \mathrm{e}^{-r(T-t)}\frac{1}{2\pi}\int_{\mathbb{R}} \widehat{g}_{T,c}(y)\Psi(t,T,c)\mathrm{e}^{(a+\mathrm{i}y)Z(t,c)}\,dy,$$

with the obvious definitions of $\Psi(t,T,c)$ and $Z(t,c)$.

We discuss the case when the logreturns of a swap are NIG distributed. We recall the market model from Sect. 6.4 with $p=0$, $n=1$, $\Upsilon_{j,c}=1$ and a Lévy process J with increments being NIG$(\alpha,\beta,\delta,\mu)$ distributed. For this model, the drift condition (see Prop. 6.4) is

$$\int_0^t A_c(u)\,du + \int_0^t d\gamma(u) + \int_0^t \int_{\mathbb{R}} \{e^z - 1 - z1_{|z|<1}\}\,\nu(dz)\,du = 0\,.$$

Using that the last two integrals in the expression above constitute the Lévy-Kintchine representation of the NIG Lévy process, we obtain from the explicit cumulant function (2.31) the following drift condition

$$\begin{aligned}\int_0^t A_c(u)\,du &= -\psi_{\mathrm{NIG}}(-\mathrm{i})t \\ &= -\mu t - \delta\left(\sqrt{\alpha^2-\beta^2} - \sqrt{\alpha^2-(\beta+1)^2}\right)t\,,\end{aligned}$$

as long as $|\beta+1| \leq \alpha$. We therefore find that

$$h(t,c) = F_c(0)\exp\left(-\mu t - \delta\left(\sqrt{\alpha^2-\beta^2} - \sqrt{\alpha^2-(\beta+1)^2}\right)t\right)\,.$$

Further, again using the explicit expression of the cumulant function of the NIG Lévy process J, we have

$$\begin{aligned}\psi(t,T;y-\mathrm{i}a) &= \psi_{\mathrm{NIG}}(y-\mathrm{i}a)(T-t) \\ &= (\mathrm{i}y+a)\mu(T-t) \\ &\quad + \delta\left(\sqrt{\alpha^2-\beta^2} - \sqrt{\alpha^2-(\beta+\mathrm{i}y+a)^2}\right)(T-t)\,,\end{aligned}$$

for $|\beta+a| \leq \alpha$. This gives us the following formula for the price of a call option

$$C(t;T,K,c) = e^{-r(T-t)}\frac{1}{2\pi}\int_{\mathbb{R}} \widehat{g}_{T,c}(y)e^{\psi_{\mathrm{NIG}}(y-\mathrm{i}a)(T-t)+(a+\mathrm{i}y)J(t)}\,dy\,.$$

This expression can be calculated using the FFT algorithm. We refer the interested reader to [Carr and Madan (1998)], where computational issues are discussed in an example using the CGMY distribution rather than the NIG.

We turn our attention to the arithmetic case, where we focus on options written on swaps only. The similar case of options on forward contracts is easily derived. Motivated from the arithmetic models in Subsect. 4.4.2, we introduce the dynamics

$$F(t,\tau_1,\tau_2) = H(t,\tau_1,\tau_2) + \sum_{k=1}^{p}\int_0^t \sigma_k(u,\tau_1,\tau_2)\,dW_k(u)$$

$$+\sum_{j=1}^{n}\int_0^t \eta_j(u,\tau_1,\tau_2)\,dJ_j(u)\,. \tag{9.19}$$

Here, H is a deterministic function, being continuous in $t \leq \tau_1$. Further, the volatility functions σ_k and η_j are also assumed to be continuous for $t \leq \tau_1$. Looking at the explicit dynamics for $F(t,\tau_1,\tau_2)$ derived from an arithmetic spot model in Prop. 4.14, we recognise the parameter functions in (9.19) as

$$\sigma_k(t,\tau_1,\tau_2) = \sum_{i=1}^{m}\sigma_{ik}(t)\int_{\tau_1}^{\tau_2} w(u,\tau_1,\tau_2)\mathrm{e}^{-\int_t^u \alpha(v)\,dv}\,du\,,$$

$$\eta_j(t,\tau_1,\tau_2) = \eta_j(t)\int_{\tau_1}^{\tau_2} w(u,\tau_1,\tau_2)\mathrm{e}^{-\int_t^u \beta(v)\,dv}\,du\,.$$

The rather long and technical expression for H can be derived from the dynamics in Prop. 4.14.

Define the process

$$Z(t,\tau_1,\tau_2) \triangleq \sum_{k=1}^{p}\int_0^t \sigma_k(u,\tau_1,\tau_2)\,dW_k(u) + \sum_{j=1}^{n}\int_0^t \eta_j(u,\tau_1,\tau_2)\,dJ_j(u)\,. \tag{9.20}$$

Introduce further a dampened version of the call payoff function with strike K at exercise time T

$$G_{T,\tau_1,\tau_2}(x) = \mathrm{e}^{-ax}\max(H(T,\tau_1,\tau_2)+x-K,0)\,, \tag{9.21}$$

with $a > 0$. We calculate the Fourier transform of this function.

Lemma 9.2 *Suppose $a > 0$. The Fourier transform of G_{T,τ_1,τ_2} in (9.21) is*

$$\widehat{G}_{T,\tau_1,\tau_2}(y) = \frac{1}{(a+iy)^2}e^{-(a+iy)(K-H(T,\tau_1,\tau_2))}\,.$$

Proof. Using the definition of the Fourier transform (9.16), we obtain

$$\begin{aligned}\widehat{G}_{T,\tau_1,\tau_2}(y) &= \int_{\mathbb{R}} \mathrm{e}^{-ax}\max(H(T,\tau_1,\tau_2)+x-K,0)\mathrm{e}^{-ixy}\,dx\\ &= \int_{K-H(T,\tau_1,\tau_2)}^{\infty} \mathrm{e}^{-(a+iy)x}(x+H(T,\tau_1,\tau_2)-K)\,dx\,.\end{aligned}$$

The result follows by a straightforward calculation. □

Based on the arithmetic dynamics in (9.19) we derive the price of an option in the following Proposition.

Proposition 9.5 *Suppose that there exist constants c_j such that for some constant $a > 0$, Condition G holds for $J_j(t)$ with $c_j \geq a \sup_{0\leq u\leq T} |\eta_j(u,\tau_1,\tau_2)|$, $j = 1,\ldots,n$. The price at time t for a call option written on a swap with delivery period over $[\tau_1,\tau_2]$ having strike price K and exercise time $\tau_1 \geq T \geq t$, is given by*

$$C(t;T,K,\tau_1,\tau_2) \\ = e^{-r(T-t)}\frac{1}{2\pi}\int_{\mathbb{R}} \widehat{G}_{T,\tau_1,\tau_2}(y)\Psi(t,T,\tau_1,\tau_2)e^{(a+iy)Z(t,\tau_1,\tau_2)}\,dy\,, \quad (9.22)$$

where

$$\ln\Psi(t,T,\tau_1,\tau_2) = \frac{1}{2}(a+iy)^2\sum_{k=1}^{p}\int_t^T \sigma_k^2(u,\tau_1,\tau_2)\,du \\ + \sum_{j=1}^{n}\psi_j(t,T;(y-ia)\eta_j(\cdot,\tau_1,\tau_2))\,.$$

Proof. The proof is analogous to the argument for Prop. 9.4. □

The price is very similar to the geometric case, except that we have a slightly different structure of the dampened payoff function G. However, the form of Ψ is the same except for some notational differences.

Note that the integrability condition in Prop. 9.5 holds as long as exponential moments exist for the jump processes $J_j(t)$. The reason being that $a > 0$ can be chosen arbitrarily small, and thus the lower bound for the constants c_j can always be made as small as desirable. We refer to [Crosby (2006)] for a thorough discussion on the use of Fourier techniques to price options in commodity markets.

The issue of hedging is delicate when the forward or swap price dynamics are modelled by jump processes. In general, and in particular for the NIG Lévy processes, the market is incomplete and the call option cannot be hedged perfectly. The theory tells us that there is an interval of possible arbitrage-free call option prices, spanned by the cheapest superhedging strategy and the dearest subhedging strategy. A super(sub)hedge is a strategy that has at least(most) the value of the option payoff at the exercise time. Since we choose the model of the forward price under a given risk-neutral probability, we effectively choose one price in this pricing interval.

For this price one can buy a partial hedge, which will leave some risk unhedged in the case of being short the call option. There is a lot of theory on how to choose partial hedges, which will reflect the risk preferences of the issuer of the option. Such partial hedges are intimately linked to risk-neutral probabilities. We refer to [Shiryaev (1999)] and [Cont and Tankov (2004)] for more on incomplete markets, jump processes and hedging.

9.2 Exotic Options

In this Section we study two classes of options particularly relevant for energy markets, namely average and spread options. Average options, also called Asian options, were traded on the Nord Pool power exchange for some time, being European call and put options written on the average electricity spot price. Spread options are much used in the market as a way to hedge price differences between energies. A typical example is the spark spread option, which is an option written on the difference between electricity and gas. At NYMEX, contracts on the crack spread, that is, the price difference of crude oil and a refined product, are traded.

We consider the arithmetic class of spot models presented in Subsect. 3.2.2, since they are analytically tractable for these types of options. The analysis in this Section is based on [Benth and Kufakunesu (2007)].

9.2.1 *Spread options*

Consider two energies (or commodities) A and B, with spot price dynamics defined as

$$S^A(t) = \Lambda^A(t) + \sum_{i=1}^{m} X_i^A(t) + \sum_{j=1}^{n} Y_j^A(t)\,, \qquad (9.23)$$

$$S^B(t) = \Lambda^B(t) + \sum_{i=1}^{m} X_i^B(t) + \sum_{j=1}^{n} Y_j^B(t)\,. \qquad (9.24)$$

We suppose that the first m factors X_i^A and X_i^B are common, in the sense that the OU processes are driven by the same jump processes, that is,

$$dX_i^A(t) = -\alpha_i^A X_i^A(t)\,dt + \sigma_i^A\,dI_i(t)\,,$$
$$dX_i^B(t) = -\alpha_i^B X_i^B(t)\,dt + \sigma_i^B\,dI_i(t)\,,$$

where $\alpha_i^A, \alpha_i^B, \sigma_i^A$ and σ_i^B are positive constants, and I_i, $i = 1, \ldots, m$ are m independent II processes. Further, we let

$$dY_j^A(t) = -\beta_j^A Y_j^A(t)\,dt + \eta_j^A\,dI_j^A(t)\,,$$
$$dY_j^B(t) = -\beta_j^B Y_j^B(t)\,dt + \eta_j^B\,dI_j^B(t)\,,$$

where all the parameters again are assumed to be positive constants and where I_j^A and I_j^B are independent II processes, $j = 1, \ldots, m$. Note that we dispense with the usual notation, since we let X be driven by jump processes and not Brownian motions. We use this convention here in order to have a clear notational separation of common and independent factors driving the spot price dynamics of the two energies.

A typical example of this model could be that energy A is electricity and B is gas. Further, since we consider arithmetic models with only jumps driving the dynamics, we have specifically in mind the class of models which yields positive price processes discussed in Subsect. 3.2.2. In the following, we restrict our attention to these models, implying that the II processes only have positive jumps. We can easily consider models driven by Brownian motions by appropriately redefining the X processes. However, then the positivity property of the spot models is lost.

The two spot price dynamics in (9.23) and (9.24) imply a correlation structure between the two energies.

Proposition 9.6 *Suppose that Condition A holds for a constant $c \geq 2$. Then the covariance between $S^A(t)$ and $S^B(t)$ is*

$$\mathit{Cov}\left[S^A(t), S^B(t)\right] = \sum_{i=1}^{m} \int_0^t \int_0^\infty z^2 \sigma_i^A \sigma_i^B e^{-(\alpha_i^A + \alpha_i^B)(t-s)}\,\ell_i(dz, ds)\,.$$

Proof. First, recall the explicit solution of X and Y from (3.3)[2]

$$X_i(t) = X_i(0)\mathrm{e}^{-\alpha_i t} + \int_0^t \sigma_i \mathrm{e}^{-\alpha_i(t-s)}\,dI_i(s)\,,$$
$$Y_j(t) = Y_j(0)\mathrm{e}^{-\beta_j t} + \int_0^t \eta_j \mathrm{e}^{-\beta_j(t-s)}\,dI_j(s)\,.$$

Imposing Condition A, we compensate I_i and I_j in order to write

$$X_i(t) = X_i(0)\mathrm{e}^{-\alpha_i t} + d_X(t) + \int_0^t \sigma_i \mathrm{e}^{-\alpha_i(t-s)}\,d\widetilde{I}_i(s)\,,$$

[2]We skip the indices A and B.

$$Y_j(t) = Y_j(0)\mathrm{e}^{-\beta_j t} + d_Y(t) + \int_0^t \eta_j \mathrm{e}^{-\beta_j(t-s)}\, d\widetilde{I}_j(s)\,,$$

where $\widetilde{I}_i$ and $\widetilde{I}_j$ are martingales, and d_X and d_Y are (deterministic) drift terms induced by the compensation. Hence, we calculate the covariance appealing to the independence of the different jump components as follows

$$\begin{aligned}\operatorname{Cov}\left[S^A(t), S^B(t)\right] &= \mathbb{E}\left[\left(\sum_{i=1}^m \int_0^t \sigma_i^A e^{-\alpha_i^A(t-s)}\, d\widetilde{I}_i(s)\right)\right.\\ &\qquad \left.\times\left(\sum_{i=1}^m \int_0^t \sigma_i^B e^{-\alpha_i^B(t-s)}\, d\widetilde{I}_i(s)\right)\right]\\ &= \sum_{i=1}^m \mathbb{E}\left[\int_0^t \sigma_i^A e^{-\alpha_i^A(t-s)}\, d\widetilde{I}_i(s) \int_0^t \sigma_i^B e^{-\alpha_i^B(t-s)}\, d\widetilde{I}_i(s)\right],\end{aligned}$$

from which the result follows. □

If the common factors I_i, $i = 1, \ldots, m$ are subordinators, we have

$$\operatorname{Cov}\left[S^A(t), S^B(t)\right] = \sum_{i=1}^m \frac{\sigma_i^A \sigma_i^B}{\alpha_i^A + \alpha_i^B}\left(1 - \mathrm{e}^{-(\alpha_i^A + \alpha_i^B)t}\right) \int_{\mathbb{R}} z^2\, \widetilde{\ell}_i(dz)\,,$$

since the compensator measures are $\ell_i(dz, ds) = \widetilde{\ell}_i(dz)\, ds$. Letting $t \to \infty$, we find the stationary covariance function

$$\operatorname{Cov}\left[S^A(t), S^B(t)\right] = \sum_{i=1}^m \frac{\sigma_i^A \sigma_i^B}{\alpha_i^A + \alpha_i^B} \int_0^\infty z^2\, \ell_i(dz)\,.$$

This can be utilised in an estimation of the spot models to actual data.

We now analyse the price of European call options written on different combinations of the two energies, like, for example, the spread. Typically, the spread between two assets is defined as

$$S^A(t) - S^B(t)\,,$$

whereas the spark spread is

$$S^A(t) - h_R S^B(t)\,,$$

with A being electricity, B gas, and the constant h_R the *heat rate* converting gas into the energy equivalent of electricity. We consider general contracts written on the linear combination

$$S(t) \triangleq a S^A(t) + b S^B(t)\,, \tag{9.25}$$

where a and b are constants. Hence, we consider options written on a basket of two energies, being slightly more general than merely restricting our attention to spreads. Consider a European option with exercise time T and payoff $f(S(T))$ at exercise. In order to apply the Fourier techniques, we restrict the class of payoff functions to those which are integrable over the whole real line, that is, $f \in L^1(\mathbb{R})$. The price $C(t)$ at time t of the option is defined (whenever $f(S(T)) \in L^1(Q)$) as

$$C(t) = \mathrm{e}^{-r(T-t)}\mathbb{E}_Q\left[f(S(T)) \mid \mathcal{F}_t\right] . \tag{9.26}$$

As usual, we consider the class of risk-neutral probabilities defined by the Esscher transform, Q^θ. We let $\theta(\cdot)$ be a bounded and continuous function on $[0, T]$, with values in $\mathbb{R}^{n+2m}$. The first n coordinate functions are denoted by $\widehat{\theta}_i$, and correspond to the Esscher transform of the common jump processes I_i. Further, we let $\widetilde{\theta}_j^A$ and $\widetilde{\theta}_j^B$ be the Esscher transforms of the jump processes I_j^A and I_j^B, respectively.

The price may be expressed in terms of the cumulant functions of the jump processes, as seen in the following Proposition.

Proposition 9.7 *For $i = 1, \ldots, m$, we assume there exist constants c_i where*

$$c_i \geq \sup_{u \leq T} \left| a\sigma_i^A e^{-\alpha_i^A(T-u)} + b\sigma_i^B e^{-\alpha_i^B(T-u)} + \widehat{\theta}(u)\right| ,$$

such that Condition G holds for I_i. Further, for $j = 1, \ldots, n$, we assume there exist constants c_j^A and c_j^B, where

$$c_j^A \geq \sup_{u \leq T} \left| a\eta_j^A e^{-\beta_j^A(T-u)} + \widetilde{\theta}_j^A(u)\right| ,$$

$$c_j^A \geq \sup_{u \leq T} \left| b\eta_j^B e^{-\beta_j^B(T-u)} + \widetilde{\theta}_j^B(u)\right| ,$$

such that Condition G holds for I_j^A and I_j^B, respectively. The price $C(t)$ of a European option with payoff $f(S(T))$ at exercise time $T \geq t$ is

$$C(t) = \frac{e^{-r(T-t)}}{2\pi} \int_{\mathbb{R}} \widehat{f}(y)\Psi(t, T, y, \theta)\, dy ,$$

where

$$\begin{aligned} \ln\Psi(t, T, y, \theta) &= iy(a\Lambda^A(T) + b\Lambda^B(T)) \\ &\quad + iy\sum_{i=1}^{m} aX_i^A(t)e^{-\alpha_i^A(T-t)} + bX_i^B(t)e^{-\alpha_i^B(T-t)} \end{aligned}$$

$$
\begin{aligned}
&+ iy \sum_{j=1}^{n} aY_j^A(t)e^{-\beta_j^A(T-t)} + bY_j^B(t)e^{-\beta_j^B(T-t)} \\
&+ \sum_{i=1}^{m} \psi_i(t,T; y(a\sigma_i^A e^{-\alpha_i^A(T-\cdot)} + b\sigma_i^B e^{-\alpha_i^B(T-\cdot)} - i\widehat{\theta}_i(\cdot))) \\
&\qquad - \psi_i(t,T; -i\widehat{\theta}_i(\cdot)) \\
&+ \sum_{j=1}^{n} \psi_j^A(t,T; y(a\eta_j^A e^{-\beta_j^A(T-\cdot)} - i\widetilde{\theta}_j^A(\cdot))) - \psi_j^A(t,T; -i\widetilde{\theta}_j^A(\cdot)) \\
&+ \sum_{j=1}^{n} \psi_j^B(t,T; y(b\eta_j^B e^{-\beta_j^B(T-\cdot)} - i\widetilde{\theta}_j^B(\cdot))) - \psi_j^B(t,T; -i\widetilde{\theta}_j^B(\cdot)) .
\end{aligned}
$$

Here, ψ_i, $i = 1, \ldots, m$ are the cumulant functions of I_i, while ψ_j^A and ψ_j^B are the cumulant functions of I_j^A and I_j^B, respectively, for $j = 1, \ldots, n$.

Proof. Using the inverse Fourier transform, we get

$$
\mathbb{E}_\theta \left[f(S(T)) \,|\, \mathcal{F}_t \right] = \frac{1}{2\pi} \int_{\mathbb{R}} \widehat{f}(y) \mathbb{E}_\theta \left[e^{iyS(T)} \,|\, \mathcal{F}_t \right] dy ,
$$

where $\widehat{f}$ is the Fourier transform of the payoff function. We need to calculate the characteristic function of $S(T)$, and show that this coincides with $\Psi(t,T,y,\theta)$ as in the Proposition.

From the explicit dynamics (see (3.3)) of the OU processes making up $S^A(T)$ and $S^B(T)$, we have

$$
\begin{aligned}
S(T) &= a\Lambda^A(T) + b\Lambda^B(T) \\
&\quad + \sum_{i=1}^{m} aX_i^A(t)e^{-\alpha_i^A(T-t)} + bX_i^B e^{-\alpha_i^B(T-t)} \\
&\quad + \sum_{j=1}^{n} aY_j^A(t)e^{-\beta_j^A(T-t)} + bY_j^B(t)e^{-\beta_j^B(T-t)} \\
&\quad + \sum_{i=1}^{m} \int_t^T (a\sigma_i^A e^{-\alpha_i^A(T-s)} + b\sigma_i^B e^{-\alpha_i^B(T-s)})\, dI_i(s) \\
&\quad + \sum_{j=1}^{n} \int_t^T a\eta_j^A e^{-\beta_j^A(T-s)}\, dI_j^A(s) + \int_t^T b\eta_j^B e^{-\beta_j^B(T-s)}\, dI_j^B(s) .
\end{aligned}
$$

Recall from Prop. 4.4 that all the involved II processes preserve the independent increment property under Q^θ, and moreover, they are mutually independent. Further, $X_i^A(t), X_i^B(t), Y_j^A(t)$ and $Y_j^B(t)$ are all $\mathcal{F}_t$-measurable.

Hence,

$$
\begin{aligned}
\ln \mathbb{E}_\theta\left[\mathrm{e}^{\mathrm{i}yS(T)} \,|\, \mathcal{F}_t\right] &= \mathrm{i}y(a\Lambda^A(T) + b\Lambda^B(T)) \\
&\quad + \mathrm{i}y\sum_{i=1}^{m} aX_i^A(t)\mathrm{e}^{-\alpha_i^A(T-t)} + bX_i^B\mathrm{e}^{-\alpha_i^B(T-t)} \\
&\quad + \sum_{j=1}^{n} aY_j^A(t)\mathrm{e}^{-\beta_j^A(T-t)} + bY_j^B(t)\mathrm{e}^{-\beta_j^B(T-t)} \\
&\quad + \sum_{i=1}^{m} \ln \mathbb{E}_\theta\left[\mathrm{e}^{\mathrm{i}y\int_t^T (a\sigma_i^A\mathrm{e}^{-\alpha_i^A(T-s)} + b\sigma_i^B\mathrm{e}^{-\alpha_i^B(T-s)})\,dI_i(s)}\right] \\
&\quad + \sum_{j=1}^{n} \ln \mathbb{E}_\theta\left[\mathrm{e}^{\mathrm{i}y\int_t^T a\eta_j^A\mathrm{e}^{-\beta_j^A(T-s)}\,dI_j^A(s)}\right] \\
&\quad + \sum_{j=1}^{n} \ln \mathbb{E}_\theta\left[\mathrm{e}^{\mathrm{i}y\int_t^T b\eta_j^B\mathrm{e}^{-\beta_j^B(T-s)}\,dI_j^B(s)}\right].
\end{aligned}
$$

By using the same argument as for Prop. 4.4, the proof is complete. □

This expression for the price of the option is suitable for the FFT method, as long as we know the cumulant functions.

We remark that the plain vanilla contracts as European call and put options are in general not included in the class of payoff functions $f(x)$ that we can directly treat here. Note in particular that the put option is not covered, which is due to the fact that the basket $S(t)$ may attain arbitrary negative values (think of a spark spread, where electricity becomes very cheap and gas prices rocket up), yielding an unbounded payoff for the holder of the option. Standard knock-out structures on calls and puts are, on the other hand, included. To allow for pricing of plain vanilla calls and puts, we dampen the payoff function by an exponential function, as suggested by [Carr and Madan (1998)], and used explicitly in Lemma 9.1 above. We leave to the interested reader to work out the details.

Also, there are contracts written on the spread between the swap (or forward) prices of two energies, rather than the spots. For instance, this is particularly relevant for gas and electricity. With the models above, it is possible to derive an explicit dynamics for the forwards and swap prices, which become arithmetic (see Subsect. 4.3.2). Hence, similar calculations as above would yield expressions for the option price in this case as well.

Of course, if both a and b are positive, then the payoff from a put option is indeed in $L^1(\mathbb{R})$. This holds true since $f(x) = \max(K - x, 0)$, and $aS^A(T) + bS^B(T)$ is positive by definition of the two positive marginal processes. Hence, f is supported on $[0, K]$, which implies integrability of the function. Call options may now be priced from the put-call parity.

As a final note, we consider modelling the spark spread of UK electricity and gas from a slightly different point of view in Sect. 9.3. There, we propose a dynamics modelling directly the observed spark spread, and price options based on this.

9.2.2 *Asian options*

We consider average-type options, or Asian options, written on an energy spot price $S(t)$ defined as in Subsect. 3.2.2

$$S(t) = \Lambda(t) + \sum_{j=1}^{n} Y_j(t)\,.$$

Here, the jump processes I_j in Y_j are increasing and $\Lambda(t)$ is the floor of the mean reversion dynamics. Suppose the option pays $f(\int_{\tau_1}^{\tau_2} S(u)\,du)$ at maturity τ_2, with $f \in L^1(\mathbb{R})$. The price $C(t)$ of the option at time t is defined (as long as $f(\int_{\tau_1}^{\tau_2} S(u)\,du) \in L^1(Q))$ as

$$C(t) = \mathrm{e}^{-r(\tau_2 - t)}\mathbb{E}_Q\left[f(\int_{\tau_1}^{\tau_2} S(u)\,du) \,|\, \mathcal{F}_t\right], \tag{9.27}$$

where we, as usual, consider risk-neutral measures Q defined by the Esscher transform. For simplicity, we restrict our attention to Asian options which are traded up to the start of the averaging period, that is, we derive a price process $C(t)$ for $t \leq \tau_1$. The price of the options is given in the following Proposition.

Proposition 9.8 *Suppose Condition G holds for constants c_j, $j = 1, \ldots, n$, with*

$$\sup_{0 \leq u \leq \tau_2} |\widetilde{\theta}_j(u)| \leq c_j\,.$$

Then the price $C(t)$ at time $t \leq \tau_1$ of an option maturing at time $\tau_2 > \tau_1$ and paying $f(\int_{\tau_1}^{\tau_2} S(u)\,du)$ is given as

$$C(t) = \frac{e^{-r(\tau_2 - t)}}{2\pi} \int_{\mathbb{R}} \widehat{f}(y)\Psi(t, \tau_1, \tau_2, y, \theta)\,dx\,.$$

Here, Ψ is defined as

$$
\begin{aligned}
&\ln\Psi(t,\tau_1,\tau_2,y,\theta)\\
&\quad = iy\int_{\tau_1}^{\tau_2}\Lambda(u)\,du + iy\sum_{j=1}^{n}\int_{\tau_1}^{\tau_2} e^{-\int_t^u \beta_j(v)\,dv}\,du\,Y_j(t)\\
&\qquad + \sum_{j=1}^{n}\psi_j(t,\tau_2;y\int_{\max(\cdot,\tau_1)}^{\tau_2}\eta_j(\cdot)e^{-\int_\cdot^u \beta_j(v)\,dv}\,du - i\widetilde{\theta}_j(\cdot))\\
&\qquad\qquad - \psi_j(t,\tau_2;-i\widetilde{\theta}_j(\cdot))\,,
\end{aligned}
$$

and ψ_j, $j=1,\dots,n$, are the cumulant functions of I_j.

Proof. Using the Fourier transform, we get

$$
\mathbb{E}_\theta\left[f(\int_{\tau_1}^{\tau_2} S(u)\,du)\,|\,\mathcal{F}_t\right] = \frac{1}{2\pi}\int_{\mathbb{R}}\widehat{f}(y)\mathbb{E}_\theta\left[e^{iy\int_{\tau_1}^{\tau_2} S(u)\,du}\,|\,\mathcal{F}_t\right]dy\,.
$$

We now calculate the conditional expectation in the expression for the inverse Fourier transform. Recall from (3.3) that with $u \geq t$ we find

$$
Y_j(u) = Y_j(t)e^{-\int_t^u \beta_j(v)\,dv} + \int_t^u \eta_j(s)e^{-\int_s^u \beta_j(v)\,dv}\,dI_j(s)\,.
$$

Thus, from the stochastic Fubini theorem

$$
\begin{aligned}
\int_{\tau_1}^{\tau_2} Y_j(u)\,du &= Y_j(t)\int_{\tau_1}^{\tau_2} e^{-\int_t^u \beta_j(v)\,dv}\,du\\
&\quad + \int_{\tau_1}^{\tau_2}\int_t^{\tau_2} 1_{[t,u]}(s)\eta_j(s)e^{-\int_s^u \beta_j(v)\,dv}\,dI_j(s)\,du\\
&= Y_j(t)\int_{\tau_1}^{\tau_2} e^{-\int_t^u \beta_j(v)\,dv}\,du\\
&\quad + \int_t^{\tau_2}\eta_j(s)\int_{\max(s,\tau_1)}^{\tau_2} e^{-\int_s^u \beta_j(v)\,dv}\,du\,dI_j(s)\,.
\end{aligned}
$$

From the $\mathcal{F}_t$-measurability of $Y_j(t)$, and the independent increment property of I_j under Q^θ, we find

$$
\begin{aligned}
&\ln\mathbb{E}_\theta\left[e^{iy\int_{\tau_1}^{\tau_2} S(u)\,du}\,|\,\mathcal{F}_t\right]\\
&\quad = iy\int_{\tau_1}^{\tau_2}\Lambda(u)\,du + iy\sum_{j=1}^{n}Y_j(t)\int_{\tau_1}^{\tau_2} e^{-\int_t^u \beta_j(v)\,dv}\,du
\end{aligned}
$$

$$+\sum_{j=1}^{n} \ln \mathbb{E}_\theta \left[\mathrm{e}^{\mathrm{i}y \int_t^{\tau_2} \eta_j(s) \int_{\max(s,\tau_1)}^{\tau_2} \mathrm{e}^{-\int_s^u \beta_j(v)\,dv}\, du\, dI_j(s)} \right].$$

By appealing to the same argument as in the proof of Prop. 4.4, we reach the conclusion of the Proposition. □

Recall that since $S(t)$ is the positive arithmetic model, we have that $S(u) \geq 0$, and therefore the average $\int_{\tau_1}^{\tau_2} S(u)\,du/(\tau_2 - \tau_1)$ is also positive. This implies that a plain vanilla put option has a payoff function $f(x) = \max(K - x/(\tau_2 - \tau_1), 0)$ which is supported on $[0, K]$. Then it follows that $f \in L^1(\mathbb{R})$. Hence, in contrast to the spread case, we can price put options by the result above. Call options may be priced by appealing to the put-call parity. Alternatively, we may use the exponential dampening of the payoff function as in [Carr and Madan (1998)].

We remark in passing that in the paper [Fusai, Marena and Roncoroni (2007)], a closed-form formula for the value of a discretely monitored Asian option written on a square root spot price process is derived. Their formula is based on the moment generating function of the spot process, and the Laplace transform can be used for numerical valuation. Empirical tests are performed on natural gas data collected at NYMEX and corn data from Chicago Board of Trade, showing an improved performance over the standard approaches with models using geometric Brownian motion.

9.3 Case Study: Valuation of spark spread options – a direct approach

In Sect. 9.2 above, we considered pricing of spread options based on a joint arithmetic model for two commodities. The approach allows for marginal modelling of each commodity, as well as correlating the two. In this Section we revisit the problem of pricing spread options, in particular, spark spread options. The idea is to model the spread between the two commodities *directly*, rather than creating a two-dimensional model. More specifically, we propose an arithmetic mean reversion model for the spread between spot electricity and gas, and show how the spark spread option can be priced by Fourier techniques.

The motivation for our approach is taken from [Carmona and Durrleman (2003)], where the authors provide a comprehensive analysis of the pricing of spread options. Their starting point is geometric models for the different commodities. In this case it is not possible, in general, to price spread options. An exception is a spread option with strike equal to zero, where Margrabe's Formula gives the price for call and put options (see [Margrabe (1978)]). [Carmona and Durrleman (2003)] argue that the difference of two lognormal random variables can be approximated reasonably well by a normal random variable. This implies analytical pricing formulas which can be used for approximating the price of a spread option written on, say, the difference of electricity and gas. More specifically, options written on the difference of two geometric Brownian motions, or two exponential OU processes driven by Brownian motions, may be approximated by a pricing formula derived from a normal distribution.

Empirical studies by [Lima (2005)] for the spread of spot electricity and gas traded in the UK show that this is far from normal. In fact, the empirical spread has a heavy-tailed distribution which is far better approximated by a NIG distribution. In [Benth and Šaltytė-Benth (2006)], these observations are taken one step further, and it is proposed to model the dynamics of the empirical spread directly. We analyse an arithmetic mean-reverting model with jumps, as studied in Subsect. 3.2.2.

The dynamics of the spread may be described reasonably well with an arithmetic process, which is simple to use when deriving prices on spread options based on forwards and swaps. The prices can be expressed in terms of characteristic functions, which easily allows for a numerical valuation based on FFT. The two-dimensional arithmetic model used in Sect. 9.2 may be difficult to fit to real data, whereas the direct approach is straightforward. Also, no approximations are involved, since we model the underlying spark spread directly, and price the options explicitly. We do not need to fit two exponential stochastic processes jointly to the two commodities, and thereafter approximate with a normal distribution. Furthermore, it is not clear what should be the right approximation in the case of an exponential jump-diffusion dynamics for the two energies. We know that a diffusion may miss the leptokurtic behaviour of power spot prices significantly, and a normal approximation of the difference seems questionable.

The disadvantage with our approach is that we lose the connection with the marginal behaviour. The pricing of spark spread options will not directly depend on any parameters which can be associated with one of the two commodities making up the spread. The dependency indirectly comes into play through the parameters of the joint model, which are estimated based on the simultaneous behaviour of the two commodities. The correlation is also not modelled directly.

In the rest of this Section, we follow the analysis in [Benth and Šaltytė-

Benth (2006)].

9.3.1 *Modelling and analysis of spark spread options*

In this Subsection we introduce, at a formal level, the spark spread options that we are going to analyse, and establish some connections related to our proposed way of modelling. We let $E(t)$ and $G(t)$ be, respectively, the electricity and gas spot price at time t. If h_R denotes the heat rate, that is, the factor converting gas prices into the units of electricity, the spark spread difference between electricity and gas is

$$S(t) = E(t) - h_R G(t)\,. \tag{9.28}$$

We call $S(t)$ simply the *spark spread*.

Let us discuss shortly the heat rate. Electricity energy is measured in units of MWh, whereas gas energy is measured in 'therm'. Now, recall from Sect. 1.2 that 1 therm corresponds to 0.029306 MWh. This leads to the following conversion from pence/therm to £/MWh.

$$1\frac{\text{pence}}{\text{therm}} \cdot \left[\frac{\text{therm}}{0.029306\text{MWh}} \cdot \frac{£}{100\text{pence}}\right] = 0.341\frac{£}{\text{MWh}}\,.$$

The heat rate h_R takes into account the relative efficiency of gas compared to electricity. Hence, we have that the heat rate is defined by

$$h_R \triangleq 0.341/\text{rate of efficiency}\,. \tag{9.29}$$

In the empirical analysis, we come back to the specification of h_R.

Consider a spark spread option written on the spread of an electricity futures and a gas futures, both with a delivery period $[\tau_1, \tau_2]$. Letting $F_{\text{el}}(t, \tau_1, \tau_2)$ and $F_{\text{gas}}(t, \tau_1, \tau_2)$ be the electricity and gas futures prices, respectively, we find the price of an option with exercise at time T with strike K to be

$$C(t) = \mathrm{e}^{-r(T-t)}\mathbb{E}_Q\left[\max\left(F_{\text{el}}(T, \tau_1, \tau_2) - h_R F_{\text{gas}}(T, \tau_1, \tau_2) - K, 0\right) \mid \mathcal{F}_t\right]\,. \tag{9.30}$$

The electricity and gas futures prices can both be represented as

$$F(t, \tau_1, \tau_2) = \mathbb{E}_Q\left[\frac{1}{\tau_2 - \tau_1}\int_{\tau_1}^{\tau_2} X(u)\,du \mid \mathcal{F}_t\right] \tag{9.31}$$

with X being either $E(t)$ or $G(t)$. We assume that there is a measure Q describing the risk preferences in both the gas and electricity markets. This can be done without loss of generality since we model both markets jointly, and the Q probability refer to the stochastic processes.

Using the definitions of $F_{\mathrm{el}}(t,\tau_1,\tau_2)$ and $F_{\mathrm{gas}}(t,\tau_1,\tau_2)$ with $w(u,\tau_1,\tau_2)=1/(\tau_1-\tau_1)$ as weight function (recall discussion in Sect. 4.1), we find

$$\begin{aligned}F_{\mathrm{spread}}(t,\tau_1,\tau_2)&=F_{\mathrm{el}}(t,\tau_1,\tau_2)-h_RF_{\mathrm{gas}}(t,\tau_1,\tau_2)\\&=\frac{1}{\tau_2-\tau_1}\mathbb{E}_Q\left[\int_{\tau_1}^{\tau_2}(E(u)-h_RG(u))\,du\,|\,\mathcal{F}_t\right]\\&=\mathbb{E}_Q\left[\frac{1}{\tau_2-\tau_1}\int_{\tau_1}^{\tau_2}S(u)\,du\,|\,\mathcal{F}_t\right].\end{aligned}\tag{9.32}$$

Thus, we see that the spark spread option can be considered as an option written on a swap contract delivering the spark spread over the period $[\tau_1,\tau_2]$. It follows from this that we can price the option based on a model for the spark spread directly, rather than modelling the electricity and gas spot prices separately. Further, we recall that we have calculated conditional expectations like (9.32) in Chapter 4 for the dynamics $S(t)$ belonging to the arithmetic class.

Let the dynamics of $S(t)$ be as in Subsect. 3.2.2, that is,

$$S(t)=\Lambda(t)+X(t)+Y(t)\,,\tag{9.33}$$

with $\Lambda(t)$ being the seasonality function, and $X(t)$ and $Y(t)$ the mean reversion processes

$$dX(t)=-\alpha X(t)\,dt+\sigma\,dB(t)\,,\tag{9.34}$$

and

$$dY(t)=-\alpha Y(t)\,dt+dI(t)\,,\tag{9.35}$$

respectively. Note that the speed of mean reversion is the same in both processes X and Y, implying that we in fact have one OU process driving the dynamics, that is,

$$d(X(t)+Y(t))=-\alpha(X(t)+Y(t))\,dt+\sigma\,dB(t)+dI(t)\,.$$

The volatility σ is a non-negative constant. Recall that this is the arithmetic model with $m=p=n=1$. We choose $I(t)$ to be a Lévy process.

In order to price options on the spark spread, we need the risk-neutral dynamics of $S(t)$. We choose, as usual, to work with the Esscher transform

Q^θ with $\theta = (\widehat{\theta}, \widetilde{\theta})$. For simplicity, we restrict our attention to constant market prices of risk. Supposing Condition G holds for a constant c so that

$$|\widetilde{\theta}| + \epsilon \leq c,$$

where $\epsilon > 0$, it follows from Prop. 4.14 that the spread swap price $F_{\text{spread}}(t, \tau_1, \tau_2)$ is

$$\begin{aligned} F_{\text{spread}}(t, \tau_1, \tau_2) &= \frac{1}{\tau_2 - \tau_1} \int_{\tau_1}^{\tau_2} \Lambda(u)\, du + \Theta(t, \tau_1, \tau_2; \theta) \\ &\quad + (S(t) - \Lambda(t)) \frac{1}{\alpha(\tau_2 - \tau_1)} \left(\mathrm{e}^{-\alpha(\tau_1 - t)} - \mathrm{e}^{-\alpha(\tau_2 - t)} \right). \end{aligned} \tag{9.36}$$

Here,

$$\begin{aligned} \Theta(t, \tau_1, \tau_2; \theta) &= \frac{1}{\alpha} \left(\sigma\widehat{\theta} + \gamma + \int_{\mathbb{R}} z \left(\mathrm{e}^{\widetilde{\theta} z} - 1_{|z|<1} \right) \widetilde{\ell}(dz) \right) \\ &\quad \times \left(1 - \frac{1}{\alpha(\tau_2 - \tau_1)} \left(\mathrm{e}^{-\alpha(\tau_1 - t)} - \mathrm{e}^{-\alpha(\tau_2 - t)} \right) \right). \end{aligned}$$

Recall that we assumed I to be a Lévy process. Then the drift $\gamma(t)$ is a constant denoted simply γ and the compensator measure becomes $\ell(dz, dt) = \widetilde{\ell}(dz)\, dt$. Thus, by using the constancy of parameters we reach the above expression for $F_{\text{spread}}(t, \tau_1, \tau_2)$. The function $\Theta(t, \tau_1, \tau_2; \theta)$ incorporates the risk adjustment in the spread swap price. We observe that since

$$1 - \frac{1}{\alpha(\tau_2 - \tau_1)} \left(\mathrm{e}^{-\alpha(\tau_1 - t)} - \mathrm{e}^{-\alpha(\tau_2 - t)} \right) > 0$$

is increasing and converging to one whenever $\tau_1 \to \infty$ and $\tau_2 - \tau_1$ is constant, the risk adjustment is increasing with increasing start of the settlement period. Hence, the market price of risk is less pronounced in the short end of the swap curve than in the long.

We now consider pricing of call options on the spread swap using the Fourier approach. First, the dynamics of $F_{\text{spread}}(t, \tau_1, \tau_2)$ is given by

$$dF_{\text{spread}}(t, \tau_1, \tau_2) = \sigma\overline{\alpha}(t, \tau_1, \tau_2)\, dB^\theta(t) + \overline{\alpha}(t, \tau_1, \tau_2) \int_{\mathbb{R}} z\, \widetilde{N}^\theta(dt, dz), \tag{9.37}$$

with

$$\overline{\alpha}(t, \tau_1, \tau_2) = \frac{1}{\alpha(\tau_2 - \tau_1)} \left(\mathrm{e}^{-\alpha(\tau_1 - t)} - \mathrm{e}^{-\alpha(\tau_2 - t)} \right).$$

From the definitions of the Esscher transform and the Lévy process $I(t)$, we have

$$\int_{\mathbb{R}} z\widetilde{N}^{\theta}(dz,dt) = dJ(t) + \left(\gamma - \int_{\mathbb{R}} z\left(e^{\widetilde{\theta}z} - 1_{|z|<1}\right)\widetilde{\ell}(dz)\right)dt\,,$$

where $J(t)$ is a Lévy process under Q^{θ} with cumulant function

$$\psi^{\widetilde{\theta}}(\lambda) = \psi(\lambda - \mathrm{i}\widetilde{\theta}) - \psi(-\mathrm{i}\widetilde{\theta})\,,$$

and ψ is the cumulant function of $I(t)$ under the market probability P. Hence,

$$F_{\text{spread}}(t,\tau_1,\tau_2) = H(t,\tau_1,\tau_2) + Z(t,\tau_1,\tau_2) \tag{9.38}$$

with

$$H(t,\tau_1,\tau_2) = F_{\text{spread}}(0,\tau_1,\tau_2) - \left(\gamma + \int_0^t\int_{\mathbb{R}} z\left(e^{\widetilde{\theta}z} - 1_{|z|<1}\right)\widetilde{\ell}(dz)\right)t\,, \tag{9.39}$$

and

$$Z(t,\tau_1,\tau_2) = \int_0^t \sigma\overline{\alpha}(s,\tau_1,\tau_2)\,dB^{\theta}(s) + \int_0^t \overline{\alpha}(s,\tau_1,\tau_2)\,dJ(s)\,. \tag{9.40}$$

Hence, we have identified the dynamics of F_{spread} as in (9.19) and (9.20), and it follows from Prop. 9.5 that the price of a call option is

$$C(t;T,K,\tau_1,\tau_2) = e^{-r(T-t)}\frac{1}{2\pi}\int_{\mathbb{R}} \widehat{G}_{T,\tau_1,\tau_2}(y)\Psi(t,T,\tau_1,\tau_2)e^{(a+\mathrm{i}y)Z(t,\tau_1,\tau_2)}\,dy\,. \tag{9.41}$$

The function $\widehat{G}$ is defined in Lemma 9.2, and $a > 0$ is the dampening factor. The function Ψ is defined in Prop. 9.5. Note that as long as $I(t)$ satisfies Condition G for some $c > 0$, the integrability condition in Prop. 9.5 is fulfilled. The reason being that $\overline{\alpha}(t,\tau_1,\tau_2)$ is bounded, and since a, the dampening factor, can be chosen arbitrarily small, the desired exponential integrability follows. Thus, having access to the cumulant function of $I(t)$, we can calculate the price dynamics of a call option written on the spark spread using the FFT technique as long as exponential moments of I exist.

We remark that applying the exponential OU models for the two marginal spot prices will not lead to any explicit dynamics for the forward price, but has to be represented as an integral. It becomes impossible to derive any analytical option prices for the spark spread swap, and Monte

Carlo methods must be used for numerical pricing. We refer to the paper [Benth and Kettler (2006)] for a modelling of spark spread using two exponential OU models joined by a copula.

9.3.2 *Empirical analysis of UK gas and electricity spread*

We want to fit the arithmetic model (9.33) for the spark spread $S(t)$ to the observed electricity and gas data from the UK. The gas spot prices are the same as analysed in Subsect. 5.1.1. In addition, we have available daily average spot prices of electricity from the England and Wales market over the same time span. Thus, after imputing the missing values in the electricity data using the technique described in Subsect. 5.1.1 for gas, we have 1,454 records of spot price data. Electricity prices are measured in £/MWh, while gas prices are quoted in pence/therm. A plot of the time series of electricity prices is depicted in Fig. 9.1.

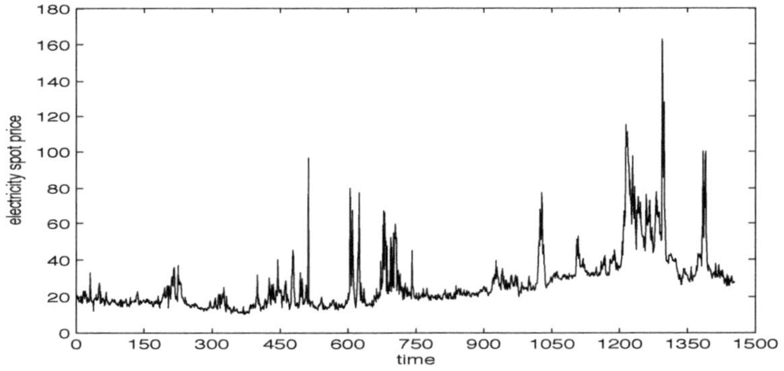

Fig. 9.1 Daily average electricity spot prices from the England and Wales market.

Denoting by $e(t)$ and $g(t)$ the daily observed electricity and gas prices at time t, we transform them into spark spread data by the formula

$$s(t) = e(t) - h_R g(t) .$$

We assume the rate of efficiency to be 40%, which from (9.29) implies a heat rate of $h_R = 0.85$. An efficiency of 40% is not unreasonable from a practical perspective (see [Eydeland and Wolyniec (2003)], where Table 2.5 on page 49 presents the efficiency of different generating units). The spark spread data are presented in Fig. 9.2. The time series shows both positive and negative price spikes, as well as long periods of smaller variations. Further, being a difference of two prices series, the spark spread attains both positive and negative values. Furthermore, it seems that there may be some seasonality present in the data set. Hence, it is not unnatural

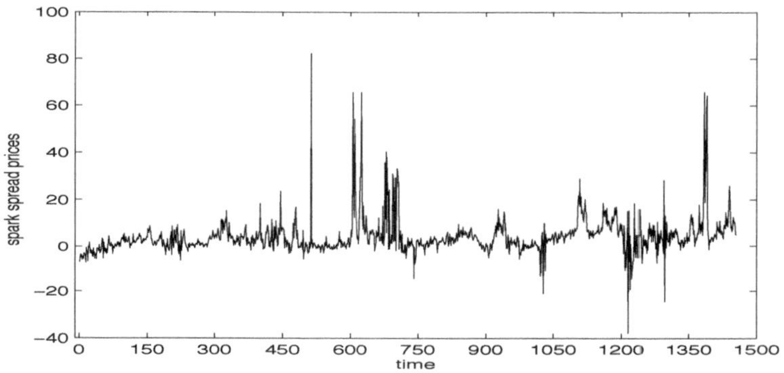

Fig. 9.2 Spark spread prices.

to propose an AR(1) model with seasonal level and mixed jump-diffusion residuals, as defined in (9.34) and (9.35). The procedure to fit this model to data follows closely the steps in Subsect. 5.1.1, and we report here only the results and some of the crucial steps.

The mean level is modelled by a trend function and four seasonal components,

$$\Lambda(t) = a_1 + a_2 t + \sum_{k=1}^{4} a_{2k+1} \cos\left(2\pi(t - a_{2k+2})/l_k\right), \tag{9.42}$$

where $l_1 = 250$, $l_2 = 52$, $l_3 = 12$, and $l_4 = 4$. This means that we consider a seasonal mean over the year, quarter, month and week. We fitted the parameters in (9.42) by using the least squares approach. The results are reported in Table 9.1 (parameters a_5, a_7 and a_9 are not significant at the 5% level). We conclude that there is an increase in the spark spread spot over the sample period, and that there are significant seasonal effects present.

The next step is to estimate the mean reversion in an AR model for the detrended and deseasonalized spark spread data. We plot the ACF of the time series in Fig. 9.3, while the partial ACF (PACF) plot is given in Fig. 9.4. Both figures indicate that we need (at least) an AR(1) model to capture the memory effect present in the data. The estimated AR(1) pa-

Table 9.1 Fitted parameters of $\Lambda(t)$

a_1	a_2	a_3	a_4	a_5
1.0618	0.0033	−2.4332	17.0352	−0.1235
a_6	a_7	a_8	a_9	a_{10}
−107.1736	0.2926	−13.3370	−0.1417	24.0090

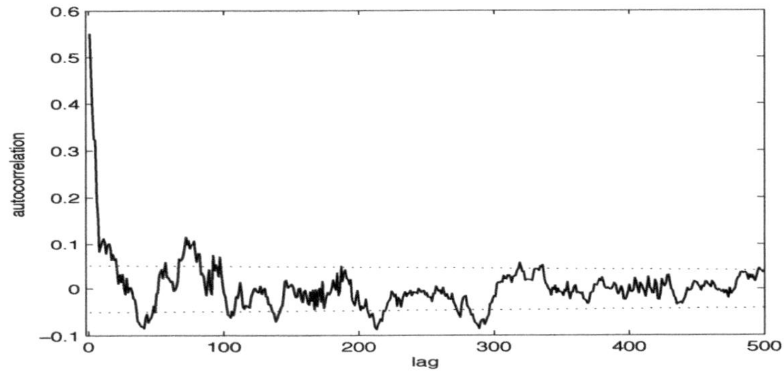

Fig. 9.3 ACF of detrended and deseasonalized spark spread data with 95% confidence interval.

rameter is 0.55, being significant at the 1% level.[3] This implies an estimate of the α parameter in (9.34) and (9.35) of

$$\widehat{\alpha} = 0.45\,, \tag{9.43}$$

implying a rather strong mean reversion in the data.

The histogram of the residuals from the autoregression with the fitted normal distribution depicted in Fig. 9.5 motivates us to go further with estimating a mixed jump-diffusion model. We clearly see that the normal distribution (the solid curve) is not at all matching the empirical distribution of the residuals, which is much more centered and with heavy tails. We use recursive filtering to identify jumps in the residuals (see Subsect. 5.1.2 for a description of recursive filtering). The results of the procedure are reported in Table 9.2. The estimated standard deviation of the filtered residuals is 2.216, which then becomes our estimate for the volatility σ,

[3]See Subsect. 5.1.1 for a comment on p-values.

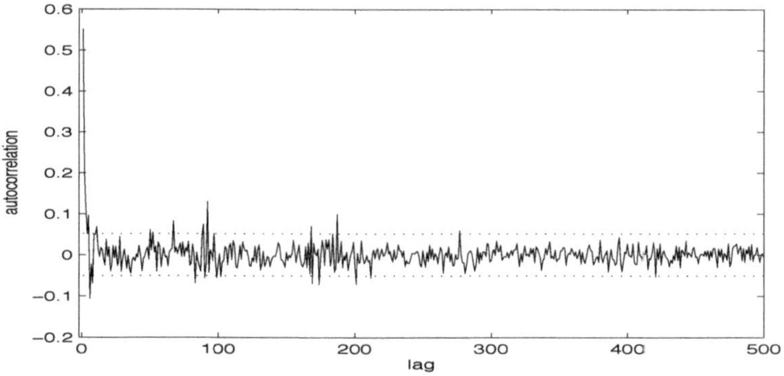

Fig. 9.4 PACF of detrended and deseasonalized spark spread prices with 95% confidence interval.

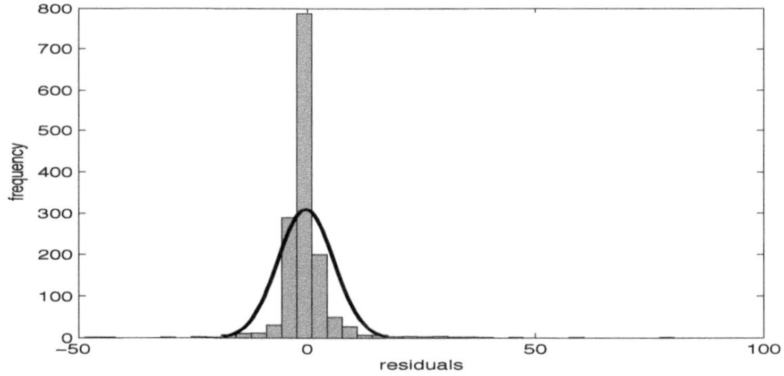

Fig. 9.5 Histogram of the residuals with normal density curve after mean level function and the autoregression effects were eliminated.

that is,

$$\widehat{\sigma} = 2.216\,. \tag{9.44}$$

To validate the normality assumption on filtered residuals implied by the Brownian motion model, we use the Kolmogorov-Smirnov statistics. This is not significant at the 5% level, meaning that the normal hypothesis for the filtered residuals cannot be rejected.

The next step is an analysis of the jumps filtered out by the recursive filtering procedure. Inspection of the histograms of the positive and nega-

Table 9.2 Summary of the recursive filtering procedure. 'Cum' and 'Std' is abbreviations for 'Cumulative' and 'Standard deviation', respectively.

Iteration	Std.	Cum. # jumps	Daily jump frequency
1	6.031	31	0.0213
2	3.524	61	0.0420
3	2.878	90	0.0619
4	2.537	104	0.0715
5	2.412	116	0.0798
6	2.310	122	0.0839
7	2.258	128	0.0880
8	2.216	128	0.0880

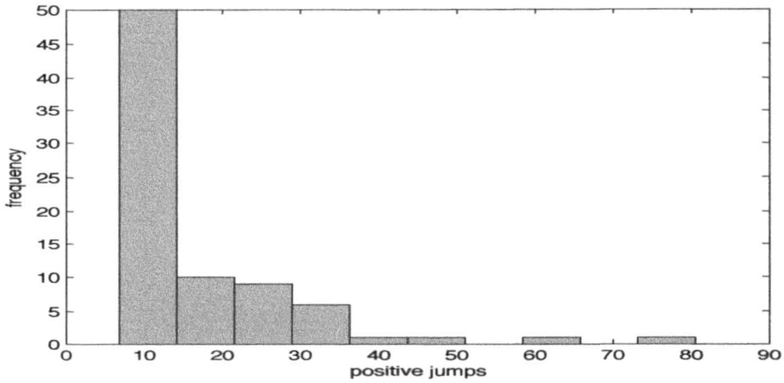

Fig. 9.6 Histogram of the filtered positive jumps.

tive jumps, plotted in Figs. 9.6 and 9.7, respectively, suggests to use a jump process $I(t)$ given in (5.5), that is,

$$I(t) = I^{+}(t) + I^{-}(t)\,,$$

and

$$I^{\pm}(t) = \sum_{i=1}^{N^{\pm}(t)} Z_i^{\pm}\,,$$

with the positive and negative jump sizes $Z_i^{\pm}$ being lognormally or exponentially distributed. The parameter estimates for the two distributions for the positive and negative jumps are reported in Table 9.3 (where the negative jumps are multiplied by minus one prior to estimation). Here, m and

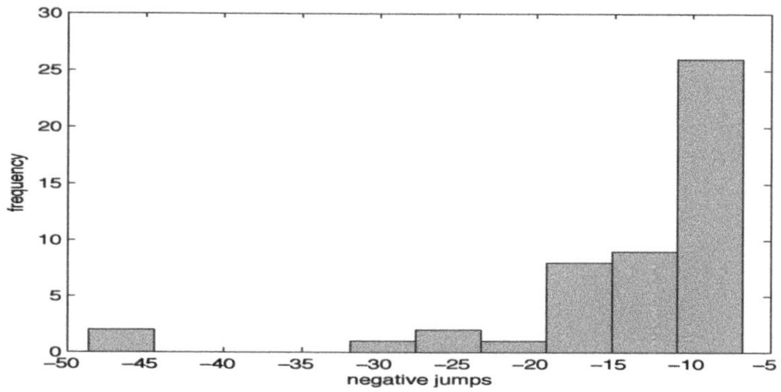

Fig. 9.7 Histogram of the filtered negative jumps.

s are the two parameters of the lognormal distribution (see (5.9)), whereas μ_J is the expectation of the exponential distribution (see (3.7)). We have

Table 9.3 Parameters of lognormal and exponential distribution

Parameter	Positive jumps	Negative (transformed) jumps
$\widehat{m}$	2.5616	2.4981
$\widehat{s}$	0.5949	0.4776
$\widehat{\mu}_J$	15.92	13.85

from the recursive filtering that the intensity of the positive jumps is

$$\widehat{\lambda}^{+} = 0.0543\,, \tag{9.45}$$

whereas the negative jump intensity is estimated to be

$$\widehat{\lambda}^{-} = 0.0337\,. \tag{9.46}$$

These estimates imply approximately 13 positive jumps and eight negative jumps on average a year.

An alternative to the jump-diffusion model above is to use, for instance, a NIG Lévy process for the residual dynamics. This approach was proposed and analysed for gas spot prices in Subsect. 5.1.1, but will not be pursued here.

Let us discuss the relation to option pricing of the proposed model above. The crucial point is if the jump model has exponential moments or not (recall discussion at the end of Subsect. 9.3.1). First, select a simple compound Poisson process as $I(t)$. In that case the cumulant function is (recall (2.22))

$$\psi(\theta) = \lambda(\mathrm{e}^{\psi_Z(\theta)} - 1)\,,$$

with λ being the jump frequency and $\psi_Z(\theta)$ the cumulant function of Z_i, the jump size. Choosing normally distributed jumps Z with mean m and variance s^2, we find that

$$\psi_Z(\theta) = \mathrm{i}\theta m - \frac{1}{2}\theta^2 s^2\,,$$

and exponential moments exist. We have specified a model which separates the positive and negative jumps, each being a compound Poisson process. We find that

$$\psi(\theta) = \lambda^+(\mathrm{e}^{\psi^+(\theta)} - 1) + \lambda^-(\mathrm{e}^{\psi^-(\theta)} - 1)\,,$$

with $\psi^{\pm}(\theta)$ being the cumulant function for the positive and negative jump sizes $Z^{\pm}$. Choosing a lognormal model for both, we can derive infinite series representations for these cumulants, however, exponential moments do not exist. Hence, selecting a lognormal jump size distribution is not feasible together with the Fourier approach. An exponential specification of the jump size distribution implies that

$$\psi_Z(\theta) = \frac{1}{\mathrm{i}\theta\mu_J - 1}\,.$$

Here, μ_J is the expectation of Z. Moreover, we see that exponential moments exist up to $1/\mu_J$, that is,

$$\mathbb{E}[\mathrm{e}^{\gamma Z}] < \infty\,,$$

as long as $\gamma < 1/\mu_J$. Thus, for exponentially distributed jumps we can apply the Fourier approach.

At the time of writing this book, there exists only OTC trading of spark spread options. As mentioned earlier, NYMEX offers a market for crack spreads. Since there is a lack of liquid prices taken from an organized market for most energy spreads, it is of importance to have available tools yielding fast and reliable prices. We believe that the above framework may serve as such a tool. From the liquid futures markets of gas and electricity, we can read off the market prices of risk. Further, our underlying stochastic model on the spark spread is estimated on historical data, which together with the market price of risk give all the required input to price the spark spread options efficiently. Reliable routines for marking-to-market and Value at Risk evaluations may be implemented based on our pricing technique.

Chapter 10

Analysis of Temperature Derivatives

In this chapter we analyse the market for temperature derivatives. We build our analysis on a stochastic dynamics for the evolution of the temperature, and propose an extension to the OU processes previously considered as spot price models. This extension is a continuous-time AR model, and we argue empirically that the daily average temperature dynamics has a memory consistent with a higher-order AR model. Furthermore, we derive prices for futures contracts typical for the CME market, including HDD and CDD futures. European options written on such futures are also analysed. We illustrate our approach using data from Stockholm, Sweden, one of the cities for which temperature dependent contracts are traded at the CME. Much of the empirical and theoretical analyses in this Chapter are taken from [Benth and Šaltytė-Benth (2007)] and [Benth, Šaltytė Benth and Koekebakker (2007)].

10.1 Some preliminaries on temperature futures

In what follows, we shall derive expression for the dynamics of futures prices based on a mean-reverting AR model for the temperature evolution. The model and the theoretical derivation of prices are most conveniently expressed in a continuous-time framework, and we restate the definitions of the different temperature indices accordingly. For the time being, we assume the temperature dynamics to be a stochastic process so that all the expressions below make sense.

The CDD (and analogously the HDD) over a measurement period $[\tau_1, \tau_2]$

is defined as (recall the discrete-time definitions in Sect. 1.3)

$$\int_{\tau_1}^{\tau_2} \max\left(T(s) - c\right)\, ds\,. \tag{10.1}$$

The CAT and PRIM indices over the same period are

$$\int_{\tau_1}^{\tau_2} T(s)\, ds \tag{10.2}$$

and

$$\frac{1}{\tau_2 - \tau_1}\int_{\tau_1}^{\tau_2} T(s)\, ds\,, \tag{10.3}$$

respectively. In these definitions, we assume that $T(t)$ is the instantaneous temperature at time t. We further assume that all contracts are settled in terms of a currency with unit one (that is, we do *not* multiply the CDD, HDD, CAT or PRIM indices with 20 GBP/USD or 250,000 JPY to convert them into money as in their respective definition in Sect. 1.3).

The buyer of a CDD futures contract will receive the amount in (10.1) at the end of the measurement period $[\tau_1, \tau_2]$. In return, the buyer pays the CDD futures price $F_{\mathrm{CDD}}(t, \tau_1, \tau_2)$ if the contract was entered at time $t \leq \tau_1$. The profit from this trade is therefore

$$\int_{\tau_1}^{\tau_2} \max\left(T(s) - c, 0\right)\, ds - F_{\mathrm{CDD}}(t, \tau_1, \tau_2)\,.$$

From arbitrage theory, the CDD futures price is given by the equation

$$0 = \mathrm{e}^{-r(\tau_2 - t)}\mathbb{E}_Q\left[\int_{\tau_1}^{\tau_2} \max\left(T(s) - c, 0\right)\, ds - F_{\mathrm{CDD}}(t, \tau_1, \tau_2) \,\middle|\, \mathcal{F}_t\right],$$

with a constant risk-free rate of return r and a risk-neutral probability Q. Since temperature (and therefore the CDD index) is not tradeable, any probability Q being equivalent to the objective probability P is a risk-neutral probability. Later, we shall specify a subclass of such probabilities via the Girsanov transform. The CDD futures price is adapted, and thus we derive it as the conditional risk-neutral expected payment from the CDD index

$$F_{\mathrm{CDD}}(t, \tau_1, \tau_2) = \mathbb{E}_Q\left[\int_{\tau_1}^{\tau_2} \max\left(T(s) - c, 0\right)\, ds \,\middle|\, \mathcal{F}_t\right]. \tag{10.4}$$

This will give an arbitrage-free dynamics for the CDD futures price. We implicitly assumed that the temperature dynamics is so that all expressions and derivations are sound.

Analogously, we find that

$$F_{\mathrm{HDD}}(t,\tau_1,\tau_2)=\mathbb{E}_Q\left[\int_{\tau_1}^{\tau_2}\max\left(c-T(s),0\right)\,ds\,|\,\mathcal{F}_t\right]. \tag{10.5}$$

Similar derivations lead us to the CAT and PRIM futures prices being

$$F_{\mathrm{CAT}}(t,\tau_1,\tau_2)=\mathbb{E}_Q\left[\int_{\tau_1}^{\tau_2}T(s)\,ds\,|\,\mathcal{F}_t\right] \tag{10.6}$$

and

$$F_{\mathrm{PRIM}}(t,\tau_1,\tau_2)=\mathbb{E}_Q\left[\frac{1}{\tau_2-\tau_1}\int_{\tau_1}^{\tau_2}T(s)\,ds\,|\,\mathcal{F}_t\right], \tag{10.7}$$

respectively. Since the PRIM futures price is simply the average of the CAT futures price, we do not investigate the former in any more detail.

We have the following useful CDD-HDD parity.

Proposition 10.1 *The CDD and HDD futures prices are linked by the relation*

$$F_{HDD}(t,\tau_1,\tau_2)=c(\tau_2-\tau_1)-F_{CAT}(t,\tau_1,\tau_2)+F_{CDD}(t,\tau_1,\tau_2). \tag{10.8}$$

Proof. The result follows by applying the equation

$$\max(c-x,0)=c-x+\max(x-c,0).$$

□

In this Chapter we focus on deriving the CDD and CAT futures prices. The Proposition above readily gives the HDD futures price as long as we know the CDD and CAT futures prices.

We recall that a Frost Day at Amsterdam airport Schiphol is defined as observed frost in the morning. More specifically, we have (with time measured in days) that

$$\begin{aligned}\mathrm{FD}(t)=1\Big(&\{T(t+7/24)\leq-3.5\}\cup\{T(t+10/24)\leq-1.5\}\\&\cup\{\{T(t+7/24)\leq-0.5\}\cap\{T(t+10/24)\leq-0.5\}\}\Big).\end{aligned}$$

The Frost Day index over a measurement period $[\tau_1,\tau_2]$ is

$$\sum_{t=\tau_1}^{\tau_2}\mathrm{FD}(t), \tag{10.9}$$

where we assume for simplicity that the currency unit is one. Using the same procedure as above, we may derive the Frost Day index futures price as

$$F_{\mathrm{FDI}}(t,\tau_1,\tau_2) = \mathbb{E}_Q\left[\sum_{s=\tau_1}^{\tau_2} \mathrm{FD}(s) \mid \mathcal{F}_t\right], \tag{10.10}$$

with $\mathrm{FD}(s)$ defined in (10.9).

Our approach to futures pricing involves specifying a temperature model together with a risk-neutral probability, and then calculate the predicted payoff from the futures contract. There exist several methodologies to assess derivatives prices on different temperature indices. We refer to [Geman (1999)], [Geman and Leonardi (2005)] and [Jewson and Brix (2005)] for detailed accounts on some established methods (including ours). [Davis (2001)] propose an approach based on marginal utility to price options on CDDs and HDDs, whereas [Platen and West (2005)] suggest an equilibrium method based on a world index for temperature derivatives valuation.

In the next Section we model the temperature dynamics $T(t)$ by a continuous-time AR stochastic process with seasonal volatility, and present a detailed empirical analysis of this model for daily average temperature data collected in Stockholm.

10.2 Modelling the dynamics of temperature

We present a class of stochastic processes generalizing the multi-factor OU models which were presented in Chapter 3. The class of models is called continuous AR (CAR) processes, since they are AR stochastic processes in continuous-time. The CAR model is a subclass of the more general CARMA (continuous autoregressive moving-average) models introduced and studied by [Brockwell and Marquardt (2005)]. Such models are particularly suitable to capture the evolution of temperature through time. We extend the models to allow for seasonality in the residual variance, and based on an empirical study on data collected in Stockholm we show the appropriateness of the models in the context of temperature derivatives.

10.2.1 *The CAR(p) model with seasonality*

Let $\mathbf{X}(t)$ be a stochastic process in $\mathbb{R}^p$ for $p \geq 1$ defined by the vectorial OU stochastic process

$$d\mathbf{X}(t) = A\mathbf{X}(t)\,dt + \mathbf{e}_p(t)\sigma(t)\,dB(t)\,, \tag{10.11}$$

where $\mathbf{e}_p$ is the pth unit vector in $\mathbb{R}^p$. Note here that we reserved the constant p to denote the order of the CAR model, and not the number of Brownian motions used in each factor of the spot price process as in Chapter 3. The number of Brownian motions is equal to *one* here. The standard deviation of the residuals is described by a function $\sigma(t) > 0$, assumed to be continuous on $[0, \mathcal{T}]$. We frequently refer to this function as the volatility of the temperature dynamics. Further, we denote by A the $p \times p$ matrix

$$A = \begin{bmatrix} 0 & 1 & 0 & \cdots & 0 \\ 0 & 0 & 1 & \cdots & 0 \\ \cdot & \cdot & \cdot & \cdot & \cdot \\ \cdot & \cdot & \cdot & \cdot & \cdot \\ 0 & 0 & 0 & 0 & 1 \\ -\alpha_p & -\alpha_{p-1} & -\alpha_{p-2} & \cdots & -\alpha_1 \end{bmatrix}. \tag{10.12}$$

Here, we suppose that α_k, $k = 1, \ldots, p$ are positive constants. We assume further that the seasonal function $\Lambda(t) : [0, \mathcal{T}] \to \mathbb{R}$ is bounded and continuously differentiable. We introduce the following CAR(p) model for the temperature dynamics

$$T(t) = \Lambda(t) + X_1(t)\,, \tag{10.13}$$

where we use the notation X_q for the qth coordinate of the vector $\mathbf{X}$. As we shall see later in Sect. 10.3, the volatility function $\sigma(t)$ will play an important role in the dynamics of the temperature. It turns out that the temperature residuals possess a variation which is a function of the season. It has a rather characteristic shape, which we will model by a truncated Fourier series.

We can represent the stochastic process $\mathbf{X}(t)$ explicitly by solving the stochastic differential equation (10.11).

Lemma 10.1 *The stochastic process* $\mathbf{X}$ *has the explicit form*

$$\mathbf{X}(s) = \exp\left(A(s-t)\right)\mathbf{x} + \int_t^s \exp\left(A(s-u)\right)\mathbf{e}_p\sigma(u)\,dB(u)\,, \tag{10.14}$$

for $s \geq t \geq 0$ *and* $\mathbf{X}(t) = \mathbf{x} \in \mathbb{R}^p$.

Proof. Follows by an application of the multidimensional Itô Formula. □

When the volatility function $\sigma(t)$ is a constant, it is known that $\mathbf{X}(t)$ is stationary as long as all the eigenvalues of the matrix A have a negative real part. This result follows from Prop. 6.2 of [Ichihara and Kunita (1974)]. When time approaches infinity, the distribution of the CAR(p) process $\mathbf{X}(t)$ converges to a normal distribution with zero expectation and a finite variance, explicitly given as

$$\int_0^\infty \exp(Ay)\mathbf{e}_p\mathbf{e}_p' \exp(A'y)\, dy\,,$$

where the notation $\mathbf{x}'$ means the transpose of the vector $\mathbf{x}$, and likewise, the transposing of the matrix A is denoted A'. Depending on the properties of the time-dependent volatility $\sigma(t)$, one may preserve the stationarity. However, when the eigenvalues of A all have a negative real part, we are sure that the process tends to have a mean equal to zero, implying that the temperature on average will coincide with the seasonal function $\Lambda(t)$.

Since our concern is to derive arbitrage-free dynamics of temperature future prices, we need the risk-neutral dynamics of temperature. Since temperature is not a tradeable asset, any equivalent probability Q is a risk-neutral probability that can serve the purpose. However, as for the spot models proposed in Chapter 3, we parametrize the set of risk-neutral measures by using the Esscher transform. However, following the theory in Sect. 4.1, the Esscher transform coincides with the Girsanov transform in the case of Brownian motion, which is the situation we have here. Hence, we use the same transform of the Brownian motion $B(t)$ as stated in Prop. 4.4, which we recall to be defined as

$$B^\theta(t) = B(t) - \int_0^t \theta(u)\, du\,,$$

for a continuous function θ on $[0, \mathcal{T}]$. Note that we skip the notation $\widehat{\theta}$, since we only deal with one function θ. The dynamics of $\mathbf{X}(t)$ under Q^θ, the probability measure which turns $B^\theta(t)$ into a Brownian motion on $[0, \mathcal{T}]$, becomes

$$d\mathbf{X}(t) = (A\mathbf{X}(t) + \mathbf{e}_p\sigma(t)\theta(t))\, dt + \mathbf{e}_p\sigma(t)\, dB^\theta(t)\,. \tag{10.15}$$

Further, by an application of the multidimensional Itô Formula, the stochastic process $\mathbf{X}(t)$ has the explicit dynamics under Q^θ defined as

$$\mathbf{X}(s) = \exp\left(A(s-t)\right)\mathbf{x} + \int_t^s \exp\left(A(s-u)\right)\mathbf{e}_p\sigma(u)\theta(u)\,du + \int_t^s \exp\left(A(s-u)\right)\mathbf{e}_p\sigma(u)\,dB^\theta(u)\,, \tag{10.16}$$

for $0 \le t \le s \le \mathcal{T}$. In the empirical study of temperature data in Sect. 10.3 we use the dynamics of $T(t)$ under the market measure P as the model, while the risk-neutral version (10.16) (or its differential (10.15)) is the appropriate model when analysing futures prices and options on these. We consider this case in Sect. 10.4.

10.2.2 *A link to time series*

We discuss the link to AR time series models. First, consider the special case of $p = 1$, where the matrix A simply becomes the constant $-\alpha_1$. The dynamics of $\mathbf{X}(t) = X_1(t)$ is then

$$dX_1(t) = -\alpha_1 X_1(t)\,dt + \sigma(t)\,dB(t)\,,$$

which we recognise as an OU process considered in Chapter 3. It is known that this process in discrete-time corresponds to an AR(1) process.

We now establish the link between a general CAR(p) process $\mathbf{X}(t)$ and an AR(p) process. First, from (10.11) we have for $q = 1, \ldots, p-1$, that

$$dX_q(t) = X_{q+1}(t)\,dt \tag{10.17}$$

and

$$dX_p(t) = -\sum_{q=1}^{p} \alpha_{p-q+1} X_q(t)\,dt + \sigma(t)\,dB(t)\,. \tag{10.18}$$

An Euler approximation (see [Kloeden and Platen (1992)]) of the stochastic differential equation (10.18) with time step one, leads to a time series $x_p(t)$, $t = 0, 1, \ldots$ of the following form

$$x_p(t+1) - x_p(t) = -\sum_{q=1}^{p} \alpha_{p-q+1} x_q(t)\,dt + \sigma(t)\epsilon(t)\,,$$

where $\epsilon(t)$, $t = 0, 1, \ldots$ are i.i.d. random variables being standard normally distributed. Similarly, we find from (10.17) that

$$x_q(t+r) - x_q(t+r-1) = x_{q+1}(t+r-1)$$

for $q = 1, \ldots, p-1$ and $r \geq 1$. Iterating this, we get the following.

Lemma 10.2 *For $q = 1, \ldots, p-1$ it holds*

$$x_{q+1}(t) = \sum_{k=0}^{q} (-1)^k b_k^q x_1(t+q-k)\,.$$

Here, the coefficients b_k^q are defined recursively as

$$b_k^q = b_{k-1}^{q-1} + b_k^{q-1}\,, k = 1, \ldots, p-1\,, q \geq 2\,,$$

and $b_0^q = b_q^q = 1$ for $q = 0, 1, \ldots, p$. Further, we have that

$$x_p(t+1) - x_p(t) = \sum_{k=0}^{p} (-1)^k b_k^p x_1(t+p-k)\,.$$

Proof. The proof goes by induction. Suppose that the result holds for $q-1$. By appealing to the induction hypothesis, we have

$$\begin{aligned} x_{q+1}(t) &= x_q(t+1) - x_q(t) \\ &= \sum_{k=0}^{q-1} (-1)^k b_k^{q-1} \left(x_1(t+q-k) - x_1(t+q-1-k) \right)\,. \end{aligned}$$

The result follows by applying the recursive definition of b_k^q. The second result follows by a similar argument. □

Inserting the expression for x_q in terms of x_1 derived in the Lemma above, we reach the following recursive expression for x_1

$$\begin{aligned} &\sum_{k=0}^{p} (-1)^k b_k^p x_1(t+p-k) = \\ &\quad - \sum_{q=1}^{p} \alpha_{p-q+1} \sum_{k=0}^{q-1} (-1)^k b_k^{q-1} x_1(t+q-1-k) + \sigma(t)\epsilon(t)\,. \end{aligned} \tag{10.19}$$

Observe that the expression includes all the terms $x_1(t+p), x_1(t+p-1), \ldots, x_1(t)$ in a linear fashion, and thus defines an AR(p) process. Hence, we have a relation where we can identify the coefficients of the CAR(p) model with those of the corresponding AR(p) model.

Example 10.1 *AR(2) model.* Let $p = 2$. Then from (10.19) we obtain

$$\begin{aligned} b_0^2 x_1(t+2) - b_1^2 x_1(t+1) + b_2^2 x_1(t) \\ = -\alpha_2 b_0^0 x_1(t) - \alpha_1 \left(b_0^1 x_1(t+1) - b_1^1 x_1(t)\right) + \sigma(t)\epsilon(t) . \end{aligned}$$

After reorganizing, we find

$$x_1(t+2) = (2-\alpha_1)x_1(t+1) + (\alpha_1 - \alpha_2 - 1)x_1(t) + \sigma(t)\epsilon(t) .$$

Temperature data is in a discrete form, and we will use AR(p) models for the time series observations. Thus, it is useful to have the above connections in order to identify the parameters in the corresponding CAR(p) model. This is the topic of the next Section, but before embarking on the empirical analysis of temperature data, we include an example on the connection between the discrete model and the continuous model for $p = 3$, since an AR(3) time series model turns out to be particularly suitable for the explaining observed temperature dynamics in Stockholm.

Example 10.2 *AR(3) model.* Let $p = 3$. Then, by applying (10.19) and reorganizing the terms, we get

$$\begin{aligned} x_1(t+3) = (3-\alpha_1)x_1(t+2) + (2\alpha_1 - \alpha_2 - 3)x_1(t+1) \\ + (\alpha_2 + 1 - \alpha_1 - \alpha_3)x_1(t) + \sigma(t)\epsilon(t) . \end{aligned}$$

We are going to utilise the explicit connection between AR(3) and CAR(3) models when analysing temperature derivatives for Stockholm.

10.3 Empirical analysis of Stockholm temperature dynamics

In this Section we study empirically the time series of daily average temperatures (DATs) observed in Stockholm, Sweden, and fit an AR(p) model to these data.

10.3.1 *Description of the data*

We have available DATs measured in degrees of Celsius from Stockholm over a period ranging from 1 January 1961 to 25 May 2006, resulting in 16,581 records. The DAT is calculated as the average of the minimum and maximum temperature during the day. The measurement on February 29 was removed from the sample in each leap year to equalise the length of all years, leading to the time series of 16,570 observations. DAT time series is plotted in Fig. 10.1 together with the fitted seasonal average described below. For illustrative purposes we only picture a snapshot of DATs from the last 10 years of the data set.

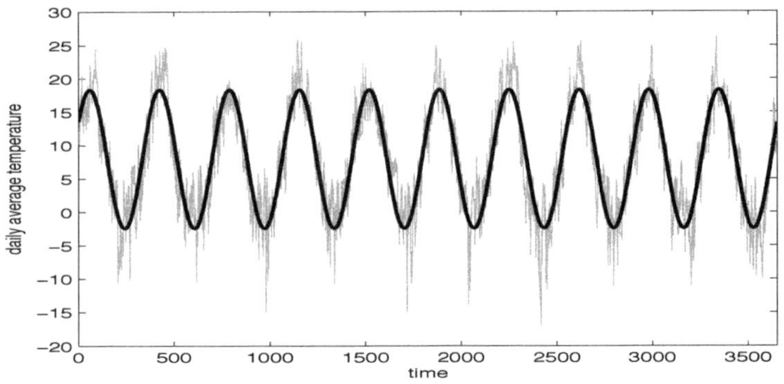

Fig. 10.1 DATs in Stockholm together with the fitted seasonal function, a snapshot of the last 10 years starting 25 May 1996.

A quick look at the histogram of the DATs in Fig. 10.2 gives us a clear indication of non-normality of data. Stockholm has rather cold winters and relatively mild summers, which is reflected in a bimodal histogram of temperatures. In addition, the data are slightly skewed to the left with a skewness coefficient being equal −0.104, and have a small negative kurtosis equal to −0.664. Figure 10.3 shows the ACF of the DATs, where we observe

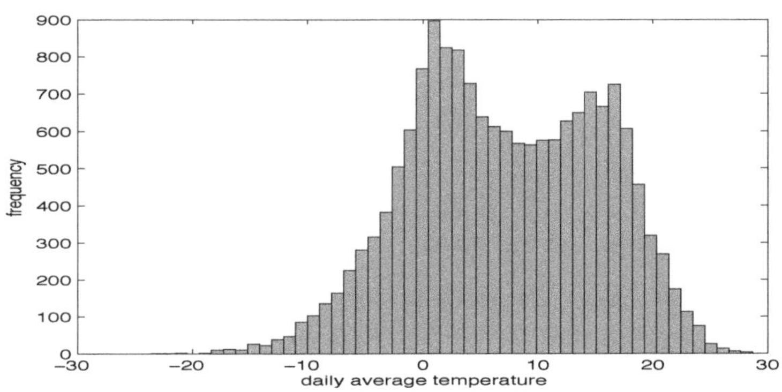

Fig. 10.2 Histogram of DATs in Stockholm.

a strong seasonal variation. This points towards a seasonal heteroskedasticity in data that we in a moment will explain by a volatility varying with the season.

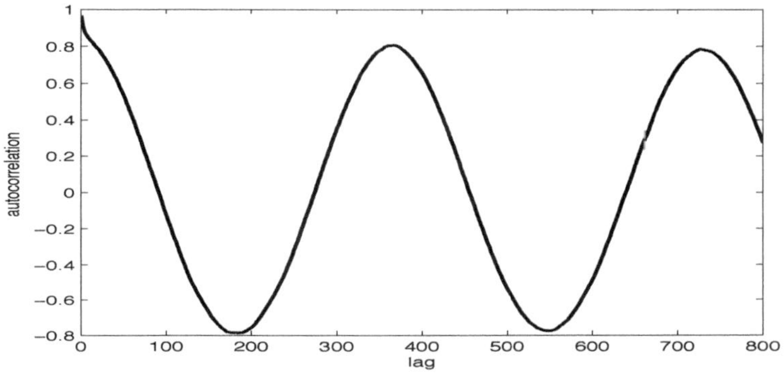

Fig. 10.3 Empirical ACF of DATs in Stockholm.

10.3.2 *Estimating the CAR(p) models*

We state the time series motivated from the CAR(p) models introduced above. Suppose that the temperature on day $i = 0, 1, 2, \ldots$ is denoted by T_i, and let

$$T_i = \Lambda_i + y_i \,, \tag{10.20}$$

where $\Lambda_i = \Lambda(i)$. The function $\Lambda(t)$ is defined by

$$\Lambda(t) = a_0 + a_1 t + a_2 \cos(2\pi(t - a_3)/365) \,, \tag{10.21}$$

where a_0 is a constant describing the average level of the temperature series, a_1 is a slope of a linear trend function, a_2 is an amplitude of the mean temperature, and a_3 defines a phase angle. Note that time is measured in days in the definition of the function $\Lambda(t)$. We now discuss briefly the background for the specification of $\Lambda(t)$ in (10.21).

In view of the many discussions of global warming (see, for example, [Rassmusson *et al.* (1993)] and [Handcock and Wallis (1994)]), one expects the presence of a trend in the temperature data. We assume it to have a linear form in $\Lambda(t)$, with a growth rate given by a_1. Urbanisation may be another source for the presence of such a trend. Further, the DAT varies naturally with the season (see Fig. 10.1). Such a cyclic pattern induced by winter and summer seasons is modelled by the cosine function in $\Lambda(t)$.

Due to natural cyclic/periodic temperature variations, which are seen on diurnal, inter-seasonal or annual time scales, temperature has a tendency to revert back to its mean over time. This mean-reverting property can be modelled by an AR(p) process, and we will show empirically that the choice $p = 3$ is the most suitable for the data under consideration. However, we also discuss the case of an AR(1) process, since this provides a natural case in view of other contributions to the literature on temperature modelling (see, for example, [Dornier and Queruel (2000)]). Hence, we suppose that the deseasonalized temperature data $T(t) - \Lambda(t)$ can be modelled by an AR(p) process y_i, with seasonally varying residuals,

$$y_{i+p} = \sum_{j=1}^{p} b_j y_{i+p-j} + \sigma_i \epsilon_i \,, \tag{10.22}$$

where ϵ_i are i.i.d. random variables following the standard normal distribution. We denote $\sigma_i = \sigma(i)$. The further empirical analysis suggests the following seasonal form of the volatility function

$$\sigma^2(t) = c_1 + \sum_{k=1}^{4} \left(c_{2k} \cos(2k\pi t/365) + c_{2k+1} \sin(2k\pi t/365)\right) . \tag{10.23}$$

In the process of estimating the parameters of our temperature model, we eliminate step by step the different model components from the data. First, we detrend and deseasonalize the observed time series, and then we fit the AR(p) process to the resulting data. The cases of $p = 1$ and $p = 3$ are considered. We show that the residuals obtained after eliminating the AR(p) process are not uncorrelated and far from being normal. The ACF of the squared residuals reveals the presence of a seasonally dependent variance, which we model by the truncated Fourier series (10.23). After the effect of estimated daily seasonal variance is removed from the data, the residuals become much closer to normal and are only slightly correlated for the few first lags. The choice of $p = 3$ gives a much better fit than $p = 1$. We now present the findings from the empirical analysis.

The seasonal function $\Lambda(t)$ in (10.21) consists of a linear trend $a_0 + a_1 t$ capturing possible influence from global warming and a cosine-term modelling the yearly seasonality level of the DAT. First, we check for the existence of a linear trend in the Stockholm data by running a simple linear regression. The obtained slope is approximately 0.00009, whereas the inter-

Table 10.1 Fitted parameters of $\Lambda(t)$

a_0	a_1	a_2	a_3
6.3759	0.0001	10.4411	−165.7591

cept is equal to 6.40. Both values are significant at the 1% level (note that the data are not normally distributed and not uncorrelated, and hence the significance level must be treated with precaution). Therefore, we have an increase in DAT over the considered period. In effect, the DAT of Stockholm has risen approximately by 1.5°C from the beginning of 1961 to the middle of 2006. This upward trend in temperature corresponds to an increase in the global mean level. Our findings for Stockholm are in line with similar observations made by [Alaton, Djehiche and Stillberger (2002)], [Campbell and Diebold (2005)], and [Šaltytė Benth, Benth and Jalinskas (2007)] for temperature data from Sweden, USA and Lithuania, respectively.

When finding the parameters of $\Lambda(t)$ in (10.21), we estimate simultaneously the linear trend and seasonal component. The parameters are fitted using the method of least squares,[1] and estimates are reported in Table 10.1. The value of R^2 after estimating $\Lambda(t)$ is equal to 80.9%, showing a reasonably good fit.

We eliminate the linear trend and seasonal component by subtracting the estimated $\Lambda(t)$ from the original observations. The ACF of the obtained residuals is plotted in Fig. 10.4. Here we clearly see a pattern indicating an autoregressive structure in the dynamics. The PACF plotted in Fig. 10.5 indicates that we need an AR(3) process to explain the evolution in the time series data. An AR(1) may be a reasonable approximation, and will correspond to an OU process in line with the models we have proposed for the spot price dynamics of energy (that is, the class of arithmetic models). Using an AR(3) model means that we base the temperature dynamics on the CAR(3) stochastic processes. We fit both models here.

10.3.2.1 *Fitting an AR(1) model*

We use a simple linear regression to estimate the parameter b_1 for the AR(1) process defined in (10.22). By regressing today's detrended and deseasonalized temperatures against those of the previous day, we find

$$b_1 = 0.8234, \tag{10.24}$$

[1] We applied the *nlinfit* procedure in MATLAB for this purpose.

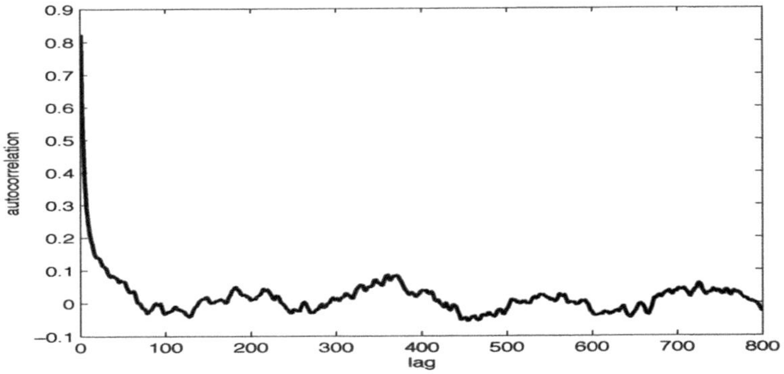

Fig. 10.4 The ACF of the residuals of DATs after linear trend and seasonal component were removed.

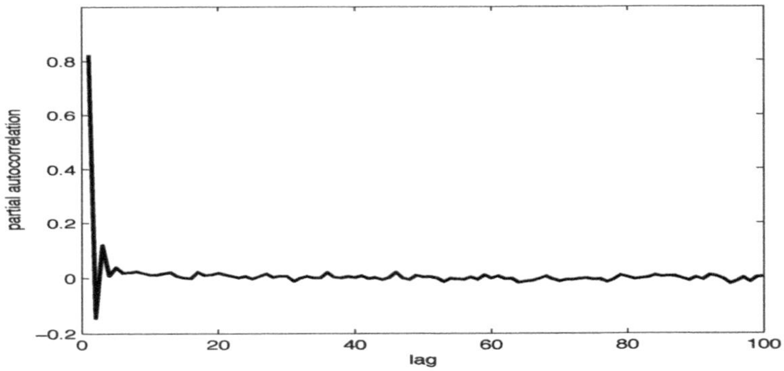

Fig. 10.5 The PACF of the residuals of DATs after linear trend and seasonal component were removed.

being significant at the 1% level. Note that p-values have to be treated with care since the assumptions of the regression model are not met. As expected, the regression constant was insignificant, and it is set to zero (in accordance with our model). The inclusion of the mean reversion process increased the value of R^2 to 93.9%.

The histogram of the obtained residuals depicted in Fig. 10.6 seems to be close to normal. However, the Kolmogorov-Smirnov statistics of 2.491 is significant at the 1% level, rejecting the normality assumption. Note that the Kolmogorov-Smirnov test is designed for use with independent data,

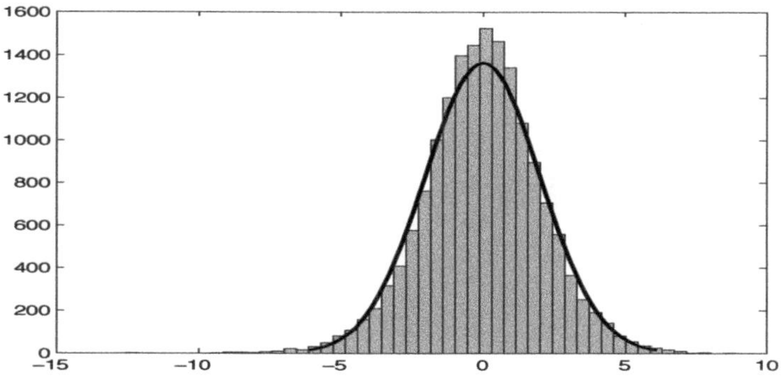

Fig. 10.6 Histogram of the residuals after linear trend, seasonal component and AR(1) process were removed.

and the p-value therefore might be unreliable.

In Fig. 10.7 we plot the residuals and squared residuals for the last 10 years. We observe a clear persistent variation (especially for squared residuals) in the noise, which is a sign of seasonal heteroskedasticity. This observation hints to the inclusion of a deterministic seasonality function when modelling the variance. Looking at the ACF for the residuals in Fig. 10.8, we see that autocorrelations for the first two lags are quite high but decrease very rapidly and vary around zero for higher lags. However, all autocorrelations up to 500 lags (and even more) are significant according to the Box-Ljung statistic. The ACF of the squared residuals in Fig. 10.9 reveals a clear seasonal pattern in DAT for Stockholm, again pointing to a seasonal heteroskedasticity of the residuals. A similar seasonal pattern in the ACF for squared residuals was observed by [Campbell and Diebold (2005)] in temperature data for several US cities. They proposed to model the temperature dynamics by an ARMA model, with a seasonal ARCH-type time series dynamics for the residuals.

Before moving on with the analysis of the seasonal variance function $\sigma(t)$, we investigate how stable (over time) the regression parameter b_1 in the AR(1) model is.

First, we estimate the regression parameter based on data from one year only, repeating this for every year until the end of data set. This yields more than 40 estimates of the regression parameter, where the average becomes 0.81 with a standard deviation equal to 0.03, indicating stable values of the yearly regression parameter. We can also investigate the seasonal structure

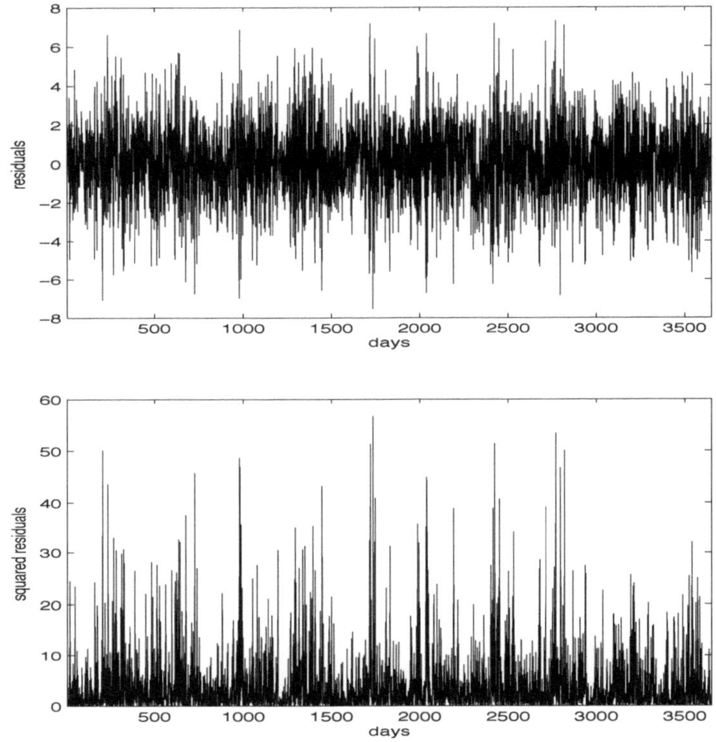

Fig. 10.7 Residuals and squared residuals of DATs after linear trend, seasonal component and AR(1) process were removed, a snapshot of the last 10 years starting 25 May 1996.

of the regression coefficient by estimating it for each month individually, that is, we consider data for January only, then for February, and so on. The results are basically the same as for the yearly study, with no clear seasonal variation. Hence, we conclude that the speed of mean reversion for the DAT observed in Stockholm is stable over time, and it is reasonable to assume it to be constant (as we do).

We continue with fitting the seasonal variance function $\sigma^2(t)$ defined in (10.23) to the obtained residuals. Let us explain how we proceed in doing this.

The daily residuals over more than 40 years are first organized into 365 groups, one group for each day of the year. This means that we let all observations on 1 January be collected into group 1, all observations on

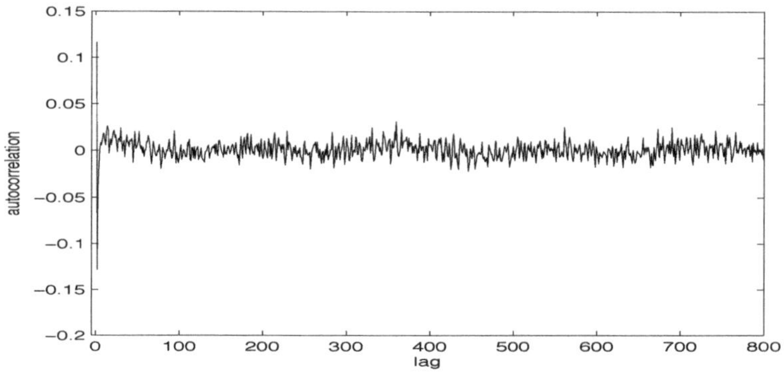

Fig. 10.8 ACF of residuals of DATs after linear trend, seasonal component and AR(1) process were removed.

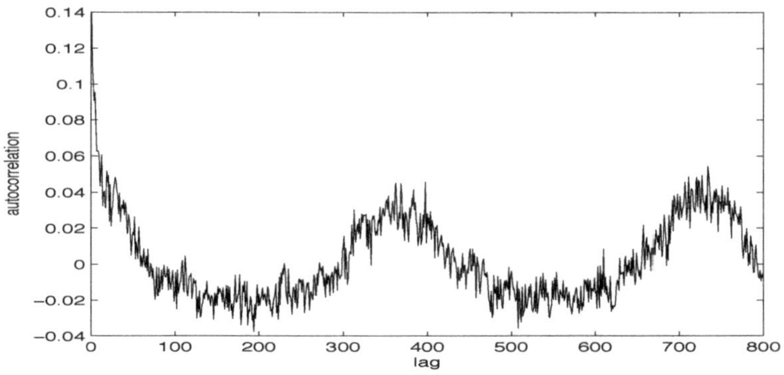

Fig. 10.9 ACF of squared residuals of DATs after linear trend, seasonal component and AR(1) process were removed.

2 January into group 2, and so on until we have grouped all the days of all years. Recall that observations on 29 February were removed in each leap year so that each year contains the same number of data, namely 365. Finding the average of the squared residuals in each group, we obtain an estimate for the expected squared residual for the day associated to the group label. This yields 365 estimates of the expected squared residuals, one for each day of the year. We model their variation by the truncated Fourier function in (10.23), where the parameters are estimated using the

Table 10.2 Fitted parameters of $\sigma^2(t)$ for the AR(1) case

c_1	c_2	c_3	c_4	c_5	c_6	c_7	c_8	c_9
4.151	1.358	0.702	0.807	−0.143	0.181	0.446	−0.054	−0.002

least squares approach.[2] The fitted parameters are presented in Table 10.2.

In Fig. 10.10 we present the empirical $\sigma^2(t)$ function together with the fitted one. It is obvious that the fluctuations in the cold season are con-

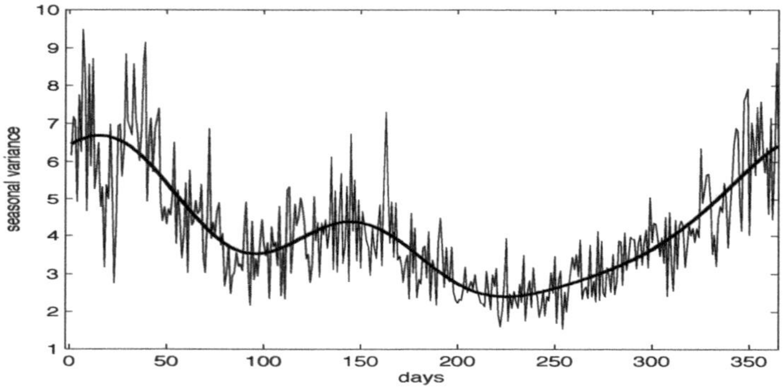

Fig. 10.10 Empirical and fitted $\sigma^2(t)$ function when AR(1) is used.

siderably higher than those during the milder seasons. Notably, there are slightly lower variations in the fall and spring than in the summer, indicating that these seasons are more stable than the summer, with the early autumn being the one with the smallest variability. Thus, we see a clear seasonal volatility effect in the temperature data. It seems that the shape of the volatility is rather characteristic for temperature observations made at many different locations. Similar seasonality effects have been observed for temperatures in Norway (see [Benth and Šaltytė-Benth (2005)]) and Lithuania (see [Šaltytė Benth, Benth and Jalinskas (2007)]). Further, [Zapranis and Alexandridis (2007)] have shown that it appears in the temperature dynamics in Paris, France, and [Rossmanith (2007)] finds it in temperature series observed at several locations in Germany.

We eliminate the influence of the seasonal variation function from the data by dividing the residuals by the square root of the fitted $\sigma^2(t)$ function.

[2]We again used the `nlinfit` function in MATLAB.

The resulting histogram in Fig. 10.11 became closer to normal compared with the one in Fig. 10.6, even though the Kolmogorov-Smirnov statistic of 1.888 still remains significant with p-value 0.002. However, considering the amount of data, even tiny departures from normality may become statistically significant. As we see from the normal plot Fig. 10.12, the reason

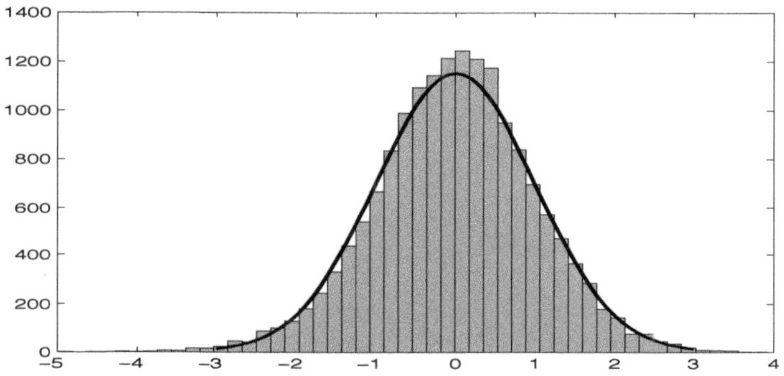

Fig. 10.11 Histogram of the residuals after linear trend, seasonal component, AR(1) process and seasonal variance were removed.

for rejecting the normal distribution can be heavier tails than normal for negative residuals. On the other hand, the normal plot indicates a really

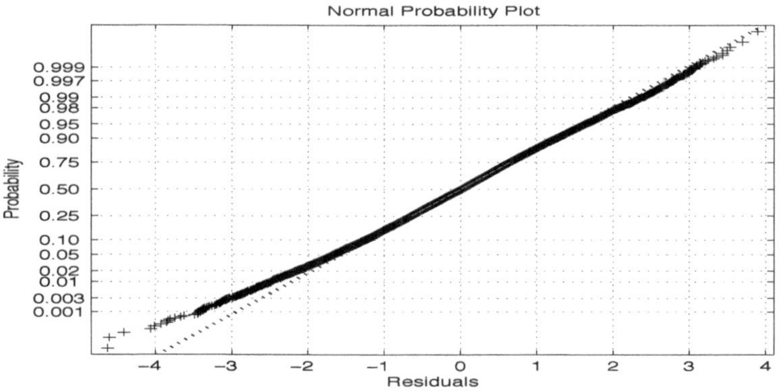

Fig. 10.12 Normal plot of the residuals after linear trend, seasonal component, AR(1) process and seasonal variance were removed.

good fit to the normal distribution elsewhere, making the choice of a Brownian motion reasonable. An alternative approach based on the GH Lévy processes is suggested in [Benth and Šaltytė-Benth (2005)].

The ACFs for residuals and squared residuals obtained after eliminating the seasonally dependent variance are plotted in Fig. 10.13 and Fig. 10.14, respectively. The ACF of the residuals basically shows that we are left with zero-mean uncorrelated noise, except for the first couple of lags where the Box-Ljung statistic remains significant. Moving our attention to the ACF for squared residuals, we observe a rapid decay in the first several lags and then a seemingly random variation around zero, clearly indicating that we managed to remove most of the seasonality in the variance. The

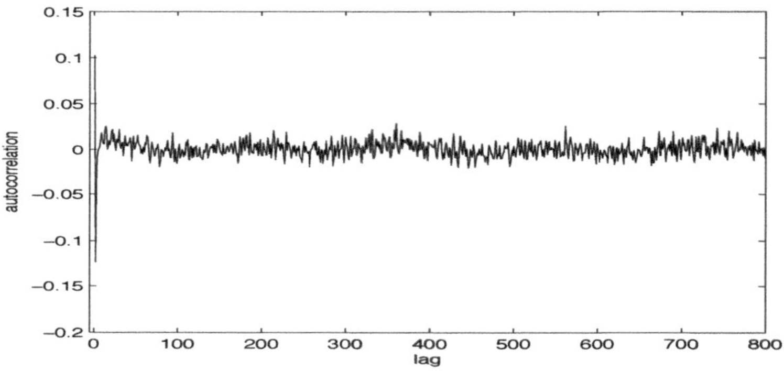

Fig. 10.13 ACF of residuals of DATs after linear trend, seasonal component, AR(1) process and seasonal variance were removed.

rapid decay in correlation for the first few lags may be modelled using a GARCH process (see [Bollerslev (1986)] for more on such processes and their properties).

10.3.2.2 *Fitting an AR(3) model*

Motivated by the PACF in Fig. 10.5, we fit an AR(3) process to the detrended and deseasonalized data. The steps fitting the AR(3) model with seasonal variance follows the sequence above closely, and we report here only main results.

We regress today's detrended and deseasonalized temperatures against the temperatures of the three previous days. The values of the regression parameters are all significant at the 1% level and reported in Table 10.3. Just as for the AR(1) case, we must treat the p-values with care. The model

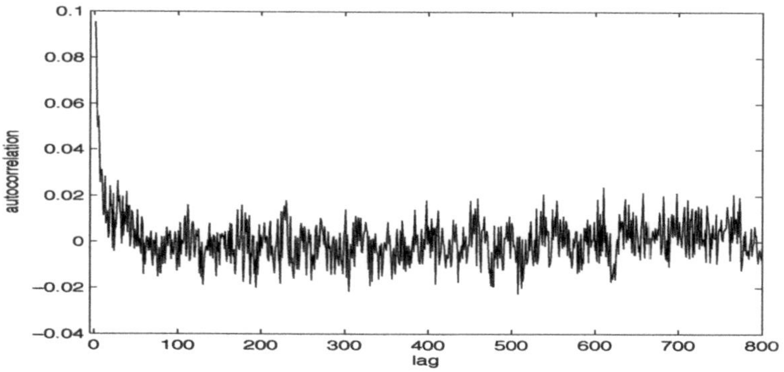

Fig. 10.14 ACF of squared residuals of DATs after linear trend, seasonal component, AR(1) process and seasonal variance were removed.

Table 10.3 Fitted regression parameters of the AR(3) process

b_1	b_2	b_3
0.957	−0.253	0.119

fit increased slightly compared to the AR(1) case, yielding $R^2 = 94.1\%$. The ACF of the residuals is presented in Fig. 10.15. The autocorrelations for the first few lags are insignficant according to the Box-Ljung statistic, and we observe a clear improvement compared with the similar ACF plot in the case of the AR(1) process (see Fig. 10.8). However, as for the AR(1) model, we still have a distinct seasonality in the ACF for the squared residuals, as seen in Fig. 10.16.

The seasonality observed in the residuals we again model with the seasonal variance function $\sigma^2(t)$ using the same approach as for the AR(1) case. The empirical and fitted seasonal variance functions are plotted in Fig. 10.17, and are similar to the AR(1) case. The estimated parameters are reported in Table 10.4. The estimates are quite close to those reported in Table 10.2.

After eliminating the seasonally dependent variance from the residuals of DAT, the autocorrelations for the obtained residuals (plotted in Fig. 10.18) became smaller, but otherwise preserved more or less the same pattern as for the AR(1) process (see Fig. 10.13 for comparison). Note

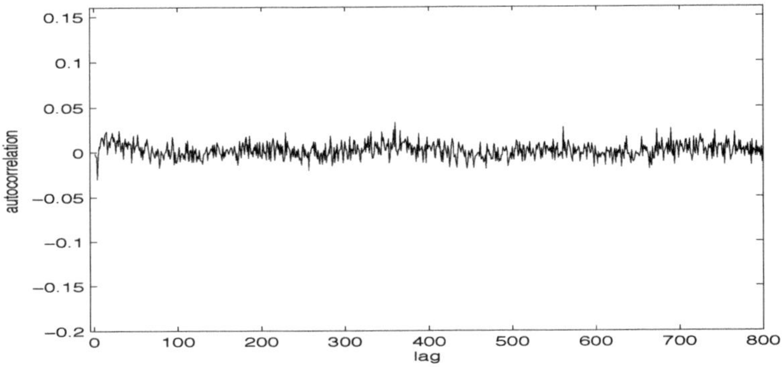

Fig. 10.15 ACF of residuals of DATs after linear trend, seasonal component and AR(3) process were removed.

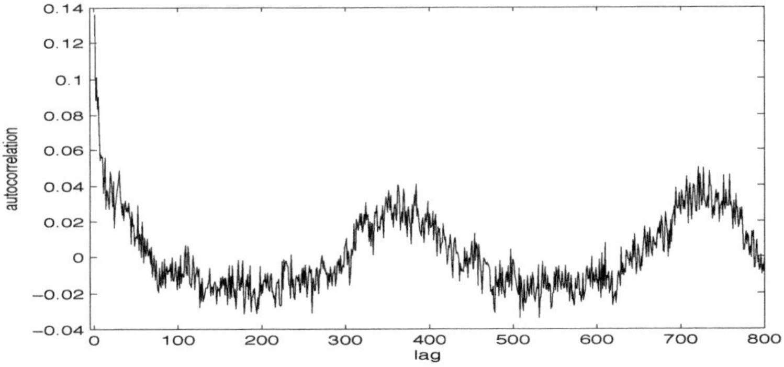

Fig. 10.16 ACF of squared residuals of DATs after linear trend, seasonal component and AR(3) process were removed.

that for the AR(3) case, only the autocorrelation at lag 1 is statistically significant. The ACF of squared residuals shown in Fig. 10.19 has the same structure as in the AR(1) case.

The average of the final residuals is −0.0022, with a standard deviation equal to 0.9997. They have a small negative skewness of −0.085 and a positive kurtosis being 0.350. Admittedly, we still did not manage to reach normality in the residuals. The histogram of the residuals looks very similar to the one for the AR(1) model (see Fig. 10.11), and the Kolmogorov-Smirnov statistics of 1.646 is significant with p-value of 0.009. Although

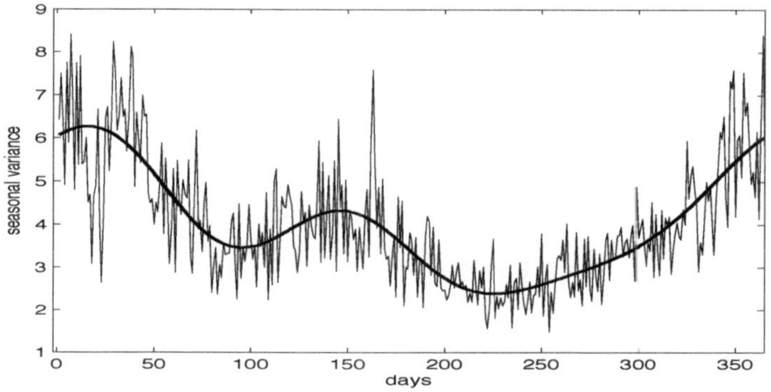

Fig. 10.17 Empirical and fitted $\sigma^2(t)$ function when AR(3) is used.

Table 10.4 Fitted parameters of $\sigma^2(t)$ for the AR(3) case

c_1	c_2	c_3	c_4	c_5	c_6	c_7	c_8	c_9
4.011	1.176	0.681	0.740	0.151	0.153	0.429	−0.042	−0.015

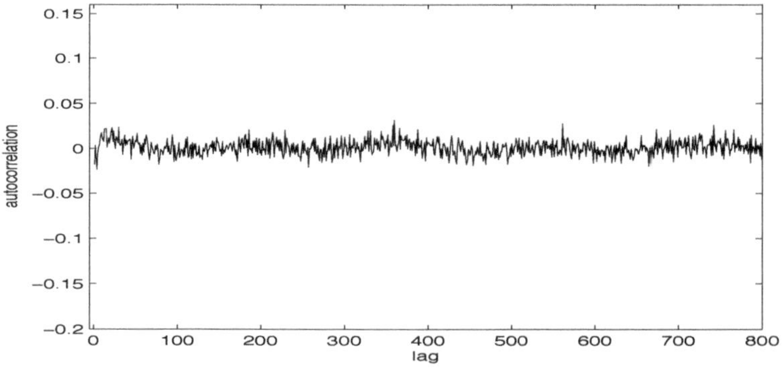

Fig. 10.18 ACF of residuals of DAT after linear trend, seasonal component, AR(3) process and seasonal variance were removed.

we reject the normality hypothesis, and thereby the argument for using the Brownian motion process for driving the noise, the histogram and normal plot are very close to those in the case of the normal distribution, except in the left tail. The error of using this distribution (that is, a Brownian motion

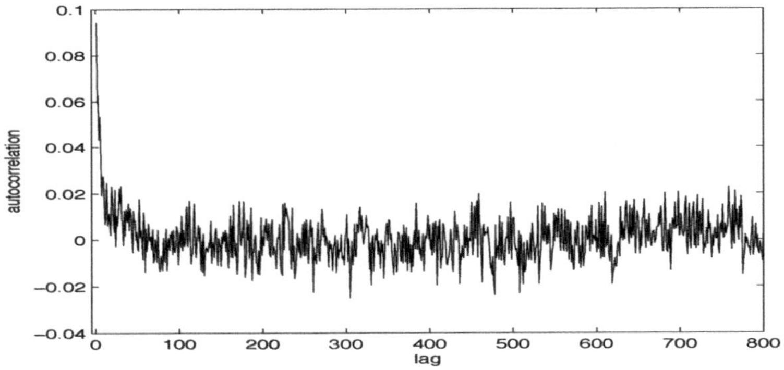

Fig. 10.19 ACF of squared residuals of DAT after linear trend, seasonal component, AR(3) process and seasonal variance were removed.

driving the dynamics) seems to be of minor importance. We conclude that the AR(3) process explains the data significantly better than the AR(1), and we give priority to this model. [Rossmanith (2007)] has shown that AR(3) is the appropriate model for the temperature dynamics in several German locations.

After establishing the shape of $\sigma^2(t)$, ML estimation can be implemented in a rather straightforward manner when appealing to the normality assumption. A more sophisticated method for estimating the coefficients of the AR model in the case of time-dependent residual variance may be weighted regression (see [Carroll and Ruppert (1988)]) or iterative reweighted regression (see [Hayman (1960)]). These approaches could of course be combined with an ML estimation of the seasonal variance function. The method which we use for estimation of the variance function $\sigma^2(t)$ may look somewhat ad hoc, however, it clearly captures the seasonality in the residuals.

10.3.2.3 *Identification of the parameters in the CAR(p) model*

The final step is to identify the corresponding parameters of the CAR(1) and CAR(3) models from the estimated parameters in the AR(1) and AR(3) models, respectively. The parameters of the seasonal mean function $\Lambda(t)$ and the variance function $\sigma^2(t)$ can be read off directly from the discrete-time model.

Recall the estimate of b_1 for the AR(1) model in (10.24). Since the AR(1) model corresponds to an OU process with speed of mean reversion

Table 10.5 Fitted regression parameters of the CAR(3) process

α_1	α_2	α_3
2.043	1.339	0.177

α_1 (the CAR(1) dynamics), we have that $b_1 = 1 - \alpha_1$. Thus, we find

$$\alpha_1 = 0.1766\,.$$

Let us turn our attention to the CAR(3) process. We find from Example 10.2 that

$$\begin{aligned} 3 - \alpha_1 &= b_1\,, \\ 2\alpha_1 - \alpha_2 - 3 &= b_2\,, \\ \alpha_2 + 1 - (\alpha_1 + \alpha_3) &= b_3\,. \end{aligned}$$

Inserting the estimates for the AR(3) model reported in Table 10.3, we derive the speeds of mean reversion in the CAR(3) model. The figures are reported in Table 10.5.

Remark that the eigenvalues of the matrix A resulting from the values of α_1, α_2 and α_3 in Table 10.5 are $\lambda_1 = -0.175$ and $\lambda_{2,3} = -0.934 \pm 0.374\,\mathrm{i}$. Thus, the matrix A satisfies the stationarity condition saying that the real parts of the eigenvalues must be negative.

10.4 Temperature derivatives pricing

In this section we investigate the dynamics of different temperature futures prices when temperature is modelled by a CAR(p) process. We concentrate on HDD, CDD and CAT futures, which constitute the three classes of futures products at CME. Along with the derivation of futures prices, we also discuss the valuation of European call and put options written on the futures. Finally, we analyse the Frost Day index futures.

10.4.1 *CAT futures*

From (10.6) we have that the CAT futures price dynamics is defined by the risk-neutral conditional expectation

$$F_{\mathrm{CAT}}(t,\tau_1,\tau_2)=\mathbb{E}_Q\left[\int_{\tau_1}^{\tau_2} T(u)\,du \,|\, \mathcal{F}_t\right].$$

We restrict our attention to risk-neutral probabilities Q^θ, and in the next Proposition we calculate the dynamics of the CAT futures with respect to this class of probabilities explicitly.

Proposition 10.2 *The CAT futures price for* $0\leq t\leq \tau_1<\tau_2$ *is given by*

$$\begin{aligned}F_{CAT}(t,\tau_1,\tau_2)&=\int_{\tau_1}^{\tau_2}\Lambda(u)\,du+\mathbf{a}(t,\tau_1,\tau_2)\mathbf{X}(t)\\&\quad+\int_t^{\tau_1}\theta(u)\sigma(u)\mathbf{a}(t,\tau_1,\tau_2)\mathbf{e}_p\,du\\&\quad+\int_{\tau_1}^{\tau_2}\theta(u)\sigma(u)\mathbf{e}_1'A^{-1}\{\exp\left(A(\tau_2-u)\right)-I_{p\times p}\}\mathbf{e}_p\,du\,,\end{aligned}$$

with $I_{p\times p}$ *being the* $p\times p$ *identity matrix and*

$$\mathbf{a}(t,\tau_1,\tau_2)=\mathbf{e}_1'A^{-1}\{\exp\left(A(\tau_2-t)\right)-\exp\left(A(\tau_1-t)\right)\}.$$

Proof. Inserting the model (10.13) for the temperature into the CAT futures definition leads to

$$F_{\mathrm{CAT}}(t,\tau_1,\tau_2)=\int_{\tau_1}^{\tau_2}\Lambda(s)\,ds+\mathbb{E}_\theta\left[\int_{\tau_1}^{\tau_2}X_1(s)\,ds\,|\,\mathcal{F}_t\right].$$

Next, we have that $X_1(u)=\mathbf{e}_1'\mathbf{X}(s)$. After commuting the integration and conditional expectation, and inserting the explicit dynamics of $\mathbf{X}(s)$ defined in (10.16), we find

$$\begin{aligned}\mathbb{E}_\theta\left[\int_{\tau_1}^{\tau_2}X_1(s)\,ds\,|\,\mathcal{F}_t\right]&=\int_{\tau_1}^{\tau_2}\mathbb{E}_\theta\left[X_1(s)\,|\,\mathcal{F}_t\right]ds\\&=\int_{\tau_1}^{\tau_2}\mathbf{e}_1'\exp\left(A(s-t)\right)\,ds\mathbf{X}(t)\\&\quad+\int_{\tau_1}^{\tau_2}\int_t^s\mathbf{e}_1'\exp\left(A(s-u)\right)\mathbf{e}_p\sigma(u)\theta(u)\,du\,ds\,.\end{aligned}$$

Consider the second integral. After using the Fubini-Tonelli Theorem, we have

$$\begin{aligned}
&\int_{\tau_1}^{\tau_2}\int_t^s \mathbf{e}_1' \exp\left(A(s-u)\right)\mathbf{e}_p\sigma(u)\theta(u)\,du\,ds \\
&\quad = \int_{\tau_1}^{\tau_2}\int_t^{\tau_2} 1_{[t,s]}(u)\mathbf{e}_1' \exp\left(A(s-u)\right)\mathbf{e}_p\sigma(u)\theta(u)\,du\,ds \\
&\quad = \int_t^{\tau_2}\int_{\tau_1}^{\tau_2} 1_{[t,s]}(u)\mathbf{e}_1' \exp\left(A(s-u)\right)\mathbf{e}_p\sigma(u)\theta(u)\,ds\,du \\
&\quad = \int_t^{\tau_1}\int_{\tau_1}^{\tau_2} \mathbf{e}_1' \exp\left(A(s-u)\right)\mathbf{e}_p\sigma(u)\theta(u)\,ds\,du \\
&\qquad + \int_{\tau_1}^{\tau_2}\int_u^{\tau_2} \mathbf{e}_1' \exp\left(A(s-u)\right)\mathbf{e}_p\sigma(u)\theta(u)\,ds\,du \\
&\quad = \int_t^{\tau_1} \sigma(u)\theta(u)\mathbf{a}(u,\tau_1,\tau_2)\,du \\
&\qquad + \int_{\tau_1}^{\tau_2} \sigma(u)\theta(u)\mathbf{e}_1'A^{-1}\{\exp\left(A(\tau_2-u)\right) - I_{p\times p}\}\mathbf{e}_p\,du\,.
\end{aligned}$$

Hence, the proposition follows. □

The futures price is given by the aggregated mean temperature over the measurement period, plus a direct dependency on $\mathbf{X}(t)$ appropriately scaled according to the mean-reverting properties of the underlying temperature dynamics. The dependency on $\mathbf{X}(t)$ means that the futures price is dependent on today's temperature $T(t)$, and all the previous days $T(t-r)$ up to lag $r \leq p$, p being the order of the CAR(p) model for temperature. The last two terms involve a smoothing of the market price of risk θ over the period from today (time t) until the end of the measurement period (time τ_2).

The dynamics of the CAT futures price under Q^θ is given in the following Proposition.

Proposition 10.3 *The Q^θ dynamics of $F_{CAT}(t,\tau_1,\tau_2)$ is*

$$dF_{CAT}(t,\tau_1,\tau_2) = \Sigma_{CAT}(t,\tau_1,\tau_2)\,,\, dB^\theta(t)$$

where

$$\Sigma_{CAT}(t,\tau_1,\tau_2) = \sigma(t)\mathbf{e}_1'A^{-1}\{\exp\left(A(\tau_2-t)\right) - \exp\left(A(\tau_1-t)\right)\}\mathbf{e}_p\,.$$

Proof. First, notice that the only coordintate of $\mathbf{X}(t)$ that has a dB^θ term is $X_p(t)$. Hence, using the fact that $F_{\mathrm{CAT}}(t,\tau_1,\tau_2)$ is a Q^θ martingale, we derive the dynamics after applying the multidimensional Itô Formula. □

Thus, we see that the CAT futures price dynamics follows, not surprisingly, an arithmetic process. This may attain negative values with positive probability as a result of the possibility of observing negative temperatures. Since the CAT futures at CME are written on temperatures in the warmer half of the year, this probability will normally be very small.

From the risk-neutral dynamics of $F_{\mathrm{CAT}}(t,\tau_1,\tau_2)$ in Prop. 10.3 it is straightforward to calculate the price of a call option written on the futures.

Proposition 10.4 *The price at time $t \leq \tau$ of a call option written on a CAT futures with strike K at exercise time $\tau \leq \tau_1$, and measurement period $[\tau_1,\tau_2]$, is*

$$C_{CAT}(t,\tau,\tau_1,\tau_2) = e^{-r(\tau-t)} \times \Big\{ (F_{CAT}(t,\tau_1,\tau_2) - K)\ \Phi(d(t,\tau,\tau_1,\tau_2)) + \int_t^\tau \Sigma^2_{CAT}(s,\tau_1,\tau_2)\,ds\,\Phi'(d(t,\tau,\tau_1,\tau_2)) \Big\},$$

where

$$d(t,\tau,\tau_1,\tau_2) = \frac{F_{CAT}(t,\tau_1,\tau_2) - K}{\sqrt{\int_t^\tau \Sigma^2_{CAT}(s,\tau_1,\tau_2)\,ds}},$$

and Φ is the cumulative standard normal distribution function.

Proof. From Prop. 10.3, we have that the Q^θ dynamics of the futures price can be written as

$$F_{\mathrm{CAT}}(\tau,\tau_1,\tau_2) = F_{\mathrm{CAT}}(t,\tau_1,\tau_2) + \int_t^\tau \Sigma_{\mathrm{CAT}}(s,\tau_1,\tau_2)\,dB^\theta(s),$$

for $0 \leq t \leq \tau \leq \tau_1 < \tau_2$. From this it follows that $F_{\mathrm{CAT}}(\tau,\tau_1,\tau_2)$ conditioned on $F_{\mathrm{CAT}}(t,\tau_1,\tau_2)$ is normally distributed, with mean $F_{\mathrm{CAT}}(t,\tau_1,\tau_2)$ and variance

$$\int_t^\tau \Sigma^2_{\mathrm{CAT}}(s,\tau_1,\tau_2)\,ds.$$

Hence, we find that

$$\begin{aligned}
&\mathbb{E}_\theta\left[\max\left(F_{\mathrm{CAT}}(\tau,\tau_1,\tau_2) - K, 0\right) \mid \mathcal{F}_t\right] \\
&\quad = \mathbb{E}_\theta\left[\max\left(F_{\mathrm{CAT}}(t,\tau_1,\tau_2) - K + \int_t^\tau \Sigma_{\mathrm{CAT}}(s,\tau_1,\tau_2)\,dB^\theta(s), 0\right) \mid \mathcal{F}_t\right]
\end{aligned}$$

$$= \mathbb{E}_\theta \left[\max \left(x - K + \int_t^\tau \Sigma_{\mathrm{CAT}}(s,\tau_1,\tau_2)\, dB^\theta(s), 0 \right) \right]_{x=F_{\mathrm{CAT}}(t,\tau_1,\tau_2)} .$$

The price C_{CAT} follows by calculating the above expectation appealing to the properties of the normal distribution. □

Note that once we have decided on a risk-neutral probability Q^θ, the market consisting of futures and options is complete. It is therefore possible to hedge the option perfectly, and the option price becomes the unique cost of replication.

The hedging strategy for this call option in the underlying CAT futures is given by the option's delta, that is, the sensitivity of the option price with respect to the underlying CAT futures price. This delta hedge ratio will give the number of CAT futures that should be held in the hedging portfolio to perfectly replicate the call.

Proposition 10.5 *The delta of the call option (or the hedge ratio), is given by*

$$\frac{\partial C_{CAT}(t,\tau,\tau_1,\tau_2)}{\partial F_{CAT}(t,\tau_1,\tau_2)} = \Phi(d(t,\tau,\tau_1,\tau_2)),$$

where the function d is defined in Prop. 10.4 and Φ is the cumulative standard normal distribution function.

Proof. The proof is an easy exercise in differentiation, and therefore omitted. □

We observe that the hedge is a real number between 0 and 1, analogous to the well-known delta hedge for call options in the Black & Scholes world. As expected, we should have close to no exposure in the CAT futures when the option is far out-of-the-money, while we should approximately be long one CAT futures in the opposite case.

10.4.2 *HDD/CDD futures*

We derive the explicit CDD futures price dynamics, and discuss issues on pricing of options. Recall from (10.4) the price of a CDD futures to be

$$F_{\mathrm{CDD}}(t,\tau_1,\tau_2) = \mathbb{E}_Q \left[\int_{\tau_1}^{\tau_2} \max\left(T(s) - c, 0\right) \mid \mathcal{F}_t \right] . \tag{10.25}$$

As for the CAT futures, we restrict our attention to Q^θ when calculating the futures price.

Proposition 10.6 *The CDD futures price is given by*

$$F_{CDD}(t,\tau_1,\tau_2)=\int_{\tau_1}^{\tau_2} v(t,s)\Psi\left(\frac{m(t,s,\mathbf{e}_1'\exp(A(s-t))\mathbf{X}(t))}{v(t,s)}\right)ds\,, \tag{10.26}$$

where

$$m(t,s,x)=\Lambda(s)-c+\int_t^s \sigma(u)\theta(u)\mathbf{e}_1'\exp(A(s-u))\mathbf{e}_p\,du+x\,,$$

$$v^2(t,s)=\int_t^s \sigma^2(u)\left(\mathbf{e}_1'\exp(A(s-u))\mathbf{e}_p\right)^2\,du$$

and $\Psi(x)=x\Phi(x)+\Phi'(x)$, *with* Φ *being the cumulative standard normal distribution function.*

Proof. After interchanging conditional expectation and integration, we get

$$\begin{aligned}F_{\mathrm{CDD}}(t,\tau_1,\tau_2)&=\mathbb{E}_\theta\left[\int_{\tau_1}^{\tau_2}\max\left(T(s)-c,0\right)\mid\mathcal{F}_t\right]\\&=\int_{\tau_1}^{\tau_2}\mathbb{E}_\theta\left[\max\left(\Lambda(s)+\mathbf{e}_1'\mathbf{X}(s)-c,0\right)\mid\mathcal{F}_t\right]ds\,.\end{aligned}$$

Under Q^θ, the explicit dynamics of $\mathbf{X}(s)$ is (see (10.16))

$$\begin{aligned}\mathbf{X}(s)=\exp(A(s-t))\mathbf{X}(t)&+\int_t^s\sigma(u)\theta(u)\exp(A(s-u))\mathbf{e}_p\,du\\&+\int_t^s\sigma(u)\exp(A(s-u))\mathbf{e}_p\,dB^\theta(u)\,,\end{aligned}$$

which therefore is normally distributed conditioned on $\mathbf{X}(t)$. It follows that $\Lambda(s)+\mathbf{e}_1'\mathbf{X}(s)-c$ is normally distributed as well, with mean $m(t,s,\mathbf{e}_1'\exp(A(s-t))\mathbf{X}(t))$ and variance $v^2(t,s)$. The Proposition then follows by standard calculations using the properties of the normal distribution. □

The dynamics of the CDD futures price is stated in the next Proposition.

Proposition 10.7 *The dynamics of* $F_{CDD}(t,\tau_1,\tau_2)$ *for* $0\le t\le\tau_1$ *under* Q^θ *is given by*

$$dF_{CDD}(t,\tau_1,\tau_2)=\sigma(t)\int_{\tau_1}^{\tau_2}\mathbf{e}_1'\exp(A(s-t))\mathbf{e}_p$$

$$\times \Phi\left(\frac{m(t,s,\mathbf{e}_1'\exp(A(s-t))\mathbf{X}(t)}{v(t,s)}\right) ds\, dB^\theta(t)\,,$$

where Φ is the cumulative standard normal distribution function.

Proof. We prove the result by applying the multidimensional Itô Formula. The CDD futures price is a Q^θ martingale, and the only terms that may contribute to a martingale part is $dX_p(t)$. Thus, after noticing that $\Psi'(x) = \Phi(x)$, a simple calculation leads to the desirable result. □

The term structure of the volatility of CDD futures is defined as

$$\Sigma_{\mathrm{CDD}}(t,\tau_1,\tau_2) \triangleq \sigma(t)\int_{\tau_1}^{\tau_2} \mathbf{e}_1'\exp(A(s-t))\mathbf{e}_p \times \Phi\left(\frac{m(t,s,\mathbf{e}_1'\exp(A(s-t))\mathbf{X}(t)}{v(t,s)}\right) ds\,. \tag{10.27}$$

We see that the volatility becomes dependent on $\mathbf{X}(t)$, which is the continuous-time analogue of today's temperature $T(t)$, and all the lagged temperatures $T(t-r)$ for $r = 1,\ldots,p$. The dependency on $\mathbf{X}(t)$ lies in the cumulative standard normal distribution function Φ. Further, we scale the temperature volatility $\sigma(t)$ by the integral over a weight function given by the regression coefficients. Observe that we recover the CAT term structure of volatility $\Sigma_{\mathrm{CAT}}(t,\tau_1,\tau_2)$ if we put $\Psi = 1$ in the above expression.

We now derive an expression for the price of a call option written on the CDD futures. It is impossible to derive an analytic expression for the price. However, we derive a formula in terms of an expectation which is suitable for simulations of the option price, using, for instance, Monte Carlo methods. Interestingly, the price is not a function of the current CDD futures price, instead it depends on the lagged temperatures.

Consider a call option on a CDD futures with measurement period $[\tau_1, \tau_2]$, where the strike price is K and the exericse time is $\tau \le \tau_1$. To simplify notation slightly in the further analysis, we introduce the function

$$\widetilde{\Psi}(t,s,x) = \Psi\left(\frac{m(t,s,x)}{v(t,s)}\right), \tag{10.28}$$

where $m(t,s,x)$ and $v(t,s)$ are defined in Prop. 10.6. The following proposition states the price of a CDD futures option at time $t \le \tau$.

Proposition 10.8 *The price at time $t \le \tau$ of a call option written on a CDD futures with measurement period $[\tau_1, \tau_2]$ and strike price K at the* *exercise time τ, is given as*

$$C_{CDD}(t,\tau,\tau_1,\tau_2) = e^{-r(\tau-t)} \times \mathbb{E}\left[\max\left(\int_{\tau_1}^{\tau_2} v(\tau,s)Z(t,s,\tau,\mathbf{x})\,ds - K, 0\right)\right]_{\mathbf{x}=\mathbf{X}(t)},$$

with the random field $Z(t,s,\tau,\mathbf{x})$ defined as follows

$$Z(t,s,\tau,\mathbf{x}) = \widetilde{\Psi}\Big(\tau, s, \mathbf{e}_1' \exp(A(s-t))\mathbf{x} + \int_t^\tau \mathbf{e}_1' \exp(A(s-u))\mathbf{e}_p\sigma(u)\theta(u)\,du + \Sigma(s,t,\tau)\,Y\Big).$$

Here,

$$\Sigma(s,t,T) = \int_t^T (\mathbf{e}_1' \exp(A(s-u))\mathbf{e}_p)^2\sigma^2(u)\,du,$$

and Y is a standard normal random variable.

Proof. The option price is by definition given as

$$C_{\mathrm{CDD}}(t,\tau,\tau_1,\tau_2) = \mathrm{e}^{-r(\tau-t)}\mathbb{E}_\theta\left[\max\left(F_{\mathrm{CDD}}(\tau,\tau_1,\tau_2) - K, 0\right) \mid \mathcal{F}_t\right].$$

We have

$$\begin{aligned} F_{\mathrm{CDD}}(\tau,\tau_1,\tau_2) &= \int_{\tau_1}^{\tau_2} v(\tau,s)\widetilde{\Psi}(\tau,s,\mathbf{e}_1' \exp(A(s-\tau))\mathbf{X}(t))\,ds \\ &= \int_{\tau_1}^{\tau_2} v(\tau,s)\widetilde{\Psi}\Big(\tau,s,\mathbf{e}_1' \exp(A(s-t))\mathbf{X}(t) \\ &\quad + \int_t^\tau \mathbf{e}_1' \exp(A(s-u))\mathbf{e}_p\sigma(u)\theta(u)\,du \\ &\quad + \int_t^\tau \mathbf{e}_1' \exp(A(s-u))\mathbf{e}_p\sigma(u)\,dB^\theta(u)\Big). \end{aligned}$$

Observe that the Itô integral inside the expectation operator is independent of $\mathcal{F}_t$ and has variance $\Sigma^2(t,s,\tau)$. Taking the conditional expectation yields the result. □

The CDD option price is a function of $\mathbf{e}_1' \exp(A(s-t))\mathbf{X}(t)$, meaning that the price is a function of the temperatures $T(t),\dots,T(t-p)$, weighted by a mean reversion dependent function over the measurement period of the futures $[\tau_1,\tau_2]$. Further, we see that the volatility, which is crucial for the price formation, is $\Sigma(t,s,\tau)$.

We now argue that $C_{\text{CDD}}(t,\tau,\tau_1,\tau_2)$ is not a function of $F_{\text{CDD}}(t,\tau_1,\tau_2)$. First, observe from the dynamics of F_{CDD} that

$$v(t,s)\widetilde{\Psi}(t,s,\mathbf{e}_1'\exp(A(s-t))\mathbf{X}(t)) = -\frac{\partial}{\partial\tau_1}F_{\text{CDD}}(t,s,\tau_2)$$

for $t \leq s$. Since $m(t,s,x)$ is linear in x, and $\Psi(x)$ is a monotonically increasing function due to the fact that $\Psi'(x) = \Phi(x)$, we have that $x \mapsto \widetilde{\Psi}(t,s,x)$ is monotonically increasing for each pair (t,s). Hence, there exists an inverse of $\widetilde{\Psi}$ for each t,s, and we derive

$$\mathbf{e}_1'\exp(A(s-t))\mathbf{X}(t) = \widetilde{\Psi}^{-1}\left(t,s,-v^{-1}(t,s)\frac{\partial}{\partial\tau_1}F_{\text{CDD}}(t,s,\tau_2)\right).$$

In conclusion, we see from Prop. 10.8 that the CDD option price will depend on the sensitivity of the CDD futures with respect to the start of the measurement period and not on the current CDD futures price.

Let us discuss the hedging strategy for a call option written on CDD futures. Considering derivatives where the price process is explicitly a function of the underlying asset, the hedging position is given as the delta of the option, as we saw in the discussion on options written on CAT futures. However, as we have just claimed, the price of an option on a CDD futures is not explicitly a function of the current CDD futures price, and therefore we cannot derive the hedging strategy straightforwardly by a differentiation. Since the market is complete, we know that there exists a hedge. From general theory of arbitrage pricing one may represent the payoff from the call option

$$C = \max\left(F_{\text{CDD}}(\tau,\tau_1,\tau_2) - K, 0\right)$$

by the Clark-Ocone Formula (see [Karatzas, Ocone and Li (1991)])

$$C = \mathbb{E}_\theta[C] + \int_0^\tau \xi(t,\tau)\,dB^\theta(t)$$

for an Itô integrable process $t \mapsto \xi(t,\tau)$, $0 \leq t \leq \tau$ which is explicitly given as

$$\xi(t,\tau) = \mathbb{E}_\theta[D_t C \mid \mathcal{F}_t]. \tag{10.29}$$

This representation involves the Malliavin derivative D_t of the random variable C. For our purposes, it is convenient to use this representation as an intermediate step to derive an expression for the hedging strategy, which will *not* involve the Malliavin derivative. The reader who wants to get familiar with the concepts of Malliavin Calculus is advised to read [Malliavin and Thalmaier (2006)]. Those, on the other hand, more interested in the results rather than the technique, can skip the next proof.

Proposition 10.9 *The Malliavin derivative of C is given as*

$$D_tC = 1\left(\int_{\tau_1}^{\tau_2} v(\tau,s)\widetilde{\Psi}(\tau,s,Z(t,\tau,\mathbf{X}(t))\,ds > K\right)$$
$$\times\,\sigma(t)\int_{\tau_1}^{\tau_2} \mathbf{e}_1'\exp(A(s-t))\mathbf{e}_p\Phi\left(\frac{m(\tau,s,Z(t,\tau,\mathbf{X}(t)))}{v(\tau,s)}\right)ds\,,$$

where

$$Z(t,\tau,\mathbf{x}) = \mathbf{e}_1'\exp(A(s-t))\mathbf{x} + \int_t^\tau \mathbf{e}_1'\exp(A(s-u))\mathbf{e}_p\sigma(u)\theta(u)\,du$$
$$+\int_t^\tau \sigma(u)\mathbf{e}_1'\exp(A(s-u))\mathbf{e}_p\,dB^\theta(u)\,,$$

and Φ the cumulative standard normal distribution function.

Proof. First, from the chain rule of the Malliavin derivative, we find

$$D_tC = 1\left(F_{\mathrm{CDD}}(\tau,\tau_1,\tau_2) > K\right)\,D_tF_{\mathrm{CDD}}(\tau,\tau_1,\tau_2)\,.$$

By the properties of the Malliavin derivative, we get

$$D_tF_{\mathrm{CDD}}(\tau,\tau_1,\tau_2) = \int_{\tau_1}^{\tau_2} v(\tau,s)D_t\widetilde{\Psi}(\tau,s,\mathbf{e}_1'\exp(A(s-\tau))\mathbf{X}(\tau)\,ds$$
$$= \int_{\tau_1}^{\tau_2} v(\tau,s)\Psi'\left(\frac{m(\tau,s,\mathbf{e}_1'\exp(A(s-\tau))\mathbf{X}(\tau))}{v(\tau,s)}\right)$$
$$\times\,\frac{1}{v(\tau,s)}D_t(\mathbf{e}_1'\exp(A(s-\tau))\mathbf{X}(\tau))\,ds\,.$$

The function $m(\tau,s,x)$ is linear in x, and $\Psi'(x) = \Phi(x)$. Moreover, from the explicit solution of $\mathbf{X}(\tau)$ in (10.16), we have

$$\mathbf{e}_1'\exp(A(s-\tau))\mathbf{X}(\tau) = \mathbf{e}_1'\exp(As)\mathbf{X}(0) + \int_0^\tau \mathbf{e}_1'\exp(A(s-u))\mathbf{e}_p\sigma(u)\theta(u)\,du$$
$$+\int_0^\tau \mathbf{e}_1'\exp(A(s-u))\mathbf{e}_p\sigma(u)\,dB^\theta(u)\,.$$

Thus,

$$D_t\mathbf{e}_1'\exp(A(s-\tau))\mathbf{X}(\tau) = \mathbf{e}_1'\exp(A(s-t))\mathbf{e}_p\sigma(t)\,,$$

which implies

$$D_t F_{\mathrm{CDD}}(\tau,\tau_1,\tau_2) = \int_{\tau_1}^{\tau_2} \mathbf{e}_1' \exp(A(s-t))\mathbf{e}_p \sigma(t) \times \Phi\left(\frac{m(\tau, s, \mathbf{e}_1'\exp(A(s-\tau))\mathbf{X}(\tau))}{v(\tau,s)}\right) ds\,.$$

Using that

$$\mathbf{X}(\tau) = \exp(A(\tau-t))\mathbf{X}(t) + \int_t^\tau \exp(A(\tau-u)\mathbf{e}_p\sigma(u)\theta(u)\,du + \int_t^\tau \exp(A(\tau-u))\mathbf{e}_p\sigma(u)\,dB^\theta(u)\,,$$

yields the result. □

We proceed to find the integrand $\xi(t,\tau)$ in (10.29).

Proposition 10.10 *The integrand process $\xi(t,\tau)$ for $0 \le t \le \tau$ in (10.29) is given as*

$$\xi(t,\tau) = \sigma(t)\mathbb{E}\left[1\left(\int_{\tau_1}^{\tau_2} v(\tau,s)\widetilde{\Psi}(\tau,s,Z)\,ds > K\right) \times \int_{\tau_1}^{\tau_2} \mathbf{e}_1'\exp(A(s-t))\mathbf{e}_p\Phi\left(\frac{m(\tau,s,Z)}{v(\tau,s)}\right) ds\right]_{\mathbf{x}=\mathbf{X}(t)}$$

for $t \le T$, where Z is a normal random variable with mean

$$\mathbf{e}_1'\exp(A(s-t))\mathbf{x} + \int_t^\tau \mathbf{e}_1'\exp(A(s-u))\mathbf{e}_p\sigma(u)\theta(u)\,du$$

and variance

$$\int_t^\tau \sigma^2(u)(\mathbf{e}_1'\exp(A(s-u))\mathbf{e}_p)^2\,du\,,$$

and Φ is the cumulative standard normal distribution function.

Proof. From the definition of ξ we have

$$\xi(t,\tau) = \mathbb{E}_{Q^\theta}\left[D_t C \,|\, \mathcal{F}_t\right]\,.$$

By using Prop. 10.9 and taking the conditional expectation we get the desired result. □

We link the integrand process $\xi(t,\tau)$ to the hedging strategy of a CDD option. Introduce the notation $h_{\text{CDD}}(t,\tau,\tau_1,\tau_2)$ as the hedging position in CDD futures with measurement period $[\tau_1,\tau_2]$ for a call option with strike K and maturity time τ. Suppose further that the volatility $\Sigma_{\text{CDD}}(t,\tau_1,\tau_2)$ is strictly positive. Then we find by applying the dynamics of $F_{\text{CDD}}(t,\tau_1,\tau_2)$ in Prop. 10.7 that

$$C = \mathbb{E}_\theta[C] + \int_0^\tau \Sigma_{\text{CDD}}^{-1}(t,\tau_1,\tau_2)\xi(t,\tau)\,dF_{\text{CDD}}(t,\tau_1,\tau_2)\,. \tag{10.30}$$

Thus, the hedging strategy is explicitly given as (with $0 \le t \le \tau$)

$$h_{\text{CDD}}(t,\tau,\tau_1,\tau_2) = \Sigma_{\text{CDD}}^{-1}(t,\tau_1,\tau_2)\xi(t,\tau)\,, \tag{10.31}$$

where $\xi(t,\tau)$ is defined in Prop. 10.10. As for the CDD option price, to calculate the hedge requires a numerical evaluation of an expectation. This expectation is easily evaluated using Monte Carlo methods.

Let us discuss the positivity of $g(t) \triangleq \mathbf{e}_1' \exp(At)\mathbf{e}_p$. Recall from (10.27) that if $g(t)$ is a positive function, it follows that the CDD volatility function Σ_{CDD} will be positive as well. From Thm. 1 (part c) in [Tsai and Chan (2005)], the function g is non-negative as long as the eigenvalues of the matrix A can be partitioned into triplets $(\lambda_i, \beta_i, \bar{\beta}_i)$, $i = 1, \ldots$, where each β_i has strictly positive imaginary part and each λ_i is real and greater than or equal to the real part of β_i. A sufficient condition is that all eigenvalues are real and negative, and a necessary condition is that there is a real eigenvalue of A being greater than or equal to the real part of all other eigenvalues of A. Further, note that when $t \to \infty$, the function $g(t)$ converges to zero whenever the eigenvalues of A are all supposed to have negative real parts. In addition, we have $g(0) = 0$. As we shall see in Subsect. 10.4.4, the function $g(t)$ is non-negative for all $t \ge 0$ when using the parameters in the CAR(3) model estimated for Stockholm.

10.4.3 *Frost Day index futures*

We discuss the pricing of the Amsterdam Frost Day index futures traded at the CME. Recall the Frost Day index futures price defined in (10.10) as

$$F_{\text{FDI}}(t,\tau_1,\tau_2) = \mathbb{E}_Q\left[\sum_{s=\tau_1}^{\tau_2} \text{FD}(s)\,|\,\mathcal{F}_t\right].$$

Before proceeding with the derivation of this price, we introduce some notation. We have that $T(s)$ conditioned on $\mathcal{F}_t$ for $s \ge t$ is normally distributed under Q^θ. The expectation and variance of $T(s)$ given $\mathcal{F}_t$ are

$$m_{s,t} \triangleq \mathbb{E}_\theta\left[T(s) \,|\, \mathcal{F}_t\right] = \Lambda(s) + \mathbf{e}_1' \exp(A(s-t))\mathbf{X}(t) + \int_t^s \mathbf{e}_1' \exp(A(s-u))\mathbf{e}_p \sigma(u)\theta(u)\,du \tag{10.32}$$

and

$$v_{s,t}^2 \triangleq \mathrm{Var}_\theta[T(s) \,|\, \mathcal{F}_t] = \int_t^s (\mathbf{e}_1 \exp(A(s-u))\mathbf{e}_p)^2 \sigma^2(u)\,du\,, \tag{10.33}$$

respectively. Since the temperatures at 7 a.m. and 10 a.m. define the Frost Day index, we need the correlation (conditioned on $\mathcal{F}_t$) between these two temperatures. It is easy to show that such a correlation is

$$\begin{aligned}\rho_{s,t} &\triangleq \mathrm{corr}[T(s+7/24), T(s+10/24) \,|\, \mathcal{F}_t] \\ &= \frac{1}{v_{s+7/24,t}v_{s+10/24,t}} \int_t^{s+7/24} \mathbf{e}_1{}' \exp(A(s+7/24-u))\mathbf{e}_p \\ &\quad \times \mathbf{e}_1 \exp(A(s+10/24-u))\mathbf{e}_p \sigma^2(u)\,du\,. \end{aligned} \tag{10.34}$$

We can now derive the price of a Frost Day index futures.

Proposition 10.11 *A Frost Day index futures price at time t with measurement period $[\tau_1, \tau_2]$ is given as*

$$\begin{aligned} F_{FDI}(t,\tau_1,\tau_2) &= \sum_{s=\tau_1}^{\tau_2} \Phi\left(\frac{-3.5 - m_{s+7/24,t}}{v_{s+7/24,t}}\right) + \sum_{s=\tau_1}^{\tau_2} \Phi\left(\frac{-1.5 - m_{s+10/24,t}}{v_{s+10/24,t}}\right) \\ &\quad + \sum_{s=\tau_1}^{\tau_2} \Phi\left(\frac{-0.5 - m_{s+7/24,t}}{v_{s+7/24,t}}, \frac{-0.5 - m_{s+10/24,t}}{v_{s+10/24,t}}, \rho_{s,t}\right) \\ &\quad - \sum_{s=\tau_1}^{\tau_2} \Phi\left(\frac{-3.5 - m_{s+7/24,t}}{v_{s+7/24,t}}, \frac{-0.5 - m_{s+10/24,t}}{v_{s+10/24,t}}, \rho_{s,t}\right) \\ &\quad - \sum_{s=\tau_1}^{\tau_2} \Phi\left(\frac{-0.5 - m_{s+7/24,t}}{v_{s+7/24,t}}, \frac{-1.5 - m_{s+10/24,t}}{v_{s+10/24,t}}, \rho_{s,t}\right) \\ &\quad + 2\sum_{s=\tau_1}^{\tau_2} \Phi\left(\frac{-3.5 - m_{s+7/24,t}}{v_{s+7/24,t}}, \frac{-1.5 - m_{s+10/24,t}}{v_{s+10/24,t}}, \rho_{s,t}\right). \end{aligned}$$

Here, $\Phi(\cdot)$ is the standard normal distribution function, while $\Phi(\cdot,\cdot,\rho)$ is the standard normal bivariate distribution function with correlation ρ. Further, $m_{s,t}, v_{s,t}^2$ and $\rho_{s,t}$ are defined in (10.32), (10.33), *and* (10.34), *respectively.*

Proof. We have that

$$\mathbb{E}_\theta \left[\mathrm{FD}(s) \,|\, \mathcal{F}_t\right] = \mathbb{E}_\theta \left[1\left(A_s \cup B_s \cup C_s\right) \,|\, \mathcal{F}_t\right] = Q^\theta \left(A_s \cup B_s \cup C_s \,|\, \mathcal{F}_t\right) ,$$

where

$$\begin{aligned} A_s &= \{T(s+7/24) \leq -3.5\} , \\ B_s &= \{T(s+7/24) \leq -1.5\} , \\ C_s &= \{T(s+7/24) \leq -0.5\} \cap \{ T(s+10/24) \leq -0.5\} . \end{aligned}$$

Using standard properties of a probability measure, we get

$$\begin{aligned} Q^\theta(A_s \cup B_s \cup C_s \,|1\, \mathcal{F}_t) &= Q^\theta(A_s \,|\, \mathcal{F}_t) + Q^\theta(B_s \,|\, \mathcal{F}_t) + Q^\theta(C_s \,|\, \mathcal{F}_t) \\ &\quad - Q^\theta(A_s \cap C_s \,|\, \mathcal{F}_t) - Q^\theta(B_s \cap C_s \,|\, \mathcal{F}_t) . \end{aligned}$$

We have that $T(s) \,|\, \mathcal{F}_t$ is normal with mean and variance defined in (10.32) and (10.33), respectively. The correlation between $T(s+7/24)$ and $T(s+10/24)$ conditioned on $\mathcal{F}_t$ is defined in (10.34). Standard calculations using the properties of the normal and bivariate normal distributions lead to the Proposition. □

To efficiently calculate the Frost Day index futures price, we can rewrite the expressions involving the standard normal distribution function introducing the error function. In Matlab, say, there are efficient algorithms to compute this function. Further, in [Drezner (1978)] a Gaussian quadrature method is proposed for the bivariate normal distribution.

We remark that the empirical analysis performed for Stockholm (in Sect. 10.3) is based on daily temperature data. In order to have a model feasible for deriving Frost Day index futures prices, we need to understand the temperature dynamics within a day, and thus need data (and models) capturing the intra-day effects of temperature. For example, there is an obvious intra-day seasonal variation since temperatures in the day are usually warmer than in the night. In addition, an intra-day seasonal volatility is expected as well.

10.4.4 *Application to futures on temperatures in Stockholm*

We end our analysis of the weather markets with a discussion of futures contracts based on our fitted model for Stockholm data. Our main emphasis will be on the volatility term structure for the CAT and CDD futures. Admittedly, the latter is not traded for Stockholm temperatures, but we are going to use the model for understanding the volatility suggesting that Stockholm is representative for the temperature dynamics in many of the cities where CME offers trading of futures.

We investigate first the function $g(t) = \mathbf{e}_1' \exp(At)\mathbf{e}_p$ which appears in the volatility of Σ_{CDD} defined in (10.27). Inserting the estimates of the CAR(3) model for Stockholm data, we find that $g(t)$ is strictly positive for $t > 0$, which is illustrated in Fig. 10.20. We conclude from this that the

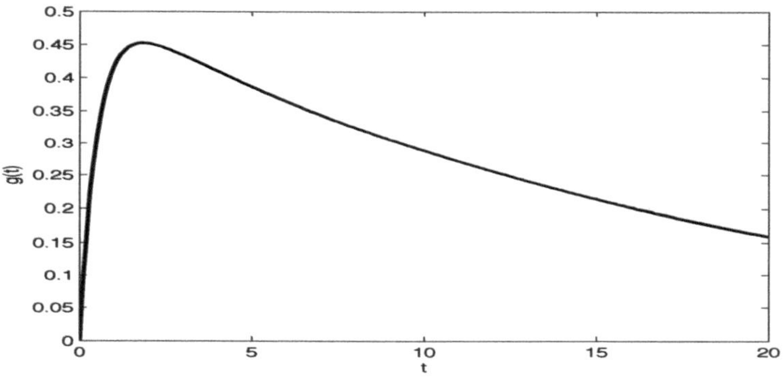

Fig. 10.20 The function $g(t)$ for the CAR(3) model for Stockholm data.

volatility of the CDD futures is positive, and that we have an expression for the hedging strategy for call (and put) options.

Let us discuss how the CDD volatility looks like. Recall that it depends on $\mathbf{X}(t)$ and is thus stochastic. To understand its shape, we restrict our attention to the following example first. Suppose that $\mathbf{X}(t) = 0$ for all t, meaning that $T(t) = \Lambda(t)$. Further, assume that the market price of risk is zero, that is $\theta = 0$. This will give us a feeling about the volatility in the case the temperature follows its mean up to the start of the CDD's measurement period. In Fig. 10.21 we plot the volatility for contracts with three different measurement periods. The graph shows the volatility the last 10 days prior to start of measurement period of the CDD, and the three curves show the volatility for measurement periods over the whole of July (top), the first two weeks (middle) and the first week (bottom) of July. The horizontal axis shows the day of the year, with 1 July being the day 182 of the year. We see that the longer the measurement period, the higher the volatility, which is rather natural keeping in mind that the longer into the future we go, the more uncertainty about the temperature

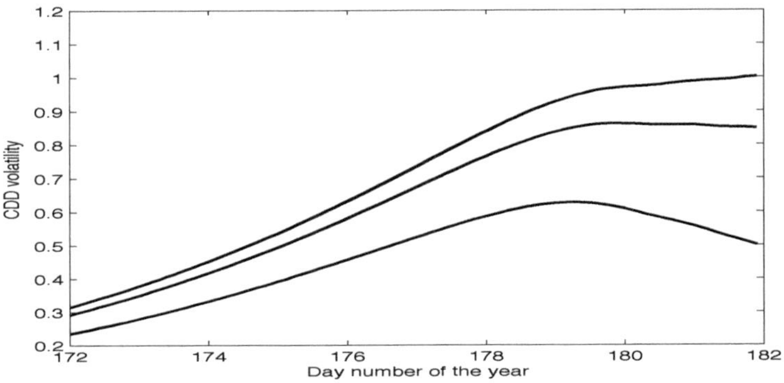

Fig. 10.21 The CDD volatility 10 days prior to start of measurement period, beginning 1 July (being day 182 of the year). The top graph shows the volatility of CDD measurement over the month of July, middle and the bottom graphs show the volatility for the first two weeks and the first week of July, respectively. The volatility is derived for the (unlikely) scenario that temperatures up to the beginning of the measurement period do not deviate from its seasonal mean $\Lambda(t)$.

we aggregate. Also, we observe a clear Samuelson effect when the volatility is increasing towards the start of the measurement period. However, this effect reverses close to maturity of the contracts. For instance, for the contract with weekly measurement the volatility starts to decrease slowly. We attribute this behaviour to the higher-order autoregression, since the memory effect captured by the CAR(3) model enables us to make good predictions about the temperature inside the measurement period when we approach the maturity of the contract. In line with this argument, the effect is less pronounced for the contract with measurement over the whole month of July. We call this the modified Samuelson effect.

In Fig. 10.22 we demonstrate the dynamics of the CDD volatility for two simulated temperature scenarios. We base our simulations of $\mathbf{X}(t)$ over the last 10 days prior to maturity on a simple Euler scheme (see [Kloeden and Platen (1992)]) with time discretization being 0.1. Note that this does not provide us with a realistic variation of the temperature over a day, because in our model we have used average daily sampled recordings which do not give us any understanding of the intra-daily temperature variations. The sampling is chosen to show the connection between temperature and CDD volatility on a time scale finer than daily. From the graph, it may not be so easy to see a connection between the temperature and the volatility. This is of course due to the complicated structure of the volatility and the ways in which the temperature affects the volatility. But we observe an increase in volatility with steadily decrease in temperature, in particular if we focus on the last part before maturity.

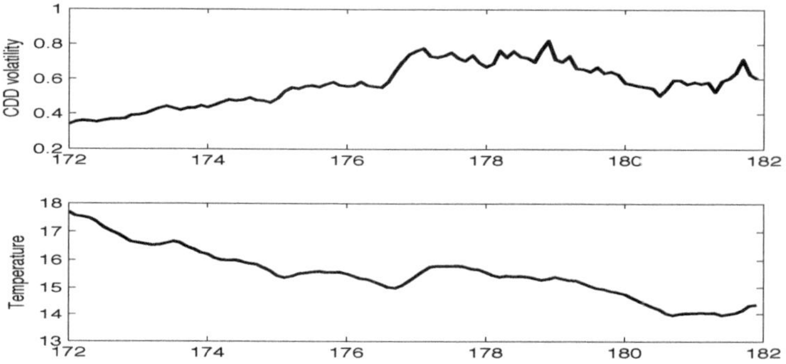

Fig. 10.22 The CDD volatility 10 days prior to start of measurement, beginning 1 July (day 182 of the year). The top graph shows the volatility for CDD measurement over the month of July. The bottom graph shows the simulated temperature for the same days.

We end with a graph showing the volatility for the CAT futures with the same measurement month of July (see Fig. 10.23). This plot again shows the three different measurement periods, starting with the whole month of July, next the first two weeks and finally the first week of July. We recover the modified Samuelson effect, manifesting strongest signs in the weekly contract. The paper [Benth, Šaltytė Benth and Koekebakker (2007)] contains further analysis on the term structure of CAT and CDD futures in the temperature market.

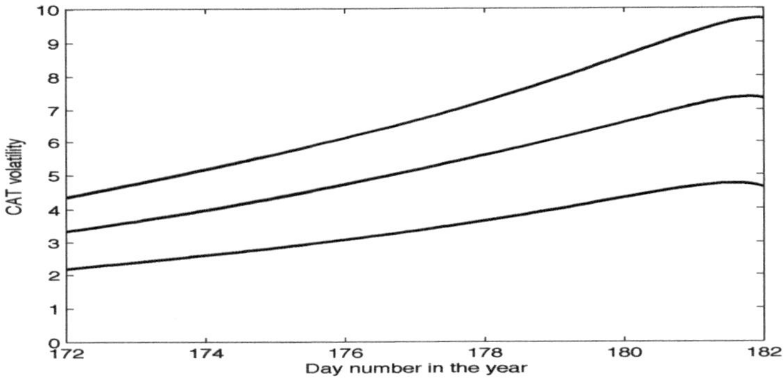

Fig. 10.23 The CAT volatility 10 days prior to start of measurement, beginning 1 July (day 182 of the year). The top graph shows the volatility of CAT measurement over the month of July, while the middle and the bottom show the volatility for the first two weeks and the first week of July, respectively.

Appendix A

List of abbreviations

ACF – *autocorrelation function*
APX – *Dutch electricity market*
AR – *autoregressive*
ARMA – *autoregressive moving average*
BIFFEX – *Baltic International Freight Futures Exchange*
Btu – *British thermal units*
CAR – *continuous autoregressive*
CARMA – *continuous autoregressive moving average*
CAT – *cumulative average temperature*
CCF – *conditional characteristic function*
CCX – *Chicago Climate Exchange*
CDD – *cooling-degree day*
CfD – *Contract for Differences*
CGMY – *Carr, Geman, Madan, Yor (distribution)*
CME – *Chicago Mercantile Exchange*
DA – *day ahead*
DAT – *daily average temperature*
ECX – *European Climate Exchange*
EEX – *European Electricity Exchange*
EUA – *EU allowances*
EU ETS – *EU emission trading system*
EUR – *Euros*
EXAA – *Energy Exchange Austria*
FFA – *forward freight agreement*
FFT – *fast Fourier transform*
FRA – *forward exchange rate agreement*
GBP – *British Pounds*

GARCH – *generalized autoregregressive conditional heteroskedastic (process)*
GH – *generalized hyperbolic (distribution)*
GIG – *generalized inverse Gaussian (distribution)*
HDD – *heating-degree days*
HJM – *Heath-Jarrow-Morton*
ICE – *Intercontinental Exchange*
IG – *inverse Gaussian (distribution)*
II – *independent increment (process)*
IMAREX – *Oslo based freight derivatives exchange*
JPY – *Japanese Yen*
LIBOR – *London interbank offer rate*
LNG – *liquified natural gas*
MWh – *Mega Watt hour*
NBP – *National Balancing Point*
NIG – *normal inverse Gaussian (distribution)*
NYMEX – *New York Mercantile Exchange*
OTC – *over-the-counter, bilateral "market"*
OU – *Ornstein-Uhlenbeck (process)*
PACF – *partial autocorrelation function*
PCA – *principal component analysis*
PJM – *Pennsylvania-New Jersey-Maryland (power market)*
PRIM – *Pacific Rim*
RCLL – *right continuous with left limits*
RT – *real time*
TSO – *transmission system operator*
UK ETS – *UK emission trading system*
UKPX – *UK Power Exchange*
USD – *US Dollars*
VG – *variance-gamma (distribution)*

Bibliography

Aadland, R. and Koekebakker, S. (2007). Forward curve dynamics in the tanker futures market. *Proceedings of the IAME-conference, Athens.* (submitted)

Abramowitz, M. and Stegun, I. A. (1972). *Handbook of Mathematical Functions.* Dover: New York

Adams, K. J. and van Deventer, D. R. (1994). Fitting yield curves and forward rate curves with maximum smoothness. *J. Fixed Income* **4**, pp. 53–62.

Alaton, P., Djehiche, B., and Stillberger, D. (2002). On modelling and pricing weather derivatives. *Appl. Math. Finance*, **9**(1), pp. 1–20.

Anderson, F. B., and Deacon, M. (1996). *Estimating and Interpreting the Yield Curve.* John Wiley and Sons.

Audet, N., Heiskanen, P., Keppo, J., and Vehvilainen, I. (2004). Modeling electricity forward curve dynamics in the Nordic market. In *Modelling Prices in Competitive Electricity Markets*, Bunn D. W. (ed), pp. 251–265, Wiley.

Barlow, M. (2002). A diffusion model for electricity prices. *Math Finance*, **12**(4), pp. 287–298.

Barlow, M., Gusev, Y., and Lai, M. (2004). Calibration of multifactor models in electricity markets. *Intern. J. Theor. Appl. Finance*, **7**(2), pp. 101–120.

Barndorff-Nielsen, O. E. (1977). Exponentially decreasing distributions for the logarithm of particle size. *Proc. R. Stat. Soc. London A*, **353**, pp. 401–419.

Barndorff-Nielsen, O. E. (1998). Processes of normal inverse Gaussian type. *Finance Stoch.*, **2**(1), pp. 41–68.

Barndorff-Nielsen, O. E. and Shephard, N. (2001). Non-Gaussian Ornstein-Uhlenbeck-based models and some of their uses in economics. *J. R. Statist. Soc. B*, **63**(2), pp. 167–241 (with discussion).

Barndorff-Nielsen, O. E. and Shephard, N. (2007). *Financial Volatility: Stochastic Volatility and Lévy Based Models.* Cambridge University Press. (to appear)

Benth, F. E., Cartea, A., and Kiesel, R. (2006). Pricing forward contracts in power markets by the certainty equivalence principle: explaining the sign of the market risk premium. To appear in *J. Banking Finance.* Available at SSRN: http://ssrn.com/abstract=941117.

Benth, F.E., Ekeland, L., Hauge, R. and Nielsen, B.F. (2003). On arbitrage-free pricing of forward contracts in energy market. *Appl. Math. Finance*, **10**, 4, pp. 325–336.

Benth, F. E., and Henriksen, P. N. (2008). Log NIG moment matching and pricing of basket options. *Preprint*, January 2008, University of Oslo.

Benth, F.E., Kallsen, J. and Meyer-Brandis, T. (2007). A non-Gaussian Ornstein-Uhlenbeck process for electricity spot price modeling and derivatives pricing. *Appl. Math. Finance*, **14**(2), pp. 153–169.

Benth, F. E., and Kettler, P. C. (2006). Dynamic copula models for the spark spread. *E-print*, 14, September 2006, Department of Mathematics, University of Oslo, Norway.

Benth, F. E., and Koekebakker, S. (2005). Stochastic modeling of financial electricity contracts. *E-print*, No. 24, September 2005, Department of Mathematics, University of Oslo. To appear in *Energy Economics*.

Benth, F. E., Koekebakker, S., and Ollmar, F. (2007). Extracting and applying smooth forward curves from average-based commodity contracts with seasonal variation. *J. Derivatives*, Fall **15**(1), pp. 52–66.

Benth, F. E., and Kufakunuseu, R. (2007). Pricing of exotic energy derivatives based on arithmetic spot models. *E-print*, No. 14, September 2007, Department of Mathematics, University of Oslo.

Benth, F.E. and Šaltytė-Benth, J. (2004). The normal inverse Gaussian distribution and spot price modelling in energy markets. *Intern. J. Theor. Appl. Finance*, **7**(2), pp. 177–192.

Benth, F.E. and Šaltytė-Benth, J. (2005). Stochastic modelling of temperature variations with a view towards weather derivatives. *Appl. Math. Finance*, **12**(1), pp. 53–85.

Benth, F.E. and Šaltytė-Benth, J. (2007). The volatility of temperature and pricing of weather derivatives. *Quantit. Finance*, **7**(5), pp. 553–561.

Benth, F.E. and Šaltytė-Benth, J. (2006). Analytical approximation for the price dynamics of spark spread options. *Stud. Nonlinear Dyn. Econom.*, **10**(3), article 8.
(Electronic publication: http://www.bepress.com/snde/vol10/iss3/art8)

Benth, F.E, Šaltytė Benth, J. and Koekebakker, S. (2007). Putting a price on temperature. *Scand. J. Stat.*, **34**, pp. 746–767.

Benz, E. and Trück, S. (2006). Modeling the price dynamics of CO2 emission allowances. Working Paper. University of Bonn.

Bessembinder, H., Coughenour, J.F., Seguin, P.J. and Smoller M.M. (1996). Is there a term structure of futures volatilities? Reevaluating the Samuelson hypothesis. *J. Derivatives*, Winter, pp. 45-58.

Bessembinder, H. and Lemmon, M. (2002). Equilibrium pricing and optimal hedging in electricity forward markets. *J. Finance*, **57**, pp. 1347-1382.

Bingham, N., and Kiesel, R. (1998). *Risk-Neutral Valuation. Pricing and Hedging of Financial Derivatives.* Springer-Verlag, Berlin Heidelberg New York.

Bjerksund, P., Rasmussen, H. and Stensland, G. (2000). Valuation and risk management in the Nordic electricity market, *Working paper, Institute of finance and management sciences*, Norwegian School of Economics and Business Administration.

Bjørk, T. (1998). *Arbitrage Theory in Continuous Time.* Oxford University Press,

Oxford.

Bjørk, T. and Landén, C. (2002). On the term structure of futures and forward prices. *Mathematical Finance - Bachelier Congress 2000*, Geman, Madan, Pliska and Ton Vorst (eds.), pp. 111–150, Springer-Verlag.

Blattberg, R. and Gonedes, N. (1974). A comparison of the stable and Student distributions as statistical models for stock prices. *J. Business*, **47**(2), pp. 244–280.

Black, F., and Scholes, M. (1973). The pricing of options and corporate liabilities. *J. Political Econ.*, **81**, pp. 637–659.

Black, F. (1976). The pricing of commodity contracts. *J Financial Econ*, **3**, pp. 167-179.

Bollerslev, T. (1986). Generalized autoregressive conditional heteroscedasticity. *J. Econometrics*, **31**, pp. 307–327.

Bølviken, E. and Benth, F. E. (2000). Quantification of risk in Norwegian stocks via the normal inverse Gaussian distribution. In *Proceedings AFIR 2000*, Tromsø, Norway, pp. 87–98.

Borak, S., Härdle, W., Trück, S. and Weron, R. (2006). Convenience yields for CO2 emission allowance futures contracts. SFB 649 Discussion Paper 2006-076. Humboldt-University of Berlin.

Börger, R., Cartea, A., Kiesel, R., and Schindlmayer, G. (2007). A multivariate commodity analysis and applications to risk management. *Preprint.* Available at SSRN: http://ssrn.com/abstract=981127

Botnen, O. J., Johannesen, A., Haugstad, A., Kroken, S., and Frøystein, O. (1992). Modelling of hydropower scheduling in a national/international context. In Broch, E., and Lysne, D. (Eds.), *Proceedings of the Second International Conference on Hydropower (Hydropower '92)*, Balkema, Rotterdam, pp. 575–584.

Brigo, D., and Liinev, J. (2005). On the distributional distance between the lognormal libor and swap market models. *Quantit. Finance*, **5**(5), pp. 433–442.

Brigo, D. and Mercurio, F. (2001). *Interest Rate Models. Theory and Practice.* Springer Verlag.

Brockwell, P. J., and Marquardt, T. (2005). Lévy-driven and fractionally integrated ARMA processes with continuous time parameter. *Statistica Sinica*, **15**, pp. 477–494.

Burger, M., Klar, B., Müller, A., and Schindlmayr, G. (2004). A spot market model for pricing derivatives in electricity markets. *Quantit. Finance*, **4**, pp. 109–122.

Campbell, S. D. and Diebold, F. X. (2005). Weather forecasting for weather derivatives. *J. Amer. Statist. Assoc.*, **100**(469), pp. 6–16.

Carmona, R., and V. Durrleman (2003). Pricing and hedging spread options. *SIAM Reviews*, **45**, pp. 627–685.

Carr, P., Geman, H., Madan, D. B., and Yor, M. (2002). The fine structure of asset returns: an empirical investigation. *J. Business*, **75**(2), pp. 305–332.

Carr, P. and Madan, D. B. (1998). Option valuation using the fast Fourier transform. *J. Comp. Finance*, **2**, pp. 61–73.

Carroll, R. J. and Ruppert, D. (1988). *Transformation and Weighting in Regres-*

sion. Chapman and Hall: New York.
Cartea, A. and Figueroa, M. G. (2005). Pricing in electricity markets: a mean reverting jump diffusion model with seasonality, *Appl. Math. Finance*, **12**(4), pp. 313–335.
Cartea, A. and Williams, Th. (2006). UK gas markets: The market price of risk and applications to multiple interruptible supply contracts. To appear in *Energy Economics*
Clewlow, L. and Strickland, C. (1999). A multifactor model for energy derivatives. *Working paper*, Quantitative Finance Research Group, University of Technology, Sydney.
Clewlow, L. and Strickland, C. (2000). *Energy Derivatives: Pricing and Risk Management*, Lacima Publications.
Cont, R., and Tankov, P. (2004). *Financial modelling with jump processes*. Chapman & Hall/CRC Financial Mathematics Series. Chapman & Hall/CRC, Boca Raton.
Cortazar, G., and Schwartz, E. S. (1994). The valuation of commodity contingent claims. *J. Derivatives*, **1**, pp. 27–39.
Crosby, J. (2005). A multi-factor jump-diffusion model for commodities. *Working Paper*.
Crosby, J. (2006). Commodity options optimised. *Risk*, **19**(5).
Davis, M. H. A. (2001). Pricing weather derivatives by marginal value. *Quantit. Finance*, **1**, pp. 305–308.
Da Prato, G. and Zabczyk, J. (1992). *Stochastic Equations in Infinite Dimensions*. Encyclopedia of Mathematics and its Applications, 44. Cambridge University Press, Cambridge.
Daskalakis, G., Psychoyios, D. and Markellos, R. N. (2006). Modeling CO2 emission allowance prices and derivatives: Evidence from the EEX. *Working Paper*. Athens University of Economics and Business.
Deng, S. (2000). Stochastic models of energy commodity prices and their applications: mean reversion with jumps and spikes, *Working paper*, PWP-073 of the Program on Workable Energy Regulation (POWER).
De Jong, F., Driessen, J., and Pelsser, A. (2004). On the Information in the interest rate term structure and option prices. *Rev. Derivatives Res.*, **7**, pp. 101–129.
Diko, P., Lawford, S., and Limpens, V. (2006). Risk premia in electricity forward prices. *Stud. Nonlinear Dyn. Econom.*, **10**(3), article 7. (Electronic publication: http://www.bepress.com/snde/vol10/iss3/art7)
Dornier, F. and Queruel, M. (2000). Caution to the wind. *Energy Power Risk Manag.*, Weather risk special report, August 30–32.
Doucet, A., de Freitas, N. and Gordon, N. (2001). *Sequential Monte Carlo Methods in Practice*. Springer Verlag.
Drezner, Z. (1978). Computation of the bivariate normal integral. *Math. Comput.*, **32**, No. 141, 277–279.
Duffie, D. (1992). *Dynamic Asset Pricing Theory*. Princeton University Press.
Dybvig, P. H. (1997). Bond and Bond Option Pricing Based on the Current Term Structure. In *Mathematics of Derivative Securities*, M. A. H. Dempster and

S. Pliska (eds.), Cambridge University Press.

Eberlein, E. and Keller, U. (1995). Hyperbolic distributions in finance. *Bernoulli*, **1**, pp. 281–299.

Eberlein, E., Keller, U. and Prause, K. (1998). New insights into smile, mispricing and value at risk measures. *J. Business*, **71**, pp. 371–406.

Eberlein, E. and Stahl, G. (2003). Both sides of a fence: A statistical and regulatory view of electricity risk. *Energy Risk*, **8**, 34–38.

Elliot, R. J., Sick, G. A. and Stein, M. (2000). Pricing electricity calls. *Preprint*, University of Alberta.

Escribano, A., Pena, J. I. and Villaplana, P. (2002). Modelling electricity prices: International evidence. *Working paper*, Universidad Carlos III de Madrid.

Esscher, F. (1932). On the probability function in the collective theory of risk. *Skandinavisk Aktuarietidskrift*, **15**, pp. 175–195.

Eydeland, A., and Wolyniec, K. (2003). *Energy and Power Risk Management.* John Wiley & Sons, Inc.

Fabra, N., von der Fehr, N-H. and Harbord, D. (2006). Designing Electricity Auctions. *RAND J. Economics*, **37**(1), pp. 23–46.

Fackler, P.L. and Tian, Y. (1999). Volatility models for commodity markets. *Working Paper*, North Carolina State University.

Fama, E. F. (1970). Efficient markets: a review of theory and empirical work. *J. Finance*, May, pp. 383–417.

Fehr, M. and Hinz J. (2006). A quantitative approach to carbon price risk modeling. *Preprint*, ETH Zentrum, Institute for Operations Research.

Financial Times, 4 October 2006. Front page, First Section: "Gas given away as surge from Norway causes price plunge".

Fleten, S. and Lemming, J. (2003). Constructing forward price curves in electricity markets, *Energy Economics*, **25**, pp. 409–424.

Folland, G. B. (1984). *Real Analysis – Modern Techniques and their Applications.* John Wiley & Sons.

Forsgren, A. (1998). A note on maximum-smoothness approximation of forward interest rates. *Technical report*, TRITA-MAT-1998-OS3, Royal Institute of Technology, Sweden.

Frestad, D. (2007). Common and unique factors influencing daily swap returns in the Nordic electricity market, 1997–2005. To appear in *Energy Econom.*

Frestad, D. (2007). Electricity forward return correlations. *Working paper*, University of Agder, Norway.

Frestad, D., Benth, F. E., and Koekebakker, S. (2007). Modelling the term structure dynamics in the Nordic electricity swap market. *Working paper*, University of Agder, Norway

Fusai, G., Marena, M., and Roncoroni, A. (2007). A note on the analytical pricing of commodity Asian-style options under discrete monitoring. *Preprint*, ESSEC Business School, France. To appear in *J. Banking Finance.*

Galloway, T.M. and Kolb, R.W. (1996). Futures prices and the maturity effect. *J. Futures Markets*, **16**, pp. 809–828.

Geman, H. (1999). *Insurance and Weather Derivatives.* RISK Books.

Geman, H. (2005). *Commodities and Commodity Derivatives.* Wiley-Finance,

John Wiley and Sons, Chichester.

Geman, H., and Leonardi, M. P. (2005). Alternative approaches to weather derivatives valuation. *Manag. Finance*, **31**(6), pp. 46–72.

Geman, H., El Karoui, N. and Rochet, J. C. (1995). Changes of numéraire, changes of probability measure and option pricing. *J Appl Probab*, **32**, pp. 385–408.

Geman, H. and Roncoroni, A. (2006). Understanding the fine structure of electricity prices. *J. Business*, **79**(3), pp. 1225–1261.

Geman, H. and Vasicek, O. (2001). Forwards and futures on non-storable commodities: the case of electricity. *RISK*, August.

Gerber, H. U., and Shiu, E. S. W. (1994). Option pricing by Esscher transforms. *Trans. Soc. Actuaries*, **46**, pp. 99–191 (with discussion).

Green, R. (2006). Risk management and the normal inverse Gaussian distribution – evidence from the Nordic electricity market. *Working paper*, Department of Economics, Lund University, Sweden.

Green, R., and Nossman, M. (2007). A jump diffusion stochastic volatility model for the electricity spot price. *Working paper*, Department of Economics, Lund University, Sweden.

Guthrie, G. and Videbeck, S. (2002). High frequency electricity spot price dynamics: An intraday markets approach, *Working Paper*, New Zealand Institute for the Study of Competition and Regulation.

Halgreen, C. (1979). Self-decomposability of the generalized inverse Gaussian and hyperbolic distributions. *Z. Wahrsch. Ver. Geb.*, **47**, pp. 13–17.

Hambly, B., Howison, S., and Kluge, T. (2007). Modelling spikes and pricing swing options in electricity markets. *Preprint*, University of Cambridge.

Handcock, M. S., and Wallis, J. R. (1994). An approach to statistical spatial-temporal modeling of meteorological fields. *J. Amer. Statist. Assoc.*, **89**(426), pp. 368–390.

Hansen, T. L., and Jensen, B. A. (2004). Energy options in a HJM framework. *WP 2004-10*, Department of Finance, Copenhagen Business School.

Harvey, A.C. (2003). *Forecasting, Structural Time Series Models and the Kalman Filter*, Cambridge University Press, Cambridge.

Hayman, B. I. (1960). Maximum likelihood estimation of genetic components of variation. *Biometrics*, **16**, pp. 369–381.

Heath, D., Jarrow, R. and Morton, A. (1992). Bond pricing and the term structure of interest rates: a new methodology for contingent claim valuation. *Econometrica*, **60**, pp. 77–105.

Henriksen, P. N. (2007). Log-normal moment matching and pricing of basket options. *Preprint*, University of Oslo.

Hinz, J. (2003). Modelling day-ahead prices. *Appl Math Finance*, **10**, 2, pp. 149–161.

Hinz, J., von Grafenstein, L., Verschuere, M., and Wilhelm, M. (2005). Pricing electricity risk by interest rate methods. *Quantit. Finance*, **5**(1), pp. 49-60.

Hjalmarsson, E. (2003). Does the Black-Scholes formula work for electricity markets? A nonparametric approach. *Working Papers in Economics*, nr. 101, Göteborg University.

Hull, J. C. (2000). *Options, Futures, and other derivatives.* Prentice Hall.
Ichihara, K. and Kunita, H. (1974). A classification of the second order elliptic operator and its probabilistic characterization. *Z. Wahrsch. Verv. Geb.*, **30**, pp. 235–254.
Ikeda, N. and Watanabe, S. (1981). *Stochastic Differential Equations and Diffusion Processes.* North-Holland/Kodansha.
Jacod, J. and Shiryaev, A. N. (1987). *Limit Theorems for Stochastic Processes.* Springer-Verlag.
James, J. and Webber, N. (2000). *Interest rate modelling.* Wiley: Chichester, England.
Jewson, S. and Brix, A. (2005). *Weather Derivatives Valuation.* Cambridge University Press.
Johnson, N., Kotz, S. and Balakrishnan, N. (1994). *Continuous Univariate Distributions.* 2nd Edition. John Wiley & Sons: p. 565.
Judd, K. L. (1998). *Numerical Methods in Economics.* The MIT Press, Cambridge, Massachusetts.
Kaaresen, K. and Husby, E. (2002). A joint state-space model for electricity spot and futures prices. *Energy Power Risk Manag.*, **7**(8).
Kallsen, J., and Tankov, P. (2006). Characterization of dependence of multidimensional Lévy processes using Lévy copulas. *J. Multivariate Analysis*, **97**, pp. 1551-1572.
Kalman, R. E. (1960). A new approach to linear filtering and prediction problems. *J. Basic Engineering* (Series D), **82**, pp. 35-45.
Karatzas, I., Ocone, D. L., and Li, J. (1991). An extension of Clark's formula. *Stoch. Stoch. Rep.*, **37**(3), pp. 215–258.
Karatzas, I. and Shreve, S. E. (1991). *Brownian Motion and Stochastic Calculus.* Springer-Verlag.
Kavussanos, M. G. and Visvikis, I. D. (2006). *Derivatives and risk management in shipping.* Witherby Publishing Limited, London.
Keller, U. (1997). Realistic modelling of financial derivatives. *PhD dissertation*, Universität Freiburg, Germany.
Kellerhals, P. (2001). Pricing electricity forwards under stochastic volatility. *Working Paper*, University of Tübingen.
Keppo, J., Audet, N., Heiskanen, P. and Vehviläinen, I. (2004). Modeling electricity forward curve dynamics in the Nordic Market. In *Modelling Prices in Competitive Electricity Markets.* Bunn, D. W. (eds.), Wiley Series in Financial Economics, pp. 251–264.
Kiesel, R., Schindlmayr, G., and Börger, R. (2006). A two-factor model for the electricity forward market. *Preprint*, University of Ulm, Germany.
Kjaer, M., and Ronn, E. (2006). Modelling natural gas futures returns. *Energy Risk*, November, pp. 60–66.
Kloeden, P. E. and Platen, E. (1992). *Numerical Solution of Stochastic Differential Equations*, Springer-Verlag: Berling Heidelberg
Knittel, C.R. and Roberts, M. R. (2005). An empirical examination of restructured electricity prices, *Energy Economics*, **27**(5), pp. 791–817.
Koekebakker, S. (2003). An arithmetic forward curve model for the electricity

market. *Working paper*, 2003:5, Agder University College, Norway.

Koekebakker, S., Aadland, R. and Sødal, S. (2007). Pricing freight rate options. *Transportation Research Part E: Logistics and Transportation Review*, **43**(5), pp 535–548.

Koekebakker, S. and Lien, G. (2004). Volatility and Price Jumps in Agricultural Futures Prices - Evidence from Wheat Options. *American J Agric Econom*, **86**(4), pp. 1018–1031.

Koekebakker, S. and Ollmar, F. (2005). Forward curve dynamics in the Nordic electricity market. *Manag. Finance*, **31**(6), pp. 74–95.

Kristiansen, T. (2004). Pricing of Contracts for Differences in the Nordic Market. *Energy Policy*, **32**, pp. 1075–1085.

Leon, A. and Rubia, A. (2004). Testing for weekly seasonal unit roots in the Spanish power pool. In *Modelling Prices in Competitive Electricity Markets*. Bunn, D. W. (eds.), Wiley Series in Financial Economics, pp. 177–189.

Lévy, P. (1965). *Processus Stochastiques et Mouvement Brownien*, 2nd Ed. Gauthier-Villars, Paris.

Levy, E. (1992). Pricing European average rate currency options. *J. Intern. Money Finance*, **11**(5), pp. 474–491.

Lillestøl, J. (2000). Risk analysis and the NIG distribution. *J. Risk*, **2**(4), pp. 41–56.

Lim, K. G., and Xiao, Q. (2002). Computing Maximum Smoothness Forward Rate Curves. *Statistics Computing*, **12**, pp. 275–279.

Lima, B. (2005). Spark spread options. *Master thesis*, Department of Mathematics, University of Oslo, Norway.

Liptser, R. S., and Shiryaev, A. N. (1977). *Statistics of Random Processes I. General Theory*. Springer-Verlag.

Litterman, R. and Sceinkman, J. (1991). Common factors affecting bond returns. *J. Fixed Income*, **1**(1), pp. 54–61.

Longstaff, F. A. and Wang, A.W. (2004). Electricity Forward Prices: A High-Frequency Empirical Analysis. *J. Finance*, **59**(4), pp. 1877–1900.

Lord, R. and Pelsser, A. A. J. (2007). Level – slope – curvature – fact or artefact? *Appl. Math. Finance*, **14**(2), pp. 105–130.

Lucia, J. and Schwartz, E. S. (2002). Electricity Prices and Power Derivatives: Evidence from the Nordic Power Exchange. *Rev. Derivatives Res.*, **5**(1), pp. 5–50.

Malliavin, P. and Thalmaier, A. (2006). *Stochastic Calculus of Variations in Mathematical Finance*. Springer-Verlag.

Marcus, M. B. (1987). ξ-radial processes and random Fourier series. *Mem. Am. Math. Soc.*, **368**.

Margrabe, W. (1978). The value of an option to exchange one asset for another. *J. Finance*, **33**, pp. 177–187.

McCulloch, J. H. (1971). Measuring the term structure of interest rates. *J. Business*, **44**, pp. 19–31.

Merton, R. (1976). Option pricing when underlying stock returns are discontinuous. *J. Finan. Econom.*, **3**, pp. 125–144.

Meyer-Brandis, T., and Tankov, P. (2007). Multi factor jump-diffusion models of electricity prices. *Working paper*, Université Paris VII, France.

Musiela, M. and Rutkowski, M. (1998). *Martingale Methods in Financial Modeling.* Springer-Verlag.

Oldfield, G. S., and Messina, R. J. (1977). Forward exchange price determination in continuous time. *J. Finan. Quantit. Anal.*, **12**(3), pp. 473–479.

Ollmar, F. (2003). An analysis of Derivatives Prices. *Ph.D. thesis*, Norwegian School of Economics and Business Administration, Norway.

Paolella, M. S. and Taschini, L. (2006). An econometric analysis of emission trading allowances. Research Paper Series 06 – 26. Swiss Finance Institute, University of Zurich.

Platen, E., and West, J. (2005). A fair pricing approach to weather derivatives. *Asian-Pac. Financial Markets*, **11**(1), pp. 23–53.

Pilipovic, D. (1998). *Energy Risk: Valuing and managing energy derivatives.* McGraw-Hill, New York.

Praetz, P. D. (1972). The distribution of share price changes. *J. Business*, **45**(1), pp. 49–55.

Prause, K. (1999). The generalized hyperbolic model: estimation, financial derivatives and risk measures. *PhD dissertation*, Universität Freiburg, Germany. Downloadable from:
`http://www.freidok.uni-freiburg.de/volltexte/15`

Protter, Ph. (1990). *Stochastic Integration and Differential Equations.* Springer-Verlag.

Raible, S. (2000). Lévy processes in finance: Theory, numerics and empirical facts. *PhD dissertation*, Universität Freiburg, Germany.

Rassmusson, E. M., Dickinson, R. E., Kutzback, J. E. and Cleveland, M. K. (1993). *Climatology.* In "Handbook of Hydrology", Maidment D.R. (editor), McGraw-Hill Professional.

Rosinski, J. (1991). On a class of infinitely divisible processes represented as mixtures of Gaussian processes. In *Stable Processes and Related Topics*, C. Cambanis, G. Samorodnitsky and M. S. Taqqu (eds.), Birkhäuser Verlag, pp. 27–41

Rossmanith, G. (2007). Das Temperaturmodell von Benth und Saltyte-Benth und dessen Anwendung auf Standorte in Deutschland. *Master thesis*, Munich University, Germany. (in German).

Rydberg, T. H. (1997). The normal inverse Gaussian Lévy process: Simulation and approximation. *Commun. Statist.-Stochastic Models*, **13**(4), pp. 887-910.

Šaltytė Benth, J., Benth, F. E., and Jalinskas P. (2007). A spatial-temporal model for temperature with seasonal variance. *J. Appl. Statistics*, **34**(7), pp. 823–841.

Samuelson, P. (1965a). Rational theory of warrant pricing. *Indust. Manag. Rev.*, **6**, pp. 13–32.

Samuelson, P. (1965b). Proof that properly anticipated prices fluctuate randomly. *Indust. Manag. Rev.*, **6**, pp. 41–44.

Saravia, C. (2003). Speculative Trading and Market Performance: The Effect of Arbitrageurs on Efficiency and Market Power in the New York Electricity

Market. *Working Paper*, No 121, Center for the Study of Energy Markets (CSEM).

Sato, K.-I. (1999). *Lévy Processes and Infinitely Divisible Distributions.* Cambridge University Press.

Schwartz, E. S. (1997). The stochastic behaviour of commodity prices: Implications for valuation and hedging. *J. Finance*, **LII**, 3, pp. 923–973.

Seifert, J., Uhrig-Homburg, M. and Wagner, M. (2006). Dynamic behavior of CO2 spot prices - theory and empirical evidence. *Working Paper*, University of Karlsruhe, Germany.

Shiryaev, A. N. (1999). *Essentials of Stochastic Finance.* World Scientific.

Singleton, K. J. (2001). Estimation of affine asset pricing models using the empricial characterstic function. *J. Econometrics*, **102**, pp. 111–141.

Skantze, P., Gubina, A. and Ilic, M. (2000). Bid-based stochastic model for electricity prices: the impact of fundamental drivers on market dynamics, *Report*, MIT Energy Laboratory.

Skantze, P. and Ilic, M. (2000). The joint dynamics of electricity spot and forward markets: implications on formulating dynamic hedging strategies, *Report*, MIT Energy Laboratory.

Steeley, J. M. (1990). Modelling the dynamics of the term structure of interest rates. *Economic Social Rev.*, **21**(4), pp. 337–361.

Stoft, S., (2002). *Power System Economics.* John Wiley and Sons, New York, NY.

Strickland L. (2002). Derivative modelling approaches used in real option valuation, Manuscript presented at Optimal Methods for Applying Real Options in the Energy Industry, London, UK, 4-5 November.

Tsai, H. and Chan, K. S. (2005). A note on non-negative continuous time processes. *J. R. Statist. Soc. B*, **67**(4), pp. 589-597.

Uhrig-Homburg, M. and Wagner, M (2006). Market dynamics and derivative instruments in the EU Emissions Trading Scheme – an early market perspective. *Manuscript*, Universität Karlsruhe, Germany.

Uhrig-Homburg, M. and Wagner, M. (2007). Forward price dynamics of CO2 emission certificates – an empirical analysis. *Manuscript*, Universität Karlsruhe, Germany.

Vehviläinen, I. (2002). Basics of electricity derivative pricing in competitive markets. *Appl. Math. Finance*, **9**, pp. 45–60.

Vehviläinen, I. and Pyykkönen, T. (2005). Stochastic factor model for the electricity spot price – the case of the Nordic market. *Energy Economics*, **27**(2), pp. 351–367.

Villaplana, P. (2004). A two-state variables model for electricity prices. *Presentation: Third World Congress of the Bachelier Finance Society, Chicago.*

Weron, R. (2005). Market price of risk implied by Asian-style electricity options. To appear in *Energy Economics.*

Weron, R. (2006). *Modeling and Forecasting Electricity Loads and Prices – A Statistical Approach.* John Wiley & Sons, Chichester.

White, H. (1982). Maximum likelihood estimation of misspecified models. *Econometrica*, **50**(1), pp. 1–25.

Wolak, F. (1997). Market design and price behavior in restructured electricity markets: an international comparison. *POWER WP-051*. University of California Energy Institute.

Yu, J. (2004). Empirical characteristic function estimation and its applications. *Econometric Reviews*, **23**(2), pp. 93–123.

Zapranis, A., and Alexandridis, A. (2007). Modeling temperature time-dependent speed of mean reversion in the context of weather derivatives pricing. *Working paper*, University of Macedonia of Economics and Social Studies, Greece.

Index

$\stackrel{d}{=}$, 75

ACF, 79
  partial, 132
active set approach, 190
Alberta Power Pool, 147
APX, 27
AR time series, 283
area price, 5
arithmetic model, 21
autocorrelation function, 79

Baltic International Freight Futures Exchange, 15
Bayes' Formula, 38
bending factor, 203
Bessel function, 51, 52
BIFFEX, 15
bivariate normal distribution function, 313
Black-76 Formula, 238
block contracts, 7
Brownian motion, 19, 39
Btu, 9
buy-and-hold strategy, 25

càdlàg, 37
Californian Power Exchange, 147
CAR process, 280
CARMA process, 280
CAT, 13, 278
CAT futures price, 279, 302
  dynamics, 303
  volatility, 307
Cauchy distribution, 54
CCX, 16
CDD, 12, 277
CDD futures price, 278, 306
  dynamics, 306
  volatility, 307
CDD-HDD parity, 279
certainty equivalent principle, 27
CfD, 7
CGMY distribution, 56, 251
characteristic function, 39, 152
Chicago Board of Trade, 262
Chicago Climate Exchange, 16
Chicago Mercantile Exchange, 11
Clark-Ocone Formula, 309
Climex, 17
CME, 11, 277
CO2 emission market, 16
coal futures, 15
compensator measure, 43
  integrability, 72, 74
complete filtered probability space, 37
complete market, 26
compound Poisson process, 49
congestion, 5
continuous AR process, 280
continuous autoregressive moving-average process, 280
Contract for Differences, 7
convenience yield, 26

cooling-degree days, 12
copulas, 179, 235, 268
covered interest rate parity, 102
cumulant function, 40
  martingale condition, 176, 177
cumulative average temperature, 13
currency risk, 100

DA market, 4
daily average temperature, 285
DAT, 285
day-ahead market, 4
delivery period, 6
delta hedge, 239
dilogarithm function, 243
Doob-Meyer decomposition theorem, 41
dry bulk, 15

ECX, 17
EEX, 2
Elbas, 5
electricity futures, 6, 18, 205, 264
  price, 30
electricity futures contract, 90
electricity futures price, 90
  two-factor model, 175
  volatility, 174
Elspot, 4
Energy Exchange Austria, 17
equivalent martingale measure, 22
error function, 314
Esscher transform, 28, 97, 161
  multivariate, 178
  original, 96
EU Allowances, 17
EU ETS, 16
EUA, 17
Euler approximation, 283
European Climate and Exchange, 17
European Power Exchange, 2
European Union Emission Trading Scheme, 16
EXAA, 17
exponential distribution, 64, 84, 99, 138

factor loading, 204, 216
FFA, 15
FFT, 158, 249, 267
flow commodity, 3
forward contract, 18, 89
Forward exchange rate agreement, 101
Forward Freight Agreement, 15
forward price, 90
  asymptotic, 108, 110, 118
  convergence, 92
  domestic currency, 103
  foreign currency, 103
  jump volatility, 112
  market dynamics, 113, 117, 162
  risk-neutral dynamics, 110, 117, 160
  smooth, 184
Fourier transform, 153, 247
  fast, 158, 249, 251
FRA, 101
freight derivatives, 15
Frost Day, 279
Frost Day index, 12, 13, 279
Frost Day index futures
  price, 313

Gamma distribution, 55, 65, 84
GARCH process, 296
gas futures, 9, 18, 91, 264
Gaussian quadrature method, 314
generalized hyperbolic distribution, 51
generalized inverse Gaussian distribution, 55
geometric Brownian motion, 19, 174
geometric model, 21
GH distribution, 51, 234
GIG distribution, 55
Girsanov transform, 28, 98, 282
global warming, 287

HDD, 12, 277
HDD futures price, 279
heat rate, 256, 264
Heath-Jarrow-Morton approach, 19

heating-degree days, 12
heavy tails, 140
Henry Hub, 8
Heren NBP DA index, 10, 130
HJM approach, 19, 31
hub, 8
  notional, 8
hyperbolic distribution, 53

ICE, 10
idiosyncratic risk, 214
IG distribution, 55
II process, 21, 38
  generating triplet, 40
  Lévy-Kintchine decomposition, 46
  pure jump, 46, 66
  random jump measure, 43
IMAREX, 15
independent increment process, 21
Intercontinental Exchange, 10
interquartile range, 130
inverse Gaussian distribution, 55
Itô isometry, 42
Itô's Formula, 48

Kalman filter, 147
Kyoto protocol, 16

Lévy copula, 179
Lévy measure, 43
  density, 83
  upper tail integral, 83
Lévy process, 20, 39
  background driving, 65
  CGMY, 56, 100
  Gamma, 56
  GH, 52, 99, 296
  hyperbolic, 53, 99
  Lévy measure, 43
  NIG, 53, 99, 140, 253
  subordinator, 39
  VG, 56
Lévy-Kintchine decomposition, 46
Lagrange multiplier, 189
Langeled pipeline, 75
Laplace transform, 73
LIBOR model, 32, 155, 172, 207
liquified natural gas, 16
LNG, 16
logarithmic returns, 20
lognormal distribution, 138, 245, 274
logreturns, 20

Malliavin derivative, 309
marginal utility, 280
Margrabe's Formula, 263
market efficiency hypothesis, 20
market model, 155, 173, 207
market price of risk, 19, 28, 145, 159, 183, 184
  jump, 95
Markov Chain Monte Carlo, 150
martingale, 38
  quadratic variation process, 41
  square integrable, 41
maximum smoothness criterion, 187
mean reversion, 18
mean-variance mixture, 54
MPS model, 183

National Balancing Point, 8
NBP, 8
negative price, 74
NEMMCO, 147
New South Wales GHG Abatement Scheme, 16
New York Mercantile Exchange, 10
Newton's algorithm, 145
NIG distribution, 53, 140, 176, 231, 250
  multivariate, 178, 235
  shape triangle, 54, 141, 231
non-storable, 26
Nord Pool, 2, 181, 203
normal backwardation, 27
normal inverse Gaussian distribution, 53
NYMEX, 10, 146, 175, 262

Omel, 4
option
  Asian, 254

average, 254
basket, 244, 257
Black-76 Formula, 238, 246
CAT futures, 304
CDD futures, 307, 309
crack spread, 254
delta, 305
delta hedge, 239, 305
real, 11
spark spread, 11, 254, 256
swap, 241
Ornstein-Uhlenbeck process, 18
OTC, 8
OU process, 18, 60
ACF, 79
CCF, 152
characteristic function, 61, 152
cumulant function, 61
Gaussian, 63
simulation, 83
stationary, 64, 282, 301
stochastic mean, 71
vectorial, 281
outliers, 130
over-the-counter market, 8

PACF, 132
Pacific Rim, 13
partial hedge, 254
particle filter, 149
PCA, 203
peak load contracts, 7
Pennsylvania-New Jersey-Maryland electricity market, 27
PJM, 27
Poisson process, 49
compensated, 49
compound, 49, 137, 274
time-inhomogeneous, 50
Poisson random measure, 160
compensator measure, 160
Powernext, 4, 17
PRIM, 13, 278
PRIM futures price, 279
principal component analysis, 203
principal component matrix, 216

QR factorisation, 190
quadratic programming, 188
quadratic variation process, 41

Radon-Nikodym derivative, 97
random field, 44
predictable, 44
random jump measure, 43, 98
compensated, 44
compensator, 98
compensator measure, 43
random variable, 37
rational expectation hypothesis, 27
real-time market, 4
recursive filtering, 136
regression
iterative reweighted, 300
weighted, 300
risk premium, 27, 210, 211
risk-free asset, 89
risk-neutral probability, 19, 22
domestic, 101
foreign, 101
RT market, 4

Samuelson effect, 111, 122
average, 126
modified, 316, 317
Scandinavian power crisis, 209
Schiphol airport, 279
Schwartz model, 20
jump, 129
seasonal floor, 85
seasonal function, 66, 78, 192
self-decomposable distribution, 65
semimartingale, 46
shifting factor, 203
sigma-algebra, 37
optional, 42
predictable, 42
snowfall, 12
spark spread, 264
spline, 188
spot freight rate, 15
spot price
arithmetic, 74, 254, 263

electricity, 6
gas, 10
geometric, 66
jump-diffusion model, 69
Lucia and Schwartz two factor model, 70
Schwartz model, 69
seasonal floor, 76
spikes, 1, 59, 68, 70, 71, 77, 85, 112, 150
stochastic Fubini Theorem, 47
stochastic integral, 42
stochastic process, 37
adapted, 37
additive, 39
càdlàg, 37
finite variation, 46
Gaussian, 63
independent increment, 38
infinite-dimensional, 171
integrable, 42
Lévy, 39
local martingale, 38
martingale, 22, 38
natural integrable increasing, 41
optional, 42
Poisson, 49
predictable, 41
RCLL, 37
semimartingale, 22, 46
stopping time, 38
storage, 26
Student t distribution, 54
subordinator, 39
swap, 6, 18, 90
basic contract, 173
market model, 173
swap model, 155
swap price, 91
asymptotic, 124
domestic currency, 104
foreign currency, 104
no-arbitrage relation, 165, 166
risk-neutral dynamics, 120, 124, 166
spread, 266
system price, 5, 212

temperature futures, 12, 18
temperature trend, 287
therm, 9
tilting factor, 203
time to delivery, 7
trading period, 6
transmission system operator, 4
TSO, 4

UK Emission Trading System, 16
UK ETS, 16
UK Power Exchange, 4
UKPX, 4
urbanisation, 287

Value at Risk, 208, 235, 275
variance-gamma distribution, 56
VG distribution, 56
volatility, 214, 224
average, 242
CAT, 307
CDD, 312
hump, 225, 227
seasonal, 224, 288, 294
smooth, 200
stochastic, 150
temperature, 281, 294

Walrasian auction, 4
weather derivatives, 11
Weber function, 52
Wiener process, 19

zonal price, 5

www.ingramcontent.com/pod-product-compliance
Ingram Content Group UK Ltd.
Pitfield, Milton Keynes, MK11 3LW, UK
UKHW022108170926
8624IPUK00024B/255

9 789812 8123